GLIMPSES OF REALITY

Episodes in the History of Science

The Octagon Room of the Royal Observatory at Greenwich, designed by Christopher Wren and set up in 1675. From an etching by Francis Place.

GLIMPSES OF REALITY

Episodes in the History of Science

Byron Wall

Department of Mathematics and Statistics
York University

Wall & Emerson, Inc.
Toronto, Ontario • Dayton, Ohio

Copyright © 2003 by Byron E. Wall

The Author

Byron E. Wall is Associate Lecturer in the Department of Mathematics and Statistics at York University in Toronto, Canada, and teaches in the Division of Natural Science of the Faculty of Pure & Applied Science. He has a Ph.D. in the History and Philosophy of Science and Technology from the University of Toronto.

All rights reserved. No part of this publication may be reproduced or transmitted in any form or by any means, electronic of mechanical, including photography, recording, or any information storage and retrieval system, without permission in writing from the publisher.

Orders for this book or requests for permission to make copies of any part of this work should be sent to:

Wall & Emerson, Inc.
Six O'Connor Drive
Toronto, Ontario, Canada M4K 2K1

Telephone: (416) 467-8685
Fax: (416) 352-5368
E-mail: wall@wallbooks.com
Web site: www.wallbooks.com

Layout and design of text and cover: Alexander Wall

National Library of Canada Cataloguing in Publication

Wall, Byron Emerson, 1943-
    Glimpses of reality : episodes in the history of science / Byron Wall.

Includes bibliographical references and index.
ISBN 0-921332-52-1

    1. Science--History. I. Title.

Q125.W34 2003   509   C2002-903920-7

Printed in Canada

TABLE OF CONTENTS

Preface ix

Part One   1
Ancient Beginnings

Chapter One. Numbers and Arithmetic 5
    Egypt 7
    Babylonia 12

Chapter Two. First Thoughts about Nature. 19
    What is the World Made Of? 20
    Change or the Lack of It 27

Chapter Three. Pythagoras and the Magic of Numbers 37
    Mathematics and Music 38
    Simple Numbers, Geometrical Shapes 40
    The Pythagorean Theorem 42
    Everything is Number 43
    The Diagonal of the Square 45

Chapter Four. Plato and the Reality of Ideas. 51
    Plato 51
    The Divided Line 53

Chapter Five. The Task of the Philosopher. 61
    The Allegory of the Cave 61
    Saving the Phenomena 69

Chapter Six. Aristotle and the World of Experience 77
    Empiricism 80
    Logic 82
    The Four Causes 83
    Aristotle's Universe 85

Chapter Seven. Axioms and Proofs 95
    Euclid 96
    The Elements of Euclid 98

Chapter Eight. Saving the Heavens. 111
    Eratosthenes and the Size of the Earth 111
    Claudius Ptolemy 117

Part Two   135
Science Emerges

Chapter Nine. Copernicus Chooses Formal Elegance over Common Sense 141
    Nicholas Copernicus 142
    The Copernican System 144
    On the Revolutions 147
    Mathematical Elegance versus Common Sense 157
    The Reception of *On the Revolutions* 161

Chapter Ten. Kepler's Celestial Harmony 165
    The Cosmographical Mystery 166
    Tycho Brahe 169
    Kepler in Prague 172
    Kepler's Laws 174

Chapter Eleven. Galileo, the Anti-Anti-Copernican 185
    Galileo and the Copernican System 187
    Galileo Reinterprets the Bible 199
    The Dialogue 201

Chapter Twelve. Galileo, Founder of Physics 209
    Niccolo Tartaglia 210
    The Law of Free Fall 211
    How, not Why 216
    Parabolic Projectiles 217
    Two New Sciences 218

Chapter Thirteen. The Res Extensa of Descartes 227
    A Jesuit-trained Soldier with a Law Degree 228
    The Principles of Philosophy 229
    Analytic Geometry 234

Chapter Fourteen. Let Newton Be! 241
    1666, the Annus Mirabilis 242
    Light 243
    Calculus 248
    The Falling Apple 253

Chapter Fifteen. *The Mathematical Principles of Natural Philosophy*. 257
    The Pieces that Newton Put Together 257
    The Lucasian Professor of Mathematicks 260
    Philosophiæ Naturalis Principia Mathematica 263

The Clockwork Universe 270
Universal Gravitation 271

Part Three 277
From Certainty to Uncertainty

Chapter Sixteen. Energy, A New Form of Being................. 283
Steam Engines 284
Heat Causes Motion, Does Motion Cause Heat? 288
The Conservation of Energy 290
Unavailable Energy 292
Thermodynamics 293
Maxwell's Demon 294
Absolute Zero 295
The Heat Death of the Universe 296

Chapter Seventeen. Electromagnetism and the Æther............ 299
Light 299
Electricity and Magnetism 305

Chapter Eighteen. Invariance and Relativity Trade Places.......... 309
Absolute Space and Time 309
The Michelson-Morley Experiment 311
Albert Einstein 317
The Special Theory of Relativity 320
General Relativity 326
Einstein = Genius 332
The Curvature of Space 334

Chapter Nineteen. Science Isn't So Sure 337
Radiation 338
Atoms are not Atomic 341
The Ultraviolet Catastrophe 343
The Quantum of Energy 345
The Bohr Atom 348
Matter Waves 351
The Uncertainty Principle 353
Schrödinger's Cat 355

Chapter Twenty. The Universe Will Go On Forever—Or It Won't .. 361
Stellar Parallax 361
Cepheid Variables 363

Glimpses of Reality

Is the Universe Finite or Infinite? 364
Edwin Hubble 367
Redshift 368
The Big Bang 371
The Cosmic Background Radiation 372
Is the Universe Open or Closed? 374

Part IV   379
What is Life?

Chapter Twenty One. The Earth and Its Inhabitants Classified......383
Theories of the Earth 384
Taxonomy 394
Lamarck's Theory of Evolution 397

Chapter Twenty Two. Evolution by Natural Selection............407
The Design Argument 410
Charles Darwin 414
Alfred Russel Wallace 423
Darwin in a Dither 425
The Book That Shook the World 426
Darwin's Attack on the Design Argument 430
Darwin on Man 438

Chapter Twenty Three. Finding the Units of Life................443
The Microscope 443
Cell Theory 446
Inheritance 449
Gregor Mendel 455
Genetics 462

Chapter Twenty Four. DNA, the Key to the Mystery.............469
DNA 472
The Structure of DNA 477
Molecular Biology 482
DNA Technology 486
The Human Genome Project 488

Index...493

PREFACE

As the title suggests, this book examines efforts made throughout the span of history to understand the natural world. This book is not comprehensive by any means. It is a selection of topics, all from the history of western science, and they are chosen as much as anything in order to make a coherent, connected story. Too often texts that survey the history of science try for completeness and end up being encyclopedias, valuable as reference works but mind numbing for those new to the subject matter.

It has been my particular goal to relate the chosen episodes in science to each other by showing how they treat the same topics, methods of study, and criteria for acceptance and rejection of explanations. I have particularly stressed the implied logic of explanatory models, beginning with the ancient philosophers and continuing throughout the book. There is much mention of Plato's "saving the phenomena," of the Ad Hoc hypothesis, of the syllogism *modus tollens*, of the mechanist model and the axiomatic system.

The intended audience for this book is a college/university class exploring the sweep of science. Such classes often satisfy a science distribution requirement for students whose interests are primarily in the humanities. Accordingly, I have avoided taking the language of science and mathematics for granted and instead have tried to explain concepts in words that might be more compactly expressed in formulas. On the other hand, I have not shied away from discussing mathematics and mathematical formulations since they are so important for understanding why certain concepts were valued more highly than others.

This text is a direct outgrowth of my classes at York University and I have written this book with my own students in mind. At York, this book will be used as one of the texts in a course that spans an entire academic year. It could be used alongside a more traditional survey text if a complete overview is desired or in combination with some other texts that explore some of these topics more deeply.

The text is divided into four parts: Part I examines the roots of scientific thinking in the ancient world, particularly ancient Greece. Part II looks at the Scientific Revolution of the 16th and 17th centuries, concentrating on an examination of the major figures. Parts III and IV divide the subject matter of science since Newton. Part III follows developments in the physical sciences, mostly physics with a glance at 20th century astronomy. Part IV concerns the world of biology as it was transformed by the theory of evolution and by genetics.

Footnotes are provided for most direct quotations and a list of further readings is appended to each chapter, but no attempt is made to document every fact or assertion made. For the most part, the facts cited are well known and generally accepted by scholars in the field. I can be held responsible for any quirks of interpretation and emphasis, though most of the views I espouse have been expressed many times by others in the field. These are, after all, some of the major events in the history of science, about which much has been written. My hope for originality lies in the comparisons and analogies I try to make among these different developments, and if this book has any value it will be in helping others to see patterns in this evolution that make science make sense.

Though my name along appears on the title page, the production of this work has been a team effort involving long hours of editing my prolix and obfuscating prose by my wife Martha, followed by my son, Alex, putting the entire output, pictures, footnotes, and text into an attractive and legible layout. When we were not going about our separate tasks on this book, we were engaging in arguments about its content. All of this often in all night sessions. Any outright errors or incomprehensible statements that remain, however, are strictly my fault.

Byron E. Wall
Toronto
August 9, 2001

GLIMPSES OF REALITY

Episodes in the History of Science

The Ancient Egyptian Temple of Karnak.

Part One

Ancient Beginnings

Science in the form we recognize today emerged in the two-hundred-year period between 1500 and 1700 CE, now called the "Scientific Revolution." It did not spring out of nothing. Indeed most of the ideas that coalesced in the scientific revolution had been around for millennia.

The great ancient civilizations had an interest in nature and some very powerful tools for studying and analyzing it. Nature had to be objectified in order to begin to think about it systematically. Abstract thinking had to be developed in order to categorize and classify ideas. Speculative ideas about the material and form of the world had to be put forward for discussion and analysis. Logic and mathematics had to be developed in order to put different insights and observations into systems of ideas. And all the effort to do these things had to be seen to be worthwhile.

These steps were not taken all at once, nor did they follow each other in an orderly and predictable manner. The first stages of abstract thinking and the development of mathematical technique took place long before ideas began to be put forth about the general makeup of the world. Then in the relatively short space of only a few hundred years, almost all the basic questions that have concerned science ever since were asked and discussed. This amazing flowering did not happen all over the world, but instead was centered on just one ancient civilization, Greece. The Greeks took an objective interest in the world and scrutinized it to the best of their abilities.

Then, just as abruptly, the push toward science ideas came to a halt and other preoccupations took their place in much of the world. During this long period those who were interested in understanding nature spent most of their time sifting through ideas already expressed hundreds of years before.

One of the problems that beset the ancient struggle to understand the world was premature success. After a few centuries of mulling over the major questions about the world, a number of standard answers emerged. These were very well-reasoned analyses based on available information. They were compelling. Unfortunately, one of the by-products of convincing answers is that people stop asking the questions. And unfortunately, as far as a scientific understanding of nature was concerned, the answers produced by the ancients were generally wrong.

Aristotle was the leading thinker of antiquity and Aristotle expressed himself convincingly on almost every subject imaginable. It took a very long time to get past Aristotle's view of the world and begin to examine it all anew. It is, of course, much more complicated than this. A vast amount of the more scientific writings of ancient times were lost for centuries and only found again when interest was rekindled during the Renaissance. As far as later Greek (Hellenistic) science is concerned, the existing works were preserved, translated, and studied extensively in the Arab world, where many lines of thought were extended in important ways. But the lively, open-ended questioning from first principles that was the hallmark of ancient Greece lay dormant for over a thousand years in the Middle Ages.

In the chapters that follow in Part One, some of those essential steps and essential questions will be reviewed: the first glimmerings of mathematical calculation; speculations about the general nature and material of the world; some attempts at grand syntheses and their pitfalls; the powerful influence of Plato and Aristotle; the triumphant emergence of reliable mathematics with Euclid; and the application of complex mathematics to solving the perceived problems of astronomy. These are only some of the vital steps taken in ancient times, but they are among the most important. Out of these issues, the Renaissance fashioned science.

Part I
Ancient Beginnings

A world map from the 15th century based upon Ptolemy's Geography from the 2nd century.

Egyptian Hieroglyphs.

Chapter One

Numbers and Arithmetic

Even the most primitive of prehistoric, pre-agricultural cultures seem to have had some system of counting. This conclusion is based on a certain amount of archaeological evidence and also on comparative evidence from some of the few remaining present-day preliterate cultures. Spoken words or gestures stood for counters from one to five, or one to ten, or to twenty. These appear to be used in interchanges between people such as bargaining, or for enumerating a list of items.

About five thousand years ago, in several places in the world, agricultural civilizations had reached a level of complexity such that there was a perceived need for keeping track of such things as possessions, obligations, organizations, rules, and other aspects of organized life. These needs were met at somewhat different times and in different places by a most extraordinary invention of human ingenuity, probably the most extraordinary invention ever: writing.

Converting spoken language to a written format has had enormous consequences for civilization, making possible the recording and preserving of details that exceed what could be easily remembered by the local group, passing on information and beliefs to future generations and to other communities, and so much more. It is not difficult for us to see the value of written language once it has been developed and begins to be used. It is more difficult to see how it is that it came about in the first place. Since it developed in different places at roughly the same time—the same era anyway—it is reasonable to suppose that at the very beginning of writing there was some common and relatively simple problem that it solved.

It has been suggested that recording ownership might have been the original "written" statement. This would be comparable to the symbolic marks used by modern cattlemen to brand their animals. The mark had to be distinctive in order to distinguish the possession of one person from that of another, but did not necessarily correspond to a spoken word. One could imagine, for example, a symbol of some sort being carved into a piece of pottery, such as an urn, before it hardened to indicate that the urn, and its contents, belonged to a certain person. This makes sense since urns and other large containers were used to hold surplus grain, which was the chief form of wealth in early agricultural societies. There is considerable evidence in the form of remains of broken pottery from these early civilizations that they did carry such markings, which probably indicated ownership.

Another kind of marking which showed up at about the same time was one that indicated quantity. How much or how many of something that existed was also an important aspect of ownership. This might have been marked on containers to indicate the amount of their contents. If the marks that indicated ownership were the earliest form of written words, the marks that designated amounts were the earliest form of writing down numbers.

Originally, written numerals were surely just the results of counting, written down. But then they began to take on a life of their own. A shepherd might see that three sheep put together with five sheep makes eight sheep, but when there were characters to represent three and five, sooner or later someone will notice that three and five of anything makes eight. That is when counters become numbers, and numbers have properties all on their own, regardless of to what they refer.

I consider numbers to be the first significant abstraction on the road to scientific thinking. A person who has noted that 3+5=8 (of anything) may also notice that 5+3=8, or that 4+4=8, or that 6+2=8, and so on. Down this road lies arithmetic. Numbers are abstractions because they can refer to anything at all that may be counted. Arithmetic is abstract reasoning because it requires thinking about matters that exist only as ideas in the mind or as representational marks in written form. My view is that only a civilization that has a well-developed system of writing numbers and has discovered rules for manipulating those numbers (arithmetic) has the chance of moving on to the issues of science: the nature of the world, the underlying structure of nature, the regular and repeatable features of the world that can be summarized in what we call natural laws.

All the great ancient civilizations had well-developed number systems and arithmetical procedures to solve complex problems. We can learn something about how complex

Chapter One
Numbers and Arithmetic

abstract systems of thought are developed by looking at some of the early attempts to solve mathematical problems, especially if we choose those which had methods different from our own.

Here, I want to discuss some of the features of number systems and arithmetical calculations of two of the great ancient civilizations of the western world, Egypt and Babylonia. By the end of the great periods of these civilizations, they both had developed mathematics to a high degree and used it to solve problems of both practical and theoretical interest. They each approached the problem of writing numbers and calculating answers to problems in different ways. Here, I will introduce only the basics of each system.

Egypt

The ancient Egyptians had a complex writing system based on pictorial representations of ordinary words. This is the system of Hieroglyphs. It began with a direct correspondence between image and object. An image of a man was the word for man. An image of a bird stood for a bird. However in time, the system was generalized into one where the hieroglyphic symbols represented sounds that could be put together into syllables and words. It was nevertheless a very complex system that was mastered only by a small select group in society, the scribes, who did all the writing. Hieroglyphic writing was slow and cumbersome and tended to be reserved for ceremonial uses, such as to display messages on temple walls. Later, a short-hand version for everyday use was developed called hieratic. Where hieroglyphics were carved or painted on stone walls, hieratic was usually written on papyrus scrolls.

The Egyptian way of writing numbers fit in with this system of writing words. They had a number system with a base of ten, just as we do. That is, they counted up to ten units, then counted by tens, then by hundreds, then thousands, and so forth. However, we distinguish units from tens and hundreds, etc., by their place in a number—the larger units are to the left of smaller. For example, the number we write as 3062 we know to mean 3 thousands, no hundreds, six tens, and two units. We have different symbols for the units from zero to nine, and then we use the same symbols again to represent tens, but written one place to the left. Moreover, because we have a symbol for zero, we can show which digit is units, which is tens, hundreds, etc. Without a symbol for zero—a place holder—it would be impossible to distinguish 3062 from 3602 or 3620; they would all be written with a 3, a 6, and a 2, in the same order.

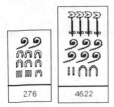
Egyptian numbers in hieroglyphs.

The Egyptians did not have a zero symbol, but then they did not need one because they had totally different symbols for units, tens, hundreds, etc. Units were written as a number of single strokes, from 1 to 9 of them. Tens were written with a figure that represented a hobble for cattle. The number 30 would be written as three hobbles. The figure for a hundred was a coil, a thousand was written as a lotus plant, ten thousand was a finger, a hundred thousand was a frog or tadpole, and a million was a god, with arms raised. Thus even without a zero, 3062 was not the same as 3602, but the down side was that for 3062 you needed three lotus plants, six hobbles, and two strokes.

Another feature of the Egyptian system is that the order in which the different levels of numbers are written is not essential. We invariably write the largest part of a number to the left and proceed to the right: thousands, hundreds, tens, and then ones. The Egyptians generally followed an order from highest to lowest, but sometimes wrote numbers from the top down, or from the right to the left or on several lines.

Ancient Egyptians wrote fractions, but curiously they only considered unit fractions (those with one in the numerator) such as 1/3, 1/5, 1/17 and so on. (An exception was made for 2/3.) A fraction was notated by placing the figure of a mouth above the number in the denominator.

Three fractions.

Later, when hieratic writing on papyrus became common another, more compact, system of writing numbers was introduced that had different symbols for 1, 2, 3, 4, and so

Chapter One
Numbers and Arithmetic

on. But then there were also separate symbols for 10, 20, 30, etc., and 200, 300, 400, and so on. There were therefore a great many symbols to be learned, but writing a number took much less space.

One of the most interesting features of Egyptian arithmetic was their system for multiplying and dividing numbers. Our own way of multiplying two numbers together (before the age of calculators!) requires knowing by heart the multiplication tables up to 9 x 9 and being able to add intermediate calculations together. The Egyptian system required only that the scribe know how to double a number (i.e., multiply by two) and add intermediate calculations. They each take about the same number of steps to reach an answer, and for many multiplication problems, the Egyptian system is more efficient.

Consider, for example, the task of multiplying 53 by 72. An Egyptian scribe might proceed as follows:

Write down one of the numbers (generally easiest to take the larger number) and beside it on the same row (and generally to the left of it), write the number 1, indicating that the number written down is equal to one of the number written at right. For simplicity, I will now switch to our own Arabic numerals.

<u>53 x 72</u>
1   72

Now double both numbers in this list and write the result on the next line, as follows:

<u>53 x 72</u>
1   72
2   144

Note that the scribe has to know what 2 x 72 is, and before we are finished will have to know what 2 times some much larger numbers are. But he does not need to know about 3 times, or 4 times, or any other times table.

You continue this process, doubling the last line and writing it below the others until there are numbers in the left-hand column that will add up to the other number, in the example case, to 53. You can always stop when the last number in the left-hand column is more than half of the number desired.

<u>53 x 72</u>

1	72
2	144
4	288
8	576
16	1152
32	2304

And this is far enough because 32 is more than half of 53. Now you find the numbers in the left-hand column that add up to 53 and mark them (it will always be possible to find numbers that work). It's easiest to find them if you work upward from the bottom.

<u>53 x 72</u>

✓	1	72
	2	144
✓	4	288
	8	576
✓	16	1152
✓	32	2304

And now, you add up the numbers in the right-hand column that are on lines with check marks.

<u>53 x 722</u>

✓	1	72	72
	2	144	
✓	4	288	288
	8	576	
✓	16	1152	1152
✓	32	2304	<u>2304</u>
			3816

Chapter One
Numbers and Arithmetic

This works because one 72 plus four 72s plus sixteen 72s plus thirty-two 72s is the same thing as fifty-three 72s.

The division process is similar, but uses the chart of doubles slightly differently. Suppose you wanted to know what 2307 divided by 115 is. This time you start with the number you are dividing into the other number and double it until the next doubling would be larger than the other number.

$$2307 \div 115$$

1	115
2	230
4	460
8	920
16	1840

And this is where you stop, because 1840 doubled is 3280, which is bigger than 2307. Now you try to find numbers in the right-hand column that add up to 2307 and put a tick mark by them. It will not always work out, so you get as close as you can without exceeding the number you want. Here, you would proceed as follows:

$$2307 \div 115$$

	1	115
	2	230
✓	4	460
	8	920
✓	16	1840

This time you tick the rows with the numbers 1840 and 460, which add to 2300. Any of the other possible numbers go higher than 2300, so they are not marked. The numbers in the left-hand column on the ticked lines are 4 and 16. Add them together to get 20. So the answer is that 2307 ÷ 115 = 20 with a remainder of 7. How the Egyptians dealt with the fraction represented by the remainder is too complicated to get into here.

Given the kind of arithmetical problems that scribes would have to solve, this was a very effective procedure. A typical multiplication problem would be measuring the area of a plot of land by multiplying its length and width. A typical division problem might be apportioning out some goods equally among the citizens. For the everyday problems of Egyptian life, this was workable.

Babylonia

A cuneiform tablet.

The ancient great civilization of Babylonia dates from about 2000 BCE, but it built upon the Sumerian and Akkadian civilizations that had occupied the same area as far back as 3500 BCE. All of these civilizations flourished in a part of the world that was called by the ancient Greeks "Mesopotamia," which means "between the rivers." The rivers in question were the Tigris and the Euphrates. Just as Egypt was made possible by the fertile waters of the Nile, Mesopotamia was a land with rich soil provided by the alluvial deposits laid down by the spring flood waters of these two rivers. Egypt and Mesopotamia provided some of the most fertile land for agriculture in the occupied world. This is the area referred to in antiquity as the Fertile Crescent. The central part of Mesopotamia, where the Sumerian, Akkadian, and later Babylonian civilizations were located, is now within the boundaries of Iraq. The Babylonian writing system derived from that originated by the Sumerians. The land had an abundance of soft clay which could be hardened if left to dry in the sun. There was also a ready supply of reeds that grew along the banks of the two rivers. The form of writing that developed consisted of making marks in a layer of wet clay, typically held in one hand, by pressing into the clay with the cut-off end of a reed. Thus, the marks made all had the shape of the stalk of the reed, a sort of an elongated triangle. Different combinations of marks imprinted at different angles had different meanings. This is called cuneiform writing. The wet clay was left to dry into a hard tablet. There are thousand and thousands of cuneiform tablets still in existence and preserved well enough to be deciphered.

The Babylonian number system is also Sumerian in origin. We may collectively speak of all of them as Mesopotamian numbers, though some differences in how they were

Chapter One
Numbers and Arithmetic

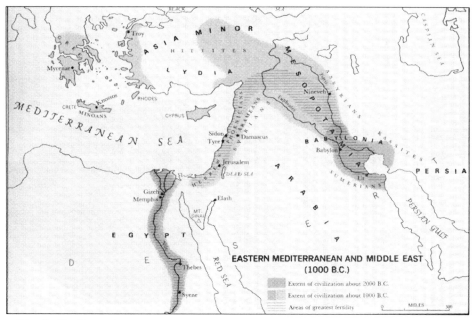

The Fertile Cresent.

written developed over time. The chief and most striking aspect of Mesopotamian numbers is that they did not use the familiar base of ten that most cultures use, including the Egyptian and our own. Instead, 60 was chosen as the base. That is, you count all the way up to 59 before going to the next higher unit. This is familiar to us because we still use this system for measuring time and angles. An hour (or an angular degree) has sixty minutes. Each minute has sixty seconds. The reason we use this peculiar system for time and angles is that we developed our way of measuring both time and angles from the Babylonians. Or, to be more precise, the Ancient Greeks used the Babylonian system and we have stuck with that.

A sexagesimal, or sixty-based, system may sound awkward, but is has a number of advantages over the more familiar ten-based system. For example, consider dividing something into exact fractional parts. An hour, for example: a half an hour is 30 minutes, a third is 20 minutes, a fourth is 15 minutes, a fifth is 12 minutes, a sixth is 10 minutes, a tenth is 6 minutes, a twelfth is 5 minutes, a fifteenth is 4 minutes, a twentieth is 3 minutes, and a thirtieth is 2 minutes. There are ten numbers that divide evenly into 60. In

Part I
Ancient Beginnings

The Babylonian numerals.

the ten-based system, only 2 and 5 go evenly into 10. For that reason, decimal fractions are often awkward affairs involving many digits..

Since we have separate symbols for all the digits from zero to nine, you might think that a sexagesimal system would be very cumbersome with sixty symbols to learn. But here is where cuneiform writing comes in. All numbers had to be made by pushing an identical stylus, made of a cross-section of a reed stalk, into a piece of wet clay. There was not much opportunity for variation. In fact, there were really only two distinct symbols: the reed pushed in vertically, and the reed pushed in horizontally. The first stood for the number one, and the second for the number ten. All other numbers were made up from combinations of these numbers. For example, the number twenty-three required two horizontal ten symbols followed by three vertical one symbols. By the time of the Babylonian civilization, a certain way of arranging the tens and ones to make up the numbers up to 59 had become standard.

Of course, the number 60 was simplicity itself, since it represented the beginning of the next higher order. Sixty was written the same as one, with a single vertical stroke, just as our number after nine is written with our number for one—and a zero after it to show it was one of the next higher category. This was a bit of a problem for the Babylonians because they did not have a symbol for zero. If you looked at the symbol for any number, you

Chapter One
Numbers and Arithmetic

could not tell for sure what the units were. To use the time analogy again, the number 45 by itself could mean either 45 seconds or 45 minutes. Generally, these problems were resolved by context. Just as we generally do not have to have it spelled out for us that if, say, a television set is priced at 230, that means $230, but a house that went for 230 would mean $230,000. Since in the sexagesimal system the next higher value represented by the same figures is sixty times the original, most times there is no confusion. That is not to say that there was *no* confusion. The lack of a zero was a severe drawback eventually.

Since the Babylonians used the same symbols over and over with different meanings, the order in which they were written was of the essence (unlike the Egyptian numbers which had different symbols for all different values). The rules were as follows: for numbers in the same order of magnitude (e.g., from 1 to 59), any ten symbols were to the left of one symbols. The number twenty-three was written as two tens, followed by three ones. Had they been written in the opposite order, the meaning would be (or at least could be) three 60s followed by two tens, in other words 180 +20 = 200. A more difficult problem is presented when the same symbol sits side by side with another having the same shape, but meaning 60 times more. For example two ten symbols side by side could mean 20, but it could also mean 10 x 60 + 10 = 610.

Though the easiest way for us to think about the Babylonian numbers is by thinking of time—hours, minutes, seconds—the system was used for more complicated numbers that may represent large or very small figures. As already mentioned, the sexagesimal system made many fractions easy to express. One fourth of 60 is 15; therefore one fourth of one would be expressed as 15/60. This would look exactly the same as the number 15. It would also look the same as the number that represented 15 times 60 or 15 times 60 x 60.

The Babylonians were very interested in astronomy and took very accurate measurements of the positions of stars and planets over a long period of time. The ancient Greeks, particularly the astronomers Hipparchus and Ptolemy, made extensive use of Babylonian astronomical records, which were, of course, expressed in their sexagesimal system. It is because these Greek astronomers used the Babylonian observations in their own systems that we still preserve the remnant of it in our way of reckoning time and angles.

For More Information

Alioto, Anthony M. *A History of Western Science,* 2nd ed. Englewood Cliffs, NJ: Prentice-Hall, 1987. Chapter 1.

Lindberg, David C. *The Beginnings of Western Science: The European Scientific Tradition in Philosophical, Religious, and Institutional Context, 600 B.C. to A.D. 1450.* Chicago: University of Chicago Press, 1992. Chapter 1.

MacLachlan, James. *Children of Prometheus: A History of Science and Technology,* 2nd ed. Toronto: Wall & Emerson, Inc., 2002. Chapter 3.

Neugebauer, Otto. *The Exact Sciences in Antiquity,* 2nd ed. New York: Dover, 1969.

O'Connor, J. J., and E. F. Robertson. "An overview of Babylonian mathematics" http://www-history.mcs.st-andrews.ac.uk/history/HistTopics/Babylonian_mathematics.html

_____. "Babylonian numerals," http://www-history.mcs.st-andrews.ac.uk/history/HistTopics/Babylonian_numerals.html

_____. "Egyptian numerals," http://www-history.mcs.st-andrews.ac.uk/history/HistTopics/Egyptian_numerals.html

_____. "Mathematics in Egyptian Papyri," http://www-history.mcs.st-andrews.ac.uk/history/HistTopics/Egyptian_papyri.html

Ronan, Colin A. Science: Its History and Development Among the World's Cultures. New York: Facts on File, 1985. Chapter 1.

Chapter One
Numbers and Arithmetic

Ruins at Ephesus.

Chapter Two

First Thoughts about Nature

The only way we know what happened in the past is if some record of it has persisted to the present day. We may be mistaken when we say that something started at a particular time and place, because whatever it is might have been preceded by other similar events of which we have no knowledge at all. Nevertheless, we work with what we have, and sometimes the historical record points to extraordinary events that appear to have been unprecedented.

Thus it is with the pre-Socratic philosophers of Ancient Greece. What we know about them is fragmentary. Almost none of their original writings have survived in the cases where they wrote down their thoughts at all. What we now have are bits and pieces of commentary and second-hand quotation in treatises written much later by other people. Nevertheless, there is a sufficient agreement among these later references that it is reasonable to conclude that they represent accurate assessments of much of what the pre-Socratics had to say.

They are called pre-Socratic philosophers because they are Greek thinkers who lived before first of the major Greek philosophers Socrates, his pupil Plato, and Plato's pupil Aristotle. In fact, much of what we know about the pre-Socratics are references in the works of Plato and Aristotle and other contemporaries of theirs.

Curiously, though ancient Greek philosophy is the foundation of scientific thinking, and while it is generally divided into two periods by the seminal work of Socrates, Socrates himself has almost no role to play in the history of science. What Socrates stood for so forcefully is the primacy of discovering truth as opposed to winning arguments. But Socrates was little interested in questions about nature. To him, we already knew what we needed to know about nature, and what we should put our minds to is how people should live. But back to the pre-Socratics.

What is so extraordinary about this handful of men who lived scattered around the Greek-speaking world over a period of nearly 200 years is that they asked questions that no one before them seems to have asked, and tried to provide answers to those questions based on reason. This is especially important for the history of science because the questions they asked were often about nature and their answers represented a groping toward a method for scientific inquiry.

And ultimately, whether these men were or were not the first to ask the questions that they did, what they did do is formulate a great many of the major problems that occupy science and, in a general way, propose all the possible answers to those questions. It is no accident that we know about these philosophers from the works of others. What each man said and became known for characterized an important viewpoint that had to be considered. It became the custom to refer to the different possible viewpoints by the person whom the later Greeks credited with first expounding each view. They, as it were, surveyed the landscape of science and marked all the major landmarks and headlands for others then to explore further.

What is the World Made Of?

The Ionians

Since Greece was not a unified state in antiquity, the term "Greeks" is applied to those peoples who spoke Greek as a native tongue and considered themselves to be Greek. They lived in scattered settlements on the mainland of what is now Greece, as well as in present day Italy, Sicily, Turkey, and other places around the Mediterranean. A group of colonies at the western edge of Anatolia or Asia Minor, which is now part of Turkey, constituted the area called Ionia. If any place on earth has a claim to be the birthplace of philosophy—at least Western philosophy—it is Ionia

Within Ionia were a number of important city-states where a considerable amount of commerce and trade was centered, some of it arriving and departing by sea into the Aegean, some of it overland from Asia Minor. The residents of Ionian towns were as likely as those anywhere to be familiar with ideas from all over the world.

Chapter Two | 21
First Thoughts about Nature

This map shows the center of the Greek world: the Aegean Sea. From the established civilizations on Crete and the Grecian Mainland, settlers spread to Ionia (the western edge of Asia Minor) and the rest of the Mediterranean.

I shall discuss briefly the lives and thoughts of several of the more important Ionian philosophers. The discussion of each is necessarily brief because so little is known about any of them.

Thales

The traditional "first philosopher" among the pre-Socratics is Thales of Miletos who lived from about 625 to 545 BCE. There is some evidence that Thales may have had Phoenician parents, which may have given him a head start in becoming aware of the rest of the world. Thales is credited with introducing the Greek world to Egyptian geometrical knowledge. He was also interested in astronomy and likely familiar with some of the extensive Babylonian work. The Greek historian Herodotus says that Thales predicted the solar eclipse of May 28, 585. The importance of this was that supposedly Thales was able to warn soldiers defending Ionia that the sun would disappear during the day but it was not to be feared, while the attacking enemy did not expect the eclipse and ran away in panic. Predicting solar eclipses is very complicated and probably Thales did not have nearly enough information to do so accurately, but it is one of the stories told about him to show what a wise fellow he was.

Thales also can lay claim to being the quintessential absent-minded professor. A story is told that as he was walking along observing the stars, he missed his footing and fell into a well.

> Theodorus, a witty and attractive Thracian servant-girl is said to have mocked Thales for falling into a well while he was observing the stars and gazing upwards; declaring that he was eager to know the things in the sky, but that what was behind him and just by his feet escaped his notice.[1]

However, Thales is also said to have exacted his revenge for this charge. Another probably wholly untrue or at least greatly exaggerated story is that Thales saw from his observations of the heavens that the coming spring was going to be a great year for the olive crop. So, long before the harvest, he went around to all the olive presses in the area and hired them out for the duration of the harvest season at the going rates. Then when the harvest came in and proved to be a bumper crop, Thales was able to lease time on the olive presses to the farmers at inflated prices and thereby make a killing. Supposedly the point of this story was that Thales wanted to demonstrate that philosophy does have practical value.

> For when they reproached him because of his poverty, as though philosophy were no use, it is said that, having observed through his study of the heavenly bodies that there would be a large olive-crop, he raised a little capital while it was still winter, and paid deposits on all the olive presses in Miletus and Chios, hiring them cheaply because no one bid against him. When the appropriate time came there was a sudden rush of requests for the presses; he then hired them out on his own terms and so made a large profit, thus demonstrating that it is easy for

---

1. From Plato, *Theatetus* 174A, quoted in Kirk and Raven, quotation 74.

Chapter Two
First Thoughts about Nature

philosophers to be rich, if they wish, but that it is not in this that they are interested.[2]

None of this would suffice to make Thales a candidate for the honour of being the "first philosopher." That rests on Thales having asked a question that seems to have been unasked before: what is the stuff of the world? The question carries with it the assumption that it makes sense to ask this, that there could be an answer we can comprehend, and that there might be ways that we can reason our way to the truth. In fact, it is the *question* that is more important than the answer. In general we can say that the importance of the pre-Socratics lies in the questions they chose to ask and the reasoning with which they reached answers. The answers themselves are not so important; in fact, by today's standards they seem silly and naïve.

In Thales' case, he asked what the primary matter of the world is from which everything else derives and came to the conclusion that it is *water*. What is really significant is that he thought it only made sense that the world, though it looked immensely varied, was ultimately and fundamentally all the same material, and that he could, through reason, find out what it was. Why did he pick water? For one thing, it was all around. He lived in a port city on the Aegean Sea. Water fell out of the sky. Water was essential for life. When water evaporated it left a residue behind, so even solid matter came out of water.

This opinion of Thales is reported by several later philosophers, who then go on to analyze the position. The most prominent of these is Aristotle. In his *Metaphysics* he discusses various views about the first principles of the world and says this about Thales:

> Most of the first philosophers thought that principles in the form of matter were the only principles of all things: for the original source of all existing things, that from which a thing first comes-into-being and into which it is finally destroyed, the substance persisting but changing in its qualities, this they declare is the element and first principle of existing things, and for this reason they consider that there is no absolute coming-to-be or passing away, on the ground that such a nature is always preserved…for there must be some natural substance, either one or more than one, from which the other things come-into-being, while it is preserved. Over the number, however, and the form of this kind of principle they do not all agree; but Thales, the founder of this type of philosophy, says that it is

---

2. Aristotle, *Politics* A11, 1259a9, quoted in Kirk and Raven, quotation 75.

water (and therefore declared that the earth is on water), perhaps taking this supposition from seeing the nurture of all things to be moist, and the warm itself coming-to-be from this and living by this (that from which they come-to-be being the principle of all things)—taking the supposition both from this and from the seeds of all things having a moist nature, water being the natural principle of moist things.[3]

Anaximander

If nothing else, Thales' speculation started a process of trying to answer this question about the fundamental nature of the world. A generation later, in the same city, Miletos, another philosopher attacked the same issue. This was Anaximander, who was probably either a pupil of Thales or his companion. He lived approximately from 611 to 547 BCE.

Anaximander studied geography and astronomy and is said to have invented the sundial and produced one of the first maps of the known world. He is also supposed to have written down his philosophical views in a book, which may have been the first Greek philosophy treatise, but it has been lost, so, like Thales, we rely on references to his ideas in the works of others. Anaximander expressed his views on a number of matters: the shape and size of the earth, the construction of the heavens, the distance of the sun and moon, the origin of the human species, the cycle of life and death, and other issues. These are interesting speculations in themselves, but here I want to focus on just one of his views, that of the primary substance of the world. This is the issue raised by Thales, who concluded that water was the material from which all else is made. Thales had chosen a visible, tangible substance with which everyone was familiar. Anaximander concluded that whatever the primary substance was, it was beyond our immediate perception.

Anaximander called the primary matter *apeiron* (in Greek, απειρων), often translated as the "boundless" or the "unlimited" or the "infinite." Anaximander believed that this apeiron was too small and rarefied to be detected by human senses but this was what underlay all material. It was an important idea that had to be considered by anyone trying to state what the ultimate material was. He concluded that something must exist that could not be perceived, because that was the only way to make sense of matters. Whether it was true or not, it was a possibility which needed to be considered. Once Anaximander expressed this view, it was no longer possible to ignore it.

---

3. Aristotle, *Metaphysics*, A3; 983b6, quoted in Kirk and Raven, quotation 87.

Chapter Two
First Thoughts about Nature

Here is one of the references to Anaximander's thought in later writing:

> Anaximander of Miletus, son of Praxiades, successor and disciple of Thales, said that the "ultimate source and first principle" as well as the primary substance is the Unlimited; he was the first to apply this name to the ultimate source. He maintained that it is neither water nor any other of the so-called elements, but is of an altogether different nature from them, in that it is unlimited [i.e., is not limited to being just this or that]. From it there arose the universe and all the worlds within it.
>
> "The Unlimited is the first-principle of things that are. It is that from which the coming-to-be [of things and qualities] takes place, and it is that into which they return when they perish, by moral necessity, giving satisfaction to one another and making reparation for their injustice, according to the order of time."
>
> Evidently since he sees the four elements changing into one another he does not think it right to identify the underlying reality with any single one of them; it must be something distinct. Coming-to-be, he holds, does not involve any alteration of basic substance; it results from the separation of opposites which the eternal motion causes.[4]

Anaximenes

Before leaving Miletos, there is one more philosopher to mention who contributed to the same debate. This is Anaximenes, possibly a student of Anaximander. Anaximenes' dates are approximately 550 to 475 BCE. Little is known about his life. His importance stems from his continuation of the debate over the fundamental stuff of the universe. Thales had named a familiar and abundant substance, water, as that from which all was made. Anaximander viewed that as too simple and literal. He opened the possibility that there was an invisible substratum, the *apeiron*, beneath the level of direct perception, that was the fundamental constituent of things.

Anaximenes' position was an interesting compromise. He picked air as the fundamental substance. Like water, it was familiar to everyone and was all around, but like the *apeiron* it was extremely rarefied and, in effect, invisible. The properties he assigned to air were similar to those that Anaximander assigned to the *apeiron*, but with one very significant difference. Since air is a tangible, identifiable part of everyday life, and since it can be perceived as wind and as heat, it is not totally inaccessible to experience.

---

4. From Simplicius Commentaria, quoted in Philip Wheelwright, ed., *The Presocratics* (Englewood Cliffs, NJ: Macmillan, 1966), p. 54-56.

Part I
Ancient Beginnings

The trouble with the totally imperceptible *apeiron* was that virtually anything could be attributed to it with no way of testing whether a claim was true. But with air, one could at least argue from the effects. Nevertheless, Anaximenes' own arguments get pretty stretched and sound silly to us. For example, he claimed that compressed air was cold while expanded air was hot. To show that this was true, he proposed the following test: Purse the lips and blow a jet of air on your hand. It will feel cool. Now, open your mouth wide and breath out onto your hand. It will feel warm. The explanation may be badly conceived and confuse one cause with another, but it does introduce a new idea into thinking about nature: an empirical test. Explanations should be supported by evidence.

Anaximenes is also notable for introducing the cosmological view that came to dominate ancient science, that the universe was a huge sphere that rotated daily. On that sphere the stars were fixed at the outer edge, the earth was at the center (immobile of course), and the planets roamed around somewhere in the space between the earth and the fixed stars.

Here are a few summaries of Anaximenes' thoughts from the works of later philosophers. The first is the basic statement of air as the fundamental substance. The second mentions the empirical "test" that demonstrated that condensed air is cold and rarefied air is hot. The third expands on the properties of air and how they account for many phenomena, and also includes his thoughts on the structure of the universe

> Anaximenes, son of Eurystratus, of Miletus, a companion of Anaximander, also says that the underlying nature is one and infinite like him, but not undefined as Anaximander said but definite, for he identifies it as air, and it differs in its substantial nature by rarity and density. Being made finer it becomes fire, being made thicker it becomes wind, then cloud, then (when thickened still more) water, then earth, then stones; and the rest come into being from these. He, too, makes motion eternal, and says that change, also comes about through it.[5]

> ... or as Anaximenes thought of old, let us leave neither the cold nor the hot as belonging to substance, but as common dispositions of matter that supervene on changes; for he says that matter which is compressed and condensed is cold, while that which is fine and "relaxed" (using this very word) is hot. Therefore, he said, the dictum is not an unreasonable one,

---

5. Theophrastus, quoted by Simplicius, quoted in Kirk and Raven, quotation number 143.

Chapter Two
First Thoughts about Nature

that man releases both warmth and cold from his mouth: for the breath is chilled by being compressed and condensed with the lips, but when the mouth is loosened the breath escapes and becomes warm through its rarity.[6]

Anaximenes…said that infinite air was the principle, from which the things that are becoming, and that are, and that shall be, and gods and things divine, all come into being, and the rest from its products. The form of air is of this kind: whenever it is most equable it is invisible to sight, but is revealed by the cold and the hot and the damp and by movement. It is always in motion: for things that change do not change unless there be movement. Through becoming denser or finer it has different appearances; for when it is dissolved into what is finer it becomes fire, while winds, again, are air that is becoming condensed, and cloud is produced from air by felting. When it is condensed still more, water is produced; with a further degree of condensation earth is produced, and when condensed as far as possible, stones. The result is that the most influential components of generation are opposites, hot and cold. …

He says that the heavenly bodies do not move under the earth, as others have supposed, but round it, just as if a felt cap turns round our head; and that the sun is hidden not by being under the earth, but through being covered by the higher parts of the earth and through its increased distance from us.[7]

Change or the Lack of It

Heraclitos

But maybe the Milesian philosophers were trying to answer the wrong question. Maybe the world is not "made of" something at all, but instead is merely a state of affairs balanced between opposites and always in tension. This view was expounded by Heraclitos, a contemporary of Anaximenes, who lived in Ephesus, a city-state about 50 km north of Miletos in Ionia. Heraclitos was not a very friendly person and had no students to carry on his work. Among the characteristics he is remembered for is a propensity to engage people in arguments to show what fools they were. Despite his lack of disciples, he exerted a great influence on Greek thinking because of a small book he wrote that was much read and quoted.

---

6. Plutarch, quoted in Kirk and Raven, quotation number 146.
7. Hippolytus, quoted in Kirk and Raven, quotation number 144 and 159.

On Heraclitos himself, Diogenes Laertius wrote:

> He grew up to be exceptionally haughty and supercilious, as is clear also from his book, in which he says: "Learning of many things does not teach intelligence; if so it would have taught Hesiod and Pythagoras, and again Xenophanes and Hecataeus."... Finally he became a misanthrope, withdrew from the world, and lived in the mountains feeding on grasses and plants.[8]

Heraclitos is best remembered for the aphorism "you can't step in the same river twice." There are various references to this in later literature. Plato wrote:

> Heraclitus somewhere says that all things are in process and nothing stays still, and likening existing things to the stream of a river he says that you would not step twice into the same river.[9]

A river is an entity that we think of as a thing and give it a name that we apply to it day after day, but in reality a river is a process of water moving downstream. The river experienced at one moment is never the same as the river experienced in another moment. More generally, any "thing" that we recognize in the world and think of as stable is constantly changing. That this is true of living things is clear, they are born, live, and die, and are never really the same from moment to moment. Stability, according to Heraclitos, is an illusion.

Heraclitos' view unnerved many Greek philosophers and caused them to think further in their efforts to explain stability and change in existence. Some of the most provocative thought came as an effort to refute Heraclitos; other important developments came from attempts to incorporate his thinking in larger systems.

Parmenides

Ionia was not always tolerant of its philosophers. Some of them had to flee for their lives when their views were seen as heretical or a threat to the prevailing social order. Such was the case with Xenophanes, who fled from Colophon in Ionia to the southern coast of Italy where he spent the remainder of his life in the Greek colony of Elea. There he became the teacher of Parmenides who was born in Elea in about 510 BCE.

---

8. From G. S. Kirk and J. E. Raven, *The Presocratic Philosophers: A Critical History with a Selection of Texts,* quotation number 193.
9. Plato, *Cratylus* 402A. Quoted in Kirk and Raven, quotation number 218.

Chapter Two
First Thoughts about Nature

Parmenides, more than any other Greek philosopher before Socrates, was committed to the notion that the ultimate good is to attain truth, and truth is to be reached through reason. Truth, or reality, was to be distinguished from appearances. Mere perception and sensation proffer appearances, but reality can only be reached through the application of reason and logic. Nature, he believed, is logical and understandable. The application of precise reasoning will lead one away from false appearances to the necessary truth even though that truth may defy common sense.

A key tool in Parmenides' logical analysis is the law of contradiction—the principle of the excluded middle. In brief, this is the assertion that a thing either is or it is not. A statement is either true or false; there is no middle ground. Applying this to the world at large, Parmenides argues that we cannot say that non-being exists. For example, it is a contradiction to speak of empty space with nothing in it, because that would imply that the "nothing" exists. Consequently, he is led to the view that the universe is "full," meaning that there is no space in it which is empty.

But this makes it very difficult to conceive of change. Nothing can change from non-being to being, because that would imply that it came from nothing. Nothing can move from place to place, because there is no empty place for anything to move into. Change is therefore impossible.

Like most pre-Socratic philosophers, Parmenides' original work is lost to us, but fortunately he was considered important enough that later writers quoted him at length, so much that we have more of Parmenides' writings passed on to us than of any other pre-Socratic. On the other hand, Parmenides wrote exclusively in hexameter verse and his attempt to put his difficult arguments into verse has made some passages exceedingly obscure. Here, for example, is the main passage in Parmenides' work that argues that everything is motionless:

> But motionless within the limits of mighty bonds, it is without beginning or end, since coming into being and perishing have been driven far away, cast out by true belief. Abiding the same in the same place it rests by itself, and so abides firm where it is; for strong Necessity holds it firm within the bonds of the limit that keeps it back on every side, because it is not lawful that what is should be unlimited; for it is not in need—if it were, it would need all. ... But since there is a furthest limit, it is bounded on every side, like the bulk of a well-rounded sphere, from the center

> equally balanced in every direction; for it needs must not be somewhat more here or somewhat less there. For neither is there that which is not, which might stop it from meeting its like, nor can what is be more here and less there than what is, since it is all inviolate; for being equal to itself on every side, it rests uniformly within its limits.[10]

This is clearly an attack on Heraclitos' position that change is all there is. Together, Heraclitos and Parmenides represent the opposite ends of a very basic problem that Greek philosophers were wrestling with: stability and change. To Heraclitos, all stability was an illusion and change was the only reality. To Parmenides it was just the reverse. Both appear to be extreme positions, but happy mediums are hard to find.

Zeno

Lest the Greek philosophers think these problems were easily resolvable, a student of Parmenides, Zeno, born in 495 BCE in Elea, showed that seemingly impeccable logic supports the view that change is impossible, no matter what ordinary experience would lead one to believe. Zeno may have moved to Athens and taught there for a time, which would have made his and Parmenides' views much more readily available.

Zeno showed the conflict between logical reasoning and common sense using paradoxes where logic points to one inescapable conclusion while common sense suggests that a different conclusion is just as inescapable. Here are three of his puzzles, along with Aristotle's attempts to solve and refute them by finding flaws in the reasoning. Aristotle's convoluted and not entirely successful arguments against Zeno serve only to show how problematic these paradoxes are.

The Stadium

Consider a stadium and a solitary athlete running from one end to the other. Will the athlete ever reach the other end of the stadium? No, says Zeno.

The reasoning is that for the athlete to reach the end of the stadium, he must first reach some point along the way, for example, halfway. Can he reach halfway? Well, before he can reach halfway, he must reach ¼ of the way. Before he can reach ¼, he must reach 1/8, and so on. Before the runner can reach any point in the direction of the end, he must

---

10. Quoted by Simplicius, reproduced in Kirk and Raven, quotations 350 and 351.

Chapter Two
First Thoughts about Nature

reach a point before that. There are an infinity of points before any point you choose that the runner must reach first. Therefore the runner can never get to the end.

Aristotle's rejoinder is as follows:

> Zeno's argument makes a false assumption in asserting that it is impossible for a thing to pass over or severally come in contact with infinite things in a finite time. For there are two senses in which length and time and generally anything continuous are called "infinite": they are called so either in respect of divisibility or in respect of their extremities. So while a thing in a finite time cannot come in contact with things quantitatively infinite, it can come in contact with things infinite in respect of divisibility: for in this sense the time itself is also infinite: and so we find that the time occupied by the passage over the infinite is not a finite but an infinite time, and the contact with the infinites is made by means of moments not finite but infinite in number.[11]

Achilles and the Tortoise

Consider a race, says Zeno, between Achilles, the fleet-footed warrior hero of Homer's Iliad, and a tortoise. Achilles is clearly a much faster runner than the ponderous tortoise. In the interest of making the race a more even contest, Achilles gives the tortoise a head start. It doesn't really matter how much of a head start, but for the sake of argument, let's say it was one-half of the distance to the finish line.

The race starts and both Achilles and the tortoise begin at the same moment, and continue moving at their same respective rates throughout the race. (For the paradox it's only necessary that they keep moving; it's not necessary that they maintain constant speeds.)

Now, we expect that Achilles will catch up to the tortoise and pass it, but Zeno shows that this cannot happen. For, before Achilles can reach the tortoise, he must first get to where the tortoise started the race, call that point T_0. When Achilles reaches point T_0, the tortoise is no longer there, since the tortoise has been moving for as long as Achilles has. The tortoise is at a new location; call it T_1. Achilles must now get to point T_1 before he can pass the tortoise, but when Achilles reaches T_1, the tortoise is at T_2. This

---

11. Aristotle, *Physics* Z2, 233a21, cited in Kirk and Raven, quotation 372.

pattern continues indefinitely. Achilles must always reach where the tortoise was, but then the tortoise, who is constantly moving, will always be ahead of that.

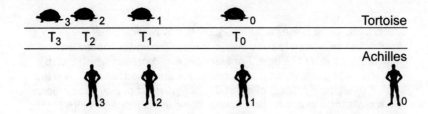

It is tempting to argue that the distances between successive tortoise positions (T_0, T_1, T_2, T_3, ...) gets smaller each time, but it is really of no significance. The tortoise will always be ahead. Yet we know from experience that fast runners can overtake slow runners. What is the message here?

Are we being duped by our senses and actually fast runners never overtake slow runners? Is there something wrong with the logic that is leading to false conclusions? Or is there something inherently inconsistent with logic itself that allows weird conclusions like this to emerge from impeccable reasoning?

Aristotle believed he could dispose of this paradox the same way as he did the first:

> The second is the so-called Achilles, and it amounts to this, that in a race the quickest runner can never overtake the slowest, since the pursuer must first reach the point whence the pursued started, so that the slower must always hold a lead. This argument is the same in principle as that which depends on bisection, though it differs from it in that the spaces with which we successively have to deal are not divided into halves.[12]

The Flying Arrow

The first two paradoxes set out to prove that continuous motion in infinitely divisible space is impossible. In the next paradox, Zeno attacks the concept of motion if time consists of indivisible moments. Suppose that an archer is shooting an arrow at a target some distance away. When the arrow is in flight, is it moving? Zeno says no.

---

12. Aristotle, *Physics* Z9, 239b30, cited in Kirk and Raven, quotation 373.

Chapter Two
First Thoughts about Nature

The arrow cannot be moving because at any moment it occupies a space equal to its own dimensions. An object is at rest when it occupies a space equal to itself. At any point in its flight, the arrow will be occupying a space equal to itself and will therefore be at rest.

Or look at this another way. At any point on the trajectory between archer and target, if the arrow is there it is there at a particular instant of time. But an instant of time has no "width" as it were, so the arrow must be at rest at that instant. Whatever position, or whatever instant you choose, the arrow is at rest. Therefore, the arrow does not move. Motion is impossible.

Aristotle's rejoinder is that this argument only works if time is composed of discrete moments:

> The third is that already given above, to the effect that the flying arrow is at rest, which result follows from the assumption that time is composed of moments: if this assumption is not granted, the conclusion will not follow.[13]

However, if you follow Aristotle here and say that time is not composed of indivisible moments, they you are back to the problems of the first two paradoxes. Zeno gave philosophers much to think about.

◆

The philosophers discussed in this chapter all lived within a two-hundred-year period and within one culture. They sprang up all of a sudden and began asking these very difficult questions about the world. These are among the issues that remain central to the scientific investigation of nature. And, these are only a few of the pre-Socratics that could have been mentioned here. Quite suddenly, serious thinkers had a lot of issues to work with and a lot of the basic groundwork of their subsequent discussion had been laid out.

---

13. Aristotle, *Physics* Z9, 239b30, cited in Kirk and Raven, quotation 374.

For More Information

Alioto, Anthony M. *A History of Western Science,* 2nd ed. Englewood Cliffs, NJ: Prentice-Hall, 1987. Chapter 2.

Clagett, Marshal. *Greek Science in Antiquity.* New York: Collier, 1963.

Kirk, G. S., and J. E. Raven. *The Presocratic Philosophers: A Critical History with a Selection of Texts.* Cambridge: Cambridge University Press, 1963.

Lindberg, David C. *The Beginnings of Western Science: The European Scientific Tradition in Philosophical, Religious, and Institutional Context, 600 B.C. to A.D. 1450.* Chicago: University of Chicago Press, 1992. Chapter 2.

Lloyd, G. E. R. *Early Greek Science: Thales to Aristotle.* London: Chatto & Windus, 1970.

_____. *Magic, Reason and Experience: Studies in the Origin and Development of Greek Science.* New York: Cambridge University Press, 1979.

Neugebauer, Otto. *The Exact Sciences in Antiquity,* 2nd ed. New York: Dover, 1969.

Ronan, Colin A. *Science: Its History and Development Among the World's Cultures.* New York: Facts on File, 1985. Chapter 2.

Wheelwright, Philip, ed. *The Presocratics.* New York: Macmillan, 1996.

Chapter Two | 35
First Thoughts about Nature

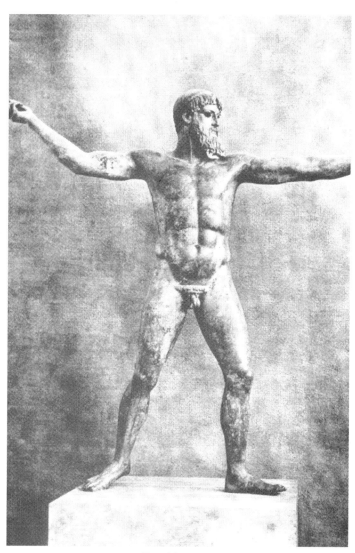

Poseidon

Chapter Three

Pythagoras and the Magic of Numbers

The pre-Socratic philosophers laid out many of the issues that arise in thinking about the world and trying to gain knowledge of it. Among these none is more vital than the question of the nature of what is real. Is reality something that is directly perceived and experienced, or is it only accessible through the mind? These are fundamental, and essentially irreconcilable and totally contrary viewpoints about nature and knowledge of nature. The pre-Socratic ferment provided some groundwork but these contrasting views were not clearly and forcibly expressed until the work of Plato and Aristotle.

Pythagoras.

Before getting to either of these two giants, there is one more pre-Socratic that must be discussed, Pythagoras of Samos. Pythagoras is important enough for the development of science that he deserves a chapter of his own. Pythagoras was born sometime between 580 and 569 BCE and died around 500. Like Xenophanes, he had to flee his homeland to escape persecution for his beliefs, and he may have spent a considerable amount of his life in hiding. Pythagoras lived before several of the pre-Socratics discussed in the previous chapter. His thought influenced a number of them, including Heraclitos, Parmenides, and Zeno. Pythagoras founded a cult of followers, a religious cult of sorts, the members of which devoted themselves to Pythagoras' teachings, gave up all personal possessions, went barefoot, wore special clothing, and had a special diet. It is impossible to distinguish the teachings of Pythagoras himself from the general beliefs of the Pythagoreans. Almost all references in later literature are to the Pythagoreans rather than to Pythagoras himself.

What makes the Pythagoreans so extraordinary is their view of number. Unlike many ancients who viewed numbers as abstractions for counting or measuring things, the Pythagoreans gave number the primary role in existence. Everything that was, was in some sense number. It is as though Pythagoras continued the discussion of Thales, Anaximander, and Anaximenes by asserting that the fundamental substance of the universe was number. But in contrast to these Milesians, number was not some kind of material. Number did take up space and define the geometry of the world. Number, in short, was the ultimate reality. It's a truly extraordinary view. Pythagoreans appeared to believe that everything, ultimately, was a geometrical configuration of numbers.

To return to Pythagoras himself, some of the stories told about him help to explain this conviction that number underlies all. The first is the relationship that Pythagoras is said to have discovered between mathematics and music.

Mathematics and Music

Music was clearly an art form. As an art form there were aesthetic conventions in Pythagoras' time as to what made pleasing sounds. The chief instruments of ancient Greece were the flute and the lyre. Greek music was composed in the "Greek" modes, as we call them today. These were arrangements of seven notes in a scale that then repeated an octave higher or lower. These modes correspond to the sequences of notes on the white keys of a modern piano, each different mode beginning on a different note. The differences between the modes were accounted for by the relative placement of "half-steps" in the scale. For example, the notes on a piano beginning on C and progressing up to the next C, which we recognize as a major scale, constituted the Ionian mode, named after the Ionian region of Greece. The sequence beginning from A and going to the next A was the Aeolian mode, our natural minor scale, from D to D was the Dorian mode, and so on. Melodies in each mode had a distinctive quality, but common to almost all of them was the relationship between the beginning note and the fourth note (the interval we call a perfect fourth) and the beginning note and the fifth note (which we call a perfect fifth). As they are today, these intervals have a special place in music and are generally considered consonant or pleasing to the ear. (The perfect fourth was considered a consonance in antiquity more than it is today.)

None of this had anything to do with mathematics or so it seemed in Pythagoras' time. But Pythagoras is said to have accidentally discovered a mathematical relationship that surprised him. In the most common version of the story, Pythagoras was walking through town one day when he passed a blacksmith's shop. The blacksmith was hammer-

Chapter Three
Pythagoras and the Magic of Numbers

ing on an iron rod that lay across his workbench or anvil. Pythagoras noticed that as the workers in the blacksmith's shop hammered on rods, a different sound was made by each hammer. Wondering why that should be so, Pythagoras weighed the different hammers and found a pattern. Boethius described this in his work *On Music (De Musica)* as follows:

> [T]hose two which gave the consonance of an octave were found to weigh in the ration 2 to 1. He took that one which was double the other and found that its weight was four-thirds the weight of a hammer with which it gave the consonance of a fourth. Again he found that this same hammer was three-halves the weight of a hammer with which it gave the consonance of a fifth. Now the two hammers to which the aforesaid hammers had been shown to bear the ration of 4 to 3 and 3 to 2, respectively, were found to bear each other the ration of 9 to 8. The fifth hammer was rejected, for it made no consonance with the others.[1]

What Pythagoras had discovered was a quantitative physical connection to the quality that is sound. And to make it more important, the experience of sound was an æsthetic one since it was related to harmonic musical intervals. Music was viewed as having mystical qualities that touched the soul. To find the pleasing musical intervals represented by simple numerical ratios of weights was a momentous discovery about the structure of the world.

In this illustration in a 12th century manuscript, Boethius is pictured strumming a monochord while Pythagoras is weighing hammers with one hand and testing the different notes obtained from each with the other hand.

These ratios held not just for vibrating hammers, but also for lengths of stretched cord (as on a lyre) or columns of vibrating air (as in a flute). In hindsight it is hard to imagine that these simple numerical ratios were not noticed before—by the makers of the

---

1. Quoted in Penelope Gouk, "The Harmonic Roots of Newtonian Science." In John Fauvel et al., *Let Newton Be! A New Perspective on His Life and Works.* (New York: Oxford University Press, 1988), p. 105.

Part I
Ancient Beginnings

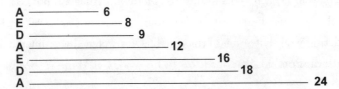

Pitch as a result of relative string length. Compare the length of each string (right) with the note produced when it is plucked.

instruments, for example. But the instruments were probably made by artisans who tuned them by ear and paid no attention to any numerical coincidences..

Pythagoras seems to have believed that he had seen into the mysteries of nature and discovered that their key was number—and fairly simple numbers at that. Once Pythagoras got the idea that numbers represented the key to understanding, he began to find numerical relationships everywhere and attribute extraordinary meaning to them.

Having discovered numerical ratios that "explained" music, Pythagoras went on to apply these same ratios to other things and thereby find great meaning in them. Consider once again the harmonic ratios for the octave and the perfect fourth: 2 to 1 and 4 to 3. These ratios are the same as the ratios 12 to 6 and 8 to 6. (And the perfect fifth can be represented by the ratios 12 to 8.) If you took, for example, a stretched string (say a violin string) of length 12 and plucked or bowed it, you would get a certain note, say, A. If you then held it down at the mid-point, so that a string of length 6 was free to vibrate, then the sound produced would be the A at the next higher octave. If instead you held the string down so that 2/3 of it, a length of 8, could vibrate, you would get the sound E, a perfect fifth above the original A. So, the numbers 12, 8, and 6, which were the respective lengths of string needed to produce these sounds, are in what Pythagoras called a harmonic progression.

Simple Numbers, Geometrical Shapes

So far this is still applying numbers to music, but for Pythagoras, he had now discovered something magical about the numbers 12, 8, and 6 taken together, and he began to look for other examples of this number pattern. He did not have to look far, because one of the simpler shapes in geometry involves just these numbers, the cube.

Chapter Three
Pythagoras and the Magic of Numbers

What is a cube? It is a regular solid having exactly 12 edges, 8 corners, and 6 faces. Therefore, according to Pythagoras, it is in "geometrical harmony." Once having started down this road, there was no stopping Pythagoras (or the Pythagoreans—we can't distinguish them here) from looking for and finding all sorts of simple numerical relationships in physical objects and attributing great significance to them.

Take for example, figurate numbers. These are numbers that can be arranged to form simple figures. By the way, the term "number" here always refers to simple, counting numbers, that is, positive integers. There are different simple figures that can be formed by taking a given number of place holders—think of marbles or some other uniform shape—and arranging them to make a figure. The number three can make a triangle, for example, so therefore three is a "triangular" number. Four counters can be arranged into a square, so therefore four is a "square" number. One can make many other regular figures by arranging counters tightly together; for example, seven can make a hexagon, and so on. Also, many other numbers can be made into the same figures. Ten is also a triangular number, so is fifteen. And we are all familiar with larger square numbers, such as nine, sixteen, twenty-five, thirty-six, and so on.

For Pythagoras, these were not coincidences; they had deep significance. Take the number ten for example. Ten is, of course, a triangular number because you can make an equilateral triangle out of ten counters that has four counters on each side. Because the figure had four on a side, it was called the tetrad, or *tetractys*.

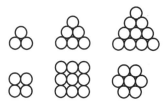

Figurate Numbers. Note the Tetrad in the upper right.

The Greek number system was based on the number ten, as were most number systems. (The Mesopotamian-Babylonian system was the chief exception.) For Pythagoras, this was not just a consequence of human beings having ten fingers; it had mystical significance. There must be something about the number ten, the *tetractys* or tetrad, that made it very special. One of the rationalizations of this attributed to the Pythagoreans is as follows:

> Ten is the very nature of number. All Greeks and all barbarians alike count up to ten, and having reached ten revert again to the unity. And again, Pythagoras maintains, the power of the number 10 lies in the number 4, the tetrad. This is the reason: if one starts at the unit (1) and adds the successive number up to 4, one will make up the number 10 (1+2+3+4=10). And if one exceeds the tetrad, one will exceed 10 too... So that the number by the unit resides in the number 10, but potentially in the number 4. And so the Pythagoreans used to invoke the Tetrad as their most binding oath: 'By him that gave to our generation the Tetractys, which contains the fount and root of eternal nature...'[2]

The *tetractys* was more than a sacred number to the Pythagoreans, it was an essential feature of the structure of the universe; they even went so far as to say that there were ten heavenly bodies, though only nine were accounted for. For their time Pythagoreans had a rather unusual notion about the structure of the universe. They did not place the Earth at the center as almost every other ancient did. (This is a further testimony to their willingness to suspend common-sense observation in favour of an arcane mystical explanation.) The Pythagorean world view was something like this:

The Earth was considered too imperfect to be at the center of the cosmos. Instead, the center was occupied by a "central fire"—the watchtower of Zeus, the chief god of the Greek heavens. Then, proceeding outward, followed the Earth, the Moon, the Sun, the five known planets (in our terminology: Mercury, Venus, Mars, Jupiter, Saturn), and ultimately, the shell containing the fixed stars. That makes nine bodies. But the *tetractys* is ten, so there must be another. They postulated that the tenth body was the *antichthon* or counter-earth, which lay on the opposite side of the central fire from the Earth. They imagined that the Earth was always turned away from the central fire (at least Greece was always turned away from it) so that it was invisible, and since the *antichthon* was across the central fire, it was also invisible. Thus, the sanctity of the number ten was preserved.

The Pythagorean Theorem

Legend has it that Pythagoras himself discovered the mathematical theorem that has been named after him, namely: that for any right triangle, squares built on the sides enclosing the right angle are equal in area to the square that can be built on the side opposite the right angle, the hypotenuse.

---

2. Aetius I. 3.8, quoted in Kirk and Raven, quotation 280.

Chapter Three
Pythagoras and the Magic of Numbers

In the case of a right triangle with sides *a* and *b* enclosing the right angle and *c* being across from the right angle, this asserts that $a^2 + b^2 = c^2$.

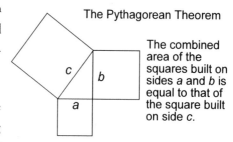

The Pythagorean Theorem

The combined area of the squares built on sides *a* and *b* is equal to that of the square built on side *c*.

Special cases of this relationship were known in both Egypt and Babylonia long before Pythagoras, but supposedly he found some sort of proof that it was generally true. What his proof was, we no longer know. But the special cases are startling enough. If a right triangle has sides of 3 and 4, the hypotenuse must be five because $3^2 + 4^2 = 5^2$. Likewise, $5^2 + 12^2 = 13^2$. These are simple numbers that underlie and determine a geometrical structure, just as the simple numerical ratios that determine musical harmonies.

Everything is Number

These marvelous numerical relationships seemed so powerful that the Pythagoreans were led to the natural conclusion, given their mystical beliefs, that number was the explanation for everything. Or, to put it more strongly, that number *was* the ultimate reality. Because numbers were thought of as small, regular, space-filling objects, the Pythagoreans could view number as the primary elements of the universe. This is another answer to the question Thales asked about the makeup of the world: it is all number. Aristotle summed up this Pythagorean outlook in his *Metaphysics* as follows:

> [S]ince by the very nature of mathematics it is numbers that stand first among the basic principles, it seemed to the Pythagoreans that they could discover in numbers, more truly than in fire or earth or water, many analogues by which to explain both existences and occurrences. For instance, they explained justice as a certain property of number [four, the square], soul and mind as another such [one, unity, the point], the "decisive moment" as another [seven]; and they gave the same kind of interpretation to virtually everything else as well. Furthermore they observed empirically that the properties and ratios of harmonious musical tones depend upon numbers. Since they found, in short, that everything else, too, in its intrinsic nature, seemed to be essentially numerical, and thus that numbers appeared to be the ultimate meaning of everything that exists, they concluded that the elements of numbers must be the elements of everything, and that the visible heavens in their entirety consist of harmony and number.

> Accordingly they collected and employed all the analogies they could find which would represent the relation of numbers and harmonies to the properties or parts of the visible heavens and even to the entire universe; and if they came up against any gaps in such analogies they would snatch at whatever additional notion they could find to bring an orderly connection into their total explanation. For example, since the decad [the tetractys, i.e. the number ten] is believed to be perfect and to embrace the essential nature of the whole system of numbers, they conclude that the number of things existing in the sky must therefore be ten; but since in actuality there are only nine that are visible, they postulate the existence of a counter-earth as the tenth....
>
> The Pythagoreans, having observed that the things presented to us in sense-perception have many attributes of number, leapt to the theory that existing things are numbers—not just in the sense that they entail numbers as something distinguishable from themselves, but in that they actually consist of numbers. And why so? On the ground that the properties of number are inherent in musical harmony, in the movements of the visible heavens, and in many other things besides.[3]

But remember that numbers meant simple, exact, positive integers. This inflexible view of numbers coupled with the insistence on a precise, geometrical correspondence with the physical world ultimately led to the undoing of the Pythagorean cult.

Commensurability and Incommensurability

To explain what happened to the Pythagoreans, it helps to understand some basic notions about measurement. To say that a right triangle has sides of three, four, and five is to say that a length that is laid down end to end fits exactly three times in one side, four times in the other side, and five times on the hypotenuse. No more, no less. That length can be called the common measure of all sides of the triangle because it fits them all an exact number of times. Depending on the triangle, the length might be, say, one centimeter, one inch, one foot, one meter, one kilometer, or two, or three, or any other amount. The requirement is that you can lay that unit off an exact number of times on each side of the triangle.

If such a measure exists, the sides of the triangle are called *commensurable*, meaning they can be measured together by the same unit. If any quantities can be shown to be

---

3. Aristotle, *Metaphysics* 985b 22 and 1090a 20, quoted in Wheelwright, *The Presocratics,* pp. 213-215.

Chapter Three | 45
Pythagoras and the Magic of Numbers

commensurable, there will be many different units that can do the job. For example, in the 3-4-5 right triangle mentioned above, if the sides are actually 30 cm by 40 cm by 50 cm, then a 10 cm unit measures each side (3 times, 4 times, 5 times respectively), but also a 5 cm unit would measure each side (6 times, 8 times, and 10 times), and a 1 cm unit, and many other units will measure it as well.

In the Pythagorean worldview, all quantities of any sort would have to be commensurable because ultimately everything was made up of little space filling units which fit exactly. Finding common measures may be difficult, but there was no reason to suppose that such measures did not exist.

The opposite of commensurability is *incommensurability*, which is a state of affairs where *no* unit can possibly measure all of the quantities. Notice that it is far easier to show that quantities are commensurable than to show that they are incommensurable. To show commensurability, you need only find a single unit that goes exactly into each quantity to be measured. To show incommensurability, you have to show that there cannot be any units that will ever fit. It's no surprise then that the Pythagoreans would assume that all quantities were commensurable even if the unit of measure could not be found.

This brings us back to the Pythagorean Theorem, the showcase of the Pythagorean worldview, the magical, powerful theorem that showed that numbers underlay everything and understanding those numerical relationships was the key to all knowledge.

The Diagonal of the Square

Far from numbers underlying all of nature, they cannot even account for so simple a figure as a square and its diagonal. Legend has it that a Pythagorean discovered that this was the case while he and his group were at sea. When he explained it to the rest of the group, they were all horrified, swore themselves to secrecy and threw the discoverer overboard! Consider that simple, regular figure, the square. If you divide a square into two equal pieces down one of the diagonals, you make two isosceles right triangles. Since they are right triangles, the Pythagorean Theorem holds and $a^2 + b^2 = c^2$. But since the sides of the triangle are both sides of a square, they are equal to each other, so $a = b$. If you substitute a for b in the basic equation then you get $a^2 + a^2 = c^2$ or $2a^2 = c^2$, and from this, trouble follows for the Pythagoreans. The reason this is trouble is that it can be shown that there are absolutely no pairs of whole numbers that can be substituted for *a* and *c* that will satisfy the equation $2a^2 = c^2$.

After this, the Pythagoreans lost their sense of certainty and the cult began to die out. But it never did die completely.

What Pythagoras started that has persisted as a minority point of view in the history of science is the notion that nature is ultimately and finally not material at all, but its fundamental existence is of a form that is completely beyond human experience and accessible only to the mind.

Incommensurability

For those interested, here is a demonstration of the incommensurability of the sides and diagonals of squares.

This will use modern notation and terminology for simplicity. We don't know how the Pythagoreans originally proved this. There is a demonstration in Euclid from several hundred years later, but it is more complex than this one.

This proof uses a common feature of ancient Greek mathematics, namely, to assume the opposite of the result and show that this leads to a contradiction. The task is to prove that there are no pairs of whole numbers a and c that can fit the equation $2a^2 = c^2$. So, begin by assuming such numbers do exist (and remember that that was an assumption).

If there are any pairs of numbers that will work, there must be a pair of numbers that are smaller than any others. (For example, if the numbers 6 and 10 satisfied the equation—which they don't—then half of each, 3 and 5, which are also whole numbers, also fit.)

Therefore, let us assume that some numbers a and c exist, and in case there are more than one set of such numbers, we will assert that a and c are the lowest possible pair that fit this equation.

So, working with the equation, we have

$2a^2 = c^2$

If this is true, then the right side, c^2 must be an even number, since it is equal to 2 times the number a^2.

Chapter Three | 47
Pythagoras and the Magic of Numbers

But c^2 is c times c. If c is an odd number, then so will c times c be an odd number. (An odd number of odd numbers is an odd number.) Since c^2 is an even number, c must also be an even number.

Any even number can be thought of as two times half of itself. So consider a number h, which is half of c. Then $c = 2h$, and $c^2 = (2h)^2 = 4h^2$.

Now, go back to the original equation and substitute for c^2 the equal number $4h^2$.

$$2a^2 = 4h^2$$

Now divide both sides of the equation by 2. The result is

$$a^2 = 2h^2$$

This looks a lot like the equation we started with. In fact we can apply the same logic as we did above and say that if a^2 is an even number, then a must also be an even number. We can therefore write it as two times a number, k, which is half of a. So, $a = 2k$, and $a^2 = 4k^2$

Now what? Substitute $4k^2$ for a^2 and you get

$$4k^2 = 2h^2$$

Okay, now you have another equation that you can divide by two. Do so and you get

$$2k^2 = h^2$$

But what is this? It is an equation in exactly the same form as the original one, $2a^2 = c^2$, but the unknowns, k and h, are exactly half of the numbers a and c. However, remember that we said that a and c would be the lowest numbers that work. How then can we have found lower numbers that satisfy the same equation? There must be something wrong. The only thing that is questionable is the assumption we made at the beginning, that such numbers as a and c actually exist. That is what must be wrong, so, using the law of contradiction in logic, if a statement is false, its direct opposite must be true.

Therefore, there are no whole numbers (i.e., integers) that satisfy the equation $2a^2 = c^2$ and that means there is no unit of measure that one can lay of an exact number of times (a times on the side and c times on the diagonal) to measure the side and diagonal of a square. Therefore, the side and diagonal of a square are incommensurable.

Note: Some people are not convinced that it was reasonable to specify that a and c were the lowest numbers that fit the criterion when we don't even know what they are. If that bothers you, there are two ways of thinking about this that may help. First, notice that one could then start all over in the demonstration with the numbers k and h and do exactly the same thing again. That is, you can show that both k and h are even numbers, and therefore there must be some even smaller numbers, say, m and n, that will also fit the equation. But they can be shown to be even too, and so on forever. That's because there weren't numbers a and c that fit to begin with. Second, think of what this means geometrically. To be commensurable, there must be some unit of measure that fits an exact number of times into the side and into the diagonal of the square. The numbers a and c are the number of times that the unit of measure fits on the side or diagonal. If k and h are half of a and c, then the unit of measure that goes with them must be twice the size of the unit used with a and c. And the unit of measure that corresponds to m and n must be four times the measure used with a and c. Since you can continue to find smaller and smaller pairs of numbers that fit the equation, you would necessarily be finding larger and larger units of measure. No matter what numbers you started with, in no time at all you would have a unit of measure that was larger than the side and diagonal themselves. Which is also absurd. Hence, it must be that the one unjustified assumption—that there were such numbers—is the thing that is wrong.

For More Information

Alioto, Anthony M. *A History of Western Science,* 2nd ed. Englewood Cliffs, NJ: Prentice-Hall, 1987. Chapter 3.

Clagett, Marshal. *Greek Science in Antiquity.* New York: Collier, 1963.

Kirk, G. S., and J. E. Raven. *The Presocratic Philosophers: A Critical History with a Selection of Texts.* Cambridge: Cambridge University Press, 1963.

Lloyd, G. E. R. *Early Greek Science: Thales to Aristotle.* London: Chatto & Windus, 1970.

_____. *Magic, Reason and Experience: Studies in the Origin and Development of Greek Science.* New York: Cambridge University Press, 1979.

Neugebauer, Otto. *The Exact Sciences in Antiquity,* 2nd ed. New York: Dover, 1969.

Ronan, Colin A. *Science: Its History and Development Among the World's Cultures.* New York: Facts on File, 1985. Chapter 2.

Wheelwright, Philip, ed. *The Presocratics.* New York: Macmillan, 1996.

Chapter Three | 49
Pythagoras and the Magic of Numbers

Ruins at Delphi, Greece.

A Roman mosaic depicting Plato's Academy.

CHAPTER FOUR

Plato and the Reality of Ideas

What made Pythagoras and the Pythagoreans so extraordinary for ancient Greek thought was the view that what was real in the world was not the material stuff of the world at all, but an abstract substratum that can only be perceived in the mind. The Pythagorean position was radical because that substratum consisted entirely of numbers. Pythagoras viewed numbers as magical entities that not only explained everything; in a sense they *were* everything. But only those who could follow the mathematics were capable of appreciating this. Everyone else was beneath contempt.

The Pythagoreans more or less fell apart after the incommensurability of the diagonal and the square was established, using the very reasoning process through which they reached their conclusions about everything else. But despite their inconsistency, they launched a way of looking at the world that has remained an important part of scientific thinking.

This view of giving pre-eminence to the abstract became much more viable in the hands of a greater and far more important thinker who valued the objects of the mind just as much as the Pythagoreans, but who was not wedded to numbers as the universal explanation.

Plato

This was Plato, born about 200 years after Pythagoras, approximately in 427 BCE, in Athens, and who died in the same city in 348. "Plato," which means "the broad," may actually have been his nickname. Some say his name was actually Aristocles. Plato was the son of a wealthy Athens family. The long Peloponnesian War between Athens and Sparta was already underway when Plato was born, and he served in the military during the last years of the war. Plato was Socrates' most notable student. In fact much of what we know about Socrates is from Plato. Socrates was charged with neglect of the gods and corruption of the young. He was tried, convicted, and sentenced to death. After his execution in 399, Plato left Athens in disgust and traveled for some years in Egypt, Sicily, and Italy. When in Italy, Plato met Pythagoreans and learned of their doctrines, gaining a particular respect for mathematics.

In about 387, Plato returned to Athens and decided to establish a school for the study of philosophy. He purchased some land containing the grove of Academos, who was a famous hero, and erected a school on the property, which he named the Academy—a name which has been synonymous with school ever since. Even now, the Academy holds the record as the longest-lived institution of learning in the world. It continued to function as a school until 529 CE when it was closed by the Byzantine emperor Justinian on the grounds that it was pagan.

Early in life, Plato had intended to embark on a political career; for a time following the Peloponnesian War he did enter Athenian politics, but after a time left, disillusioned. His conviction that his fellow citizens were not fit to rule was even more confirmed by the execution of Socrates. When he set up the Academy, Plato had intended to train young men to become statesmen. For Plato, the proper training for politics was a thorough grounding in all aspects of philosophy. Partly perhaps because of his sojourn in Italy and contact with the Pythagoreans, Plato had become convinced that no true understanding of anything important could be achieved by anyone who did not have a deep understanding of mathematics and mathematical reasoning.

So, his Academy had, as it were, an entrance requirement: every incoming student must already be competent in mathematics. Over the entrance of the Academy, Plato had the following words inscribed:

> Let no one who does not know geometry enter here.

This edict says much about Plato's view of the world. Knowledge was not to be had without the kind of reasoning that mathematics brought, and it was necessary to gain those reasoning skills by the study of mathematics itself, regardless of the eventual subject matter to be understood.

Chapter Four
Plato and the Reality of Ideas

Most of what we know about Plato's thought is from his writings in the form of dialogues. These thirty or so works are written in the form of conversations among a small number of people discussing a wide variety of philosophical topics. All of the characters in the dialogues were real people, philosophers and statesmen, mostly, who represented different viewpoints. These were not necessarily people who were alive in Plato's lifetime. They were just characters from which to expound contrasting views. The main speaker in most of the dialogues is Socrates, Plato's teacher. In fact, much of what we know about

Socrates.

Socrates is from Plato's dialogues, but since Plato exercised considerable dramatic license, it is hard to know whether the historical Socrates would have espoused the views that Plato gives him to say in the dialogues. The general consensus is that in Plato's early dialogues, the character Socrates presents the views of the historical Socrates, but in the later dialogues, Socrates is more of a stand-in for Plato himself.

The subject matter of the dialogues is very broad: knowledge, ethics, politics, justice, physics, astronomy, mathematics, and many other issues. For the development of science, Plato's particular views about nature are less important than his concept of reality and of knowledge. These were most cogently expressed in two relatively short passages in his masterful work, *The Republic*, written later in his life. The first of these is a discussion of reality and appearances, of perception and knowledge, and provides a hierarchy of all things that are. This is the metaphor of the Divided Line, which is reproduced below. The second addresses the stages on the road to knowledge and the tasks facing the philosopher. This is the Allegory of the Cave. This appears in the next chapter.

The Divided Line

In Book VI of *The Republic*, Plato describes a vertical line divided into four sections which together represent and categorize all that is. The line is first divided into two main sections, which are specified as being unequal. Though Plato does not say so exactly, the upper portion of the line is taken to be larger than the lower portion. The lower section represents the "visible" world or the "sensible" world, meaning that it contains all things that one can see or perceive with the senses. The upper portion represents the "intelligible" world, containing all things that one can apprehend only with the mind.

Then each of the major sections are divided again, in the same proportion as the main divisions. (This in fact will make the two middle subdivisions equal to each other.) The visible world (the lower main section) has in *its* lower subsection images, reflections, shadows, and so on. In other words, things that aren't really there at all, but are perceptions caused by objects. The upper subsection of the visible world contains those objects: things of all sorts, animals, whatever has a physical presence in the world. The point of this division is to establish that there is a different level of being in the two subsections. We can speak of the shadows and so forth, but they don't exist other than as a perception, while the objects of the world are tangible.

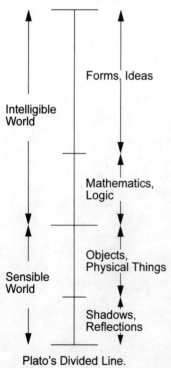

Plato's Divided Line.

The upper part of the line, the intelligible world, also has two divisions. The lower part of it contains the theorems of mathematics and logic—what we might call valid arguments. These are examples of correct reasoning from stated premises. Consider a geometrical object, a triangle. A triangle is a closed figure composed of three straight line segments that touch end-to-end. The line segments have no width whatsoever, are perfectly straight, and meet in a point. We have never seen a triangle, though we can draw representations of it. A triangle drawn on a chalkboard, for example, has sides that are lines of chalk marks wide enough to be seen, are doubtless not perfectly straight, and try as we might, we cannot get the line segments to meet exactly in points. But when we look at the representation, we can think of the idea of the triangle—the Form of triangle. Likewise, Plato views all objects in the visible world as being representations of

Chapter Four | 55
Plato and the Reality of Ideas

something that exists only in the intelligible world. Even so, the ideal triangle is dependent upon the definitions of line, angle, figure, and so forth that are given assumptions. The uppermost part of the Divided Line contains the Forms of the ultimate aspects of existence, including the forms of beauty, justice, and the good.

Corresponding to each section of the Divided Line are degrees of certainty and truth. In the lower main section, only opinion is possible because the senses must be relied upon and they can be wrong. In the upper section, knowledge is possible because reason is the means through which these entities are grasped. Then, in the subsections there are divisions too. In the lowest subsection, the section of images and shadows, what is perceived is not even material. There is much room for error. In the next subsection, the objects and things can be perceived directly and verified by different means, so one can have much stronger convictions about them. This is still not knowledge.

The difference between the two upper sections is that in the lower section, the reasoning is all in one direction. The theorems of mathematics, for example, are contingent upon assumed premises, which are not examined. Only in the very uppermost section, the realm of the Forms, is true knowledge possible through the process of reason alone.

The following is the text of the Divided Line discussion from Book VI of *The Republic*. It is a conversation between Socrates, who does most of the talking, and Glaucon, who is led along point by point.

> Socrates: You have to imagine, then, that there are two ruling powers, and that one of them is set over the intellectual world, the other over the visible. I do not say heaven, lest you should fancy that I am playing upon the name. May I suppose that you have this distinction of the visible and intelligible fixed in your mind?
>
> Glaucon: I have.
>
> Socrates: Now take a line which has been cut into two unequal parts and divide each of them again in the same proportion, and suppose the two main divisions to answer, one to the visible and the other to the intelligible, and then compare the subdivisions in respect of their clearness and want of clearness, and you will find that the first section in the sphere of the visible consists of images. And by images I mean, in the first place, shadows, and in the second place, reflections in water and in solid, smooth and polished bodies and the like: Do you understand?

Glaucon: Yes, I understand.

Socrates: Imagine, now, the other section, of which this is only the resemblance, to include the animals which we see, and everything that grows or is made.

Glaucon: Very good.

Socrates: Would you not admit that both the sections of this division have different degrees of truth, and that the copy is to the original as the sphere of opinion is to the sphere of knowledge?

Glaucon: Most undoubtedly.

Socrates: Next proceed to consider the manner in which the sphere of the intellectual is to be divided.

Glaucon: In what manner?

Socrates: Thus: There are two subdivisions, in the lower of which the soul uses the figures given by the former division as images; the enquiry can only be hypothetical, and instead of going upwards to a principle descends to the other end; in the higher of the two, the soul passes out of hypotheses, and goes up to a principle which is above hypotheses, making no use of images as in the former case, but proceeding only in and through the ideas themselves.

Glaucon: I do not quite understand your meaning.

Socrates: Then I will try again; you will understand me better when I have made some preliminary remarks. You are aware that students of geometry, arithmetic, and the kindred sciences assume the odd and the even and the figures and three kinds of angles and the like in their several branches of science; these are their hypotheses, which they and everybody are supposed to know, and therefore they do not deign to give any account of them either to themselves or others; but they begin with them, and go on until they arrive at last, and in a consistent manner, at their conclusions?

Glaucon: Yes, I know.

Socrates: And do you not know also that although they make use of the visible forms and reason about them, they are thinking not of these, but of the ideas which they resemble; not of the figures which they draw, but of the absolute square and the absolute diameter, and so on, the forms which they draw or make, and which have shadows and reflections in water of their own, are converted by them into images, but they are really seeking to behold the things themselves, which can only be seen with the eye of the mind?

Chapter Four
Plato and the Reality of Ideas

Glaucon: That is true.

Socrates: And of this kind I spoke as the intelligible, although in the search after it the soul is compelled to use hypotheses; not ascending to a first principle, because she is unable to rise above the region of hypothesis, but employing the objects of which the shadows below are resemblances in their turn as images, they having in relation to the shadows and reflections of them a greater distinctness, and therefore a higher value.

Glaucon: I understand that you are speaking of the province of geometry and the sister arts.

Socrates: And when I speak of the other division of the intelligible, you will understand me to speak of that other sort of knowledge which reason herself attains by the power of dialectic, using the hypotheses not as first principles, but openly as hypotheses, that is to say, as steps and points of departure into a world which is above hypotheses, in order that one may soar beyond them to the first principle of the whole; and clinging to this and then to that which depends on this, by successive steps she descends again without the aid of any sensible object, from ideas through ideas and in ideas she ends.

Glaucon: I understand you, he replied; not perfectly, for you seem to me to be describing a task which is really tremendous; but at any rate, I understand you to say that knowledge and being, which the science of dialectic contemplates, are clearer than the notions of the arts, as they are termed, which proceed from hypotheses only: these are also contemplated by the understanding, and not by the senses: yet, because they start from hypotheses and do not ascend to a principle, those who contemplate them appear to you not to exercise the higher reason upon them, although when a first principle is added to them they are cognizable by the higher reason. And the habit which is concerned with geometry and the cognate sciences I suppose that you would term understanding and not reason, as being intermediate between opinion and reason.

Socrates: You have quite conceived my meaning; and now, corresponding to these four divisions, let there be four faculties in the soul, intelligence answering to the highest, reason to the second, belief (or conviction) to the third, and perception of shadows or illusion to the last, and let there be a scale of them, and let us suppose that the several faculties have clearness in the same degree that their objects have truth.

Glaucon: I understand and give my assent, and accept your argument.[1]

---

1. Plato, *The Republic*, translated by Benjamin Jowett, Book VI 509-513.

Part I
Ancient Beginnings

For More Information

Alioto, Anthony M. *A History of Western Science,* 2nd ed. Englewood Cliffs, NJ: Prentice-Hall, 1987. Chapter 3.

Clagett, Marshal. *Greek Science in Antiquity.* New York: Collier, 1963.

Lindberg, David C. *The Beginnings of Western Science: The European Scientific Tradition in Philosophical, Religious, and Institutional Context, 600 B.C. to A.D. 1450.* Chicago: University of Chicago Press, 1992. Chapter 2.

Lloyd, G. E. R. *Early Greek Science: Thales to Aristotle.* London: Chatto & Windus, 1970.

O'Connor, J. J. and E. F. Robertson, "Plato." http://www-history.mcs.st-andrews.ac.uk/history/Mathematicians/Plato.html

Plato, *The Republic,* Benjamin Jowett, trans. New York: Modern Library.

Ronan, Colin A. *Science: Its History and Development Among the World's Cultures.* New York: Facts on File, 1985. Chapter 2.

Chapter Four
Plato and the Reality of Ideas

| 59

The ancient Stoa in Athens, reconstructed.

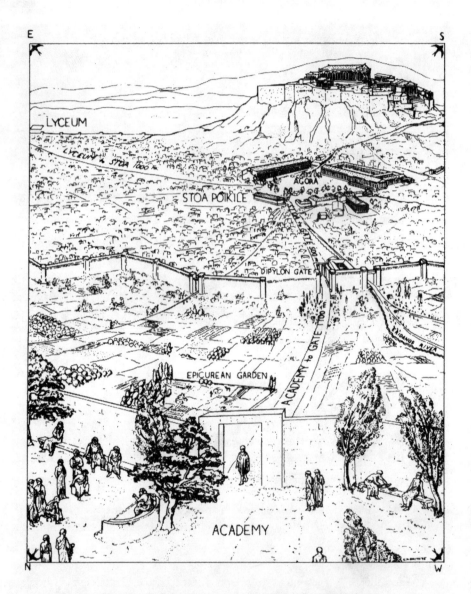

An artist's conception of Plato's Academy situated outside the walls of ancient Athens. Aristotle's Lyceum is indicated off in the distance.

CHAPTER FIVE

The Task of the Philosopher

Plato's purpose in founding the Academy was to train young men to become statesmen. But to Plato, that required becoming well-educated philosophers first. Much of what Plato saw as the proper education of future statesmen and their civic responsibilities applied just as well to philosophers who focused on the study of scientific topics, as well as those who would become political leaders. In the Book VII of *The Republic*, Plato discussed the training, development, and responsibilities of the philosopher-statesmen in very general terms by means of as a fable, the famous allegory of the cave, which is probably the best known of all of Plato's creations.

What makes the allegory so important for science is that it sets out responsibilities for the philosopher-scientist to reveal and show to the world the relevance of any discoveries and insights into the workings of nature.

The Allegory of the Cave

In the allegory, a prisoner is held bound in a cave along with other prisoners. They are all seated in a row and their legs and necks are so chained that they can only look forward. In front of them is a featureless wall on which shadows are thrown from objects behind them which they cannot see. Thus, their experience is limited to what Plato characterized as the lowest part of the Divided Line. Moreover, since they have been in the cave since childhood, they know nothing else, and have no reason to think that the shadows before them are not reality.

Behind them is a large fire, which produces the light for the shadows, though they are unaware of that. Between the prisoners and the fire there is a walkway with a low wall along it. People are passing along the walkway, carrying a variety of objects. The prisoners will see only the shadows of the objects being carried. If the people walking behind them are talking, the prisoners will hear only an echo off the wall before them. Therefore, they will believe those shadows to be reality.

If a prisoner is released and made to turn around and see the actual cause of the shadows, he is likely to believe that they are the illusions and the shadows are the reality, for that is his entire experience. Note here that Plato shows how experience can be a faulty guide. Suppose further that the prisoner is led out of the cave into the world outside and sees things as they are. In time, and after much resistance, the prisoner will come to realize that the world outside is the real one and much more desirable than the world he knew in the cave. This of course would be analogous to the ascent to the upper region of the Divided Line.

But, having shown that the prisoner is better off out of the cave—and the philosopher is better off in the upper part of the Divided Line—Plato says that he must be forced to go back down into the cave and instruct the other prisoners in what he has learned. He must explain to them that shadows are mere illusions and that there is a world beyond that they can only reach through the mind. This is his responsibility. Here is the text of the Allegory, from Book VII.

> Socrates: And now, let me show in a figure how far our nature is enlightened or unenlightened:, Behold! human beings living in an underground den, which has a mouth open towards the light and reaching all along the den; here they have been from their childhood, and have their legs and necks chained so that they cannot move, and can only see before them, being prevented by the chains from turning round their heads. Above and behind them a fire is blazing at a distance, and between the fire and the prisoners there is a raised way; and you will see, if you look, a low wall built along the way, like the screen which marionette players have in front of them, over which they show the puppets.
>
> Glaucon: I see.
>
> Socrates: And do you see, men passing along the wall carrying all sorts of vessels, and statues and figures of animals made of wood and stone and various materials, which appear over the wall? Some of them are talking, others silent.
>
> Glaucon: You have shown me a strange image, and they are strange prisoners.

Chapter Five
The Task of the Philosopher

Socrates: Like ourselves, and they see only their own shadows, or the shadows of one another, which the fire throws on the opposite wall of the cave?

Glaucon: True; how could they see anything but the shadows if they were never allowed to move their heads?

Socrates: And of the objects which are being carried in like manner they would only see the shadows?

Glaucon: Yes.

Socrates: And if they were able to converse with one another, would they not suppose that they were naming what was actually before them?

And suppose further that the prison had an echo which came from the other side, would they not be sure to fancy, when one of the passers-by spoke that the voice which they heard came from the passing shadow?

Glaucon: No question.

Socrates: To them, the truth would be literally nothing but the shadows of the images.

Glaucon: That is certain.

Socrates: And now look again, and see what will naturally follow if the prisoners are released and disabused of their error. At first, when any of them is liberated and compelled suddenly to stand up and turn his neck round and walk and look towards the light, he will suffer sharp pains; the glare will distress him, and he will be unable to see the realities of which in his former state he had seen the shadows; and then conceive some one saying to him, that what he saw before was an illusion, but that now, when he is approaching nearer to being and his eye is turned towards more real existence, he has a clearer vision, what will be his reply? And you may further imagine that his instructor is pointing. And when to the objects as they pass and requiring him to name them, will he not be perplexed? Will he not fancy that the shadows which he formerly saw are truer than the objects which are now shown to him?

Glaucon: Far truer.

Socrates: And if he is compelled to look straight at the light, will he not have a pain in his eyes which will make him turn away to take refuge in the objects of vision which he can see, and which he will conceive to be in reality clearer than the things which are now being shown to him?

Glaucon: True.

Socrates: And suppose once more, that he is reluctantly dragged up a steep and rugged ascent, and held fast until he is forced into the presence of the sun himself, is he not likely to be pained and irritated? When he approaches the light his eyes will be dazzled, and he will not be able to see anything at all of what are now called realities?

Glaucon: Not all in a moment.

Socrates: He will require to grow accustomed to the sight of the upper world. And first he will see the shadows best, next the reflections of men and other objects in the water, and then the objects themselves; then he will gaze upon the light of the moon and the stars and the spangled heaven; and he will see the sky and the stars by night better than the sun or the light of the sun by day?

Glaucon: Certainly.

Socrates: Last of all he will be able to see the sun, and not mere reflections of him in the water, but he will see him in his own proper place, and not in another; and he will contemplate him as he is.

Glaucon: Certainly.

Socrates: He will then proceed to argue that this is he who gives the season and the years, and is the guardian of all that is in the visible world, and in a certain way the cause of all things which he and his fellows have been accustomed to behold?

Glaucon: Clearly, he would first see the sun and then reason about it.

Socrates: And when he remembered his old habitation, and the wisdom of the den and his fellow-prisoners, do you not suppose that he would felicitate himself on the change, and pity them?

Glaucon: Certainly, he would.

Socrates: And if they were in the habit of conferring honors among themselves on those who were quickest to observe the passing shadows and to remark which of them went before, and which followed after, and which were together; and who were therefore best able to draw conclusions as to the future, do you think that he would care for such honors and glories, or envy the possessors of them? Would he not say with Homer, "Better to be the poor servant of a poor master," and to endure anything, rather than think as they do and live after their manner?

Glaucon: Yes, I think that he would rather suffer anything than entertain these false notions and live in this miserable manner.

Chapter Five
The Task of the Philosopher | 65

Socrates: Imagine once more, such a one coming suddenly out of the sun to be replaced in his old situation; would he not be certain to have his eyes full of darkness?

Glaucon: To be sure.

Socrates: And if there were a contest, and he had to compete in measuring the shadows with the prisoners who had never moved out of the den, while his sight was still weak, and before his eyes had become steady (and the time which would be needed to acquire this new habit of sight might be very considerable), would he not be ridiculous? Men would say of him that up he went and down he came without his eyes; and that it was better not even to think of ascending; and if any one tried to loose another and lead him up to the light, let them only catch the offender, and they would put him to death.

Glaucon: No question.

Socrates: This entire allegory, you may now append, dear Glaucon, to the previous argument; the prison-house is the world of sight, the light of the fire is the sun, and you will not misapprehend me if you interpret the journey upwards to be the ascent of the soul into the intellectual world according to my poor belief, which, at your desire, I have expressed, whether rightly or wrongly God knows. But, whether true or false, my opinion is that in the world of knowledge the idea of good appears last of all, and is seen only with an effort; and, when seen, is also inferred to be the universal author of all things beautiful and right, parent of light and of the lord of light in this visible world, and the immediate source of reason and truth in the intellectual; and that this is the power upon which he who would act rationally either in public or private life must have his eye fixed.

Glaucon: I agree, as far as I am able to understand you.

Socrates: Moreover, you must not wonder that those who attain to this beatific vision are unwilling to descend to human affairs; for their souls are ever hastening into the upper world where they desire to dwell; which desire of theirs is very natural, if our allegory may be trusted.

Glaucon: Yes, very natural.

Socrates: And is there anything surprising in one who passes from divine contemplations to the evil state of man, misbehaving himself in a ridiculous manner; if, while his eyes are blinking and before he has become accustomed to the surrounding darkness, he is compelled to fight in courts of law, or in other places, about the images or the shadows of images of justice, and is endeavoring to meet the conceptions of those who have never yet seen absolute justice?

Glaucon: Anything but surprising.

Socrates: Any one who has common sense will remember that the bewilderments of the eyes are of two kinds, and arise from two causes, either from coming out of the light or from going into the light, which is true of the mind's eye, quite as much as of the bodily eye; and he who remembers this when he sees any one whose vision is perplexed and weak, will not be too ready to laugh; he will first ask whether that soul of man has come out of the brighter life, and is unable to see because unaccustomed to the dark, or having turned from darkness to the day is dazzled by excess of light. And he will count the one happy in his condition and state of being, and he will pity the other; or, if he has a mind to laugh at the soul which comes from below into the light, there will be more reason in this than in the laugh which greets him who returns from above out of the light into the den.

Glaucon: That, is a very just distinction.

Socrates: But then, if I am right, certain professors of education must be wrong when they say that they can put a knowledge into the soul which was not there before, like sight into blind eyes?

Glaucon: They undoubtedly say this.

Socrates: Whereas, our argument shows that the power and capacity of learning exists in the soul already; and that just as the eye was unable to turn from darkness to light without the whole body, so too the instrument of knowledge can only by the movement of the whole soul be turned from the world of becoming into that of being, and learn by degrees to endure the sight of being, and of the brightest and best of being, or in other words, of the good.

Glaucon: Very true.

Socrates: And must there not be some art which will effect conversion in the easiest and quickest manner; not implanting the faculty of sight, for that exists already, but has been turned in the wrong direction, and is looking away from the truth?

Glaucon: Yes, such an art may be presumed.

Socrates: And whereas the other so-called virtues of the soul seem to be akin to bodily qualities, for even when they are not originally innate they can be implanted later by habit and exercise, the virtue of wisdom more than anything else contains a divine element which always remains, and by this conversion is rendered useful and profitable; or, on the other hand, hurtful and useless.

Chapter Five | 67
The Task of the Philosopher

Did you never observe the narrow intelligence flashing from the keen eye of a clever rogue, how eager he is, how clearly his paltry soul sees the way to his end; he is the reverse of blind, but his keen eye-sight is forced into the service of evil, and he is mischievous in proportion to his cleverness?

Glaucon: Very true.

Socrates: But what if there had been a circumcision of such natures in the days of their youth; and they had been severed from those sensual pleasures, such as eating and drinking, which, like leaden weights, were attached to them at their birth, and which drag them down and turn the vision of their souls upon the things that are below, if, I say, they had been released from these impediments and turned in the opposite direction, the very same faculty in them would have seen the truth as keenly as they see what their eyes are turned to now.

Glaucon: Very likely.

Socrates: Yes, and there is another thing which is likely, or rather a necessary inference from what has preceded, that neither the uneducated and uninformed of the truth, nor yet those who never make an end of their education, will be able educated ministers of State; not the former, because they have no single aim of duty which is the rule of all their actions, private as well as public; nor the latter, because they will not act at all except upon compulsion, fancying that they are already dwelling apart in the islands of the blest.

Glaucon: Very true.

Socrates: Then, the business of us who are the founders of the State will be to compel the best minds to attain that knowledge which we have already shown to be the greatest of all, they must continue to ascend until they arrive at the good; but when they have ascended and seen enough we must not allow them to do as they do now.

Glaucon: What do you mean?

Socrates: I mean that they remain in the upper world: but this must not be allowed; they must be made to descend again among the prisoners in the den, and partake of their labors and honors, whether they are worth having or not.

Glaucon: But is not this unjust? Ought we to give them a worse life, when they might have a better?

Socrates: You have again forgotten, my friend, the intention of the legislator, who did not aim at making any one class in the State happy above the rest; the happiness was to be in the whole State, and he held the citizens together by persuasion and necessity, making them benefactors of the State, and therefore benefactors of one another; to this end he created them, not to please themselves, but to be his instruments in binding up the State.

Glaucon: True, I had forgotten.

Socrates: Observe, Glaucon, that there will be no injustice in compelling our philosophers to have a care and providence of others; we shall explain to them that in other States, men of their class are not obliged to share in the toils of politics: and this is reasonable, for they grow up at their own sweet will, and the government would rather not have them. Being self-taught, they cannot be expected to show any gratitude for a culture which they have never received. But we have brought you into the world to be rulers of the hive, kings of yourselves and of the other citizens, and have educated you far better and more perfectly than they have been educated, and you are better able to share in the double duty. That is why each of you, when his turn comes, must go down to the general underground abode, and get the habit of seeing in the dark. When you have acquired the habit, you will see ten thousand times better than the inhabitants of the den, and you will know what the several images are, and what they represent, because you have seen the beautiful and just and good in their truth. And thus our State, which is also yours will be a reality, and not a dream only, and will be administered in a spirit unlike that of other States, in which men fight with one another about shadows only and are distracted in the struggle for power, which in their eyes is a great good. Whereas the truth is that the State in which the rulers are most reluctant to govern is always the best and most quietly governed, and the State in which they are most eager, the worst.

Glaucon: Quite true.

Socrates: And will our pupils, when they hear this, refuse to take their turn at the toils of State, when they are allowed to spend the greater part of their time with one another in the heavenly light?

Glaucon: Impossible, for they are just men, and the commands which we impose upon them are just; there can be no doubt that every one of them will take office as a stern necessity, and not after the fashion of our present rulers of State.

Socrates: Yes, my friend, and there lies the point. You must contrive for your future rulers another and a better life than that of a ruler, and then you may have a well-ordered State; for only in the State which offers this, will they rule who are truly rich, not in silver and gold, but in virtue and wisdom, which are the true bless-

Chapter Five
The Task of the Philosopher

> ings of life. Whereas if they go to the administration of public affairs, poor and hungering after their own private advantage, thinking that hence they are to snatch the chief good, order there can never be; for they will be fighting about office, and the civil and domestic broils which thus arise will be the ruin of the rulers themselves and of the whole State.
>
> Glaucon: Most true.
>
> Socrates: And the only life which looks down upon the life of political ambition is that of true philosophy. Do you know of any other?
>
> Glaucon: Indeed, I do not.
>
> Socrates: And those who govern ought not to be lovers of the task? For, if they are, there will be rival lovers, and they will fight.
>
> Glaucon: No question.
>
> Socrates: Who then are those whom we shall compel to be guardians? Surely they will be the men who are wisest about affairs of the state. [1]

Saving the Phenomena

Because the Allegory of the Cave occurs in *The Republic*, the practical applications mentioned concern government. Only those who have come to understand the intelligible world are fit leaders, but to be leaders, they must go again among the general population who have not experienced the intelligible world and explain it to them.

This has very important ramifications for the history of science. A philosopher who discovers something important about the world is not encouraged to keep it secret, but instead is obliged to find ways of explaining it to others and indeed to convince others of its correctness. But Plato views explaining reality to the uneducated as a general responsibility of all philosophers. Here he departs completely from the precepts of the Pythagoreans, who considered only those who could appreciate the finer points of philosophy to be worth troubling with. Plato definitely saw it as a burden of the philosopher to show others the way.

---

1. Plato, *The Republic,* translated by Benjamin Jowett, Book VII 514-521.

Likewise, a philosopher who discovers the rational explanation for something that is generally held to be mysterious should show the world what the reasonable explanation is. This is known as "saving the phenomena" or "saving the appearances."

To "save" phenomena is to show how some puzzle of the world can be explained by reasonable causes. A trivial example would be to explain that day follows night again and again because the sun, which is bright enough to produce daylight, goes around the earth every 24 hours. A slightly more sophisticated "saving" of phenomena would be to explain the annual cycle of the seasons by pointing out that the sun rises and sets at different places on the horizon each day and does so in a cycle through the year. In the winter, the sun rises and sets farther to the south (if you are in the northern hemisphere) than it does in the summer. The result is that the path across the sky is shorter and the hours of daylight are shorter in the winter. Fewer hours of daylight make for colder days. Note that the explanation raises as many questions as it answers. Why are hours of daylight associated with temperature? Why does the sun rise at different points on the horizon? Why does the sun cross the sky at all? Nevertheless, the explanation does give a place to start and points to the next questions. This way, a process is begun that can lead to understanding.

Plato's admonition to philosophers to save the phenomena is the foundation of a program of research leading to general knowledge. It gave the impetus for philosophers to look at nature with a critical eye, find explanations using rational arguments, and then explain how those explanations account for mysteries of nature that had been nothing but enigmas until then.

The Problem of the Planets

It is no coincidence that the examples of saving the phenomena that I mentioned above have to do with one of the heavenly bodies. The regular and repeating events in the sky were among the greatest mysteries pondered by ancient peoples. Night after night, the same patterns occur again and again. Day after day, the sun rises in the east and sets in the west. Year after year, the seasons follow in the same order. All early civilizations had some explanations of these events, but usually the explanations were mythic and religious and did not admit of rational analysis. The heavenly motions stood among the top priorities of phenomena that needed saving in order to show that philosophy has value and relevance.

Chapter Five
The Task of the Philosopher

Explanations of the sort given above were common enough by Plato's time, though perhaps not widely understood. But there was another astronomical problem that defied analysis. It would be a great coup to show how this problem yielded to a rational explanation. This was the "problem of the planets."

What is a planet? That's problem number one. A planet is a heavenly body that looks like a star. But unlike all the other stars that appeared fixed in place in the sky relative to each other, planets move around, appearing in one constellation for a time and then moving on to another. That's in fact what the word planet means: it's Greek for "wanderer." A planet is a star that wanders.

What the stars are themselves and why they move at all is a mystery, but one can say, as several pre-Socratic philosophers did, that the stars are all fixed to a huge sphere at the edge of the universe and that sphere rotates completely around once a day. It may raise other questions, but at least it accounts for the daily motion. But the planets don't behave this way.

To take a simple case, the sun moves along through the constellations during the year, reaching about the same place one year later. (This, by the way, is judged by noting the stars that are visible just before dawn and just after dusk on different days.) You can explain this by saying that the sun moves along a great circle in the sky at about one degree a day, and this whole circle is carried around in the daily motion of the stars. A similar explanation can be given for the moon. Both the sun and the moon were considered planets by the ancients, but these were not the problem.

It was the other "planets" that defied explanation: Mercury, Venus, Mars, Jupiter, and Saturn. They not only traveled around daily with the stars and also worked their way through the constellations more slowly, like the sun; they also changed directions. Sometimes they moved in an easterly direction against the stars; sometimes they appeared to stop and just move with the stars; sometimes they reversed direction and moved west. How can this be explained as due to natural causes?

The "problem of the planets" was to show how the planetary motions that were seen from Earth were actually some sort of understandable process. Now, here we run into a mindset that dominated ancient Greek thinking. Because the motions in the heavens repeated themselves endlessly, it seemed only reasonable to philosophers to think that their motions must be in a circle. A circle was a closed path, so it never ended. But in addition to that, a perfect circle is the same everywhere, so that even if a heavenly body moved, it only moved on something that was changeless. Therefore, it was sort of eternal. Therefore, the task became one of showing how all planetary motions were in fact combinations of circular motions, since circular motions were "philosophical."

This must have been one of the problems that was considered at the Academy. It would be a perfect test of the philosopher's methods to find a way to account for the planetary motions as combinations of circular motions and thereby "save" them. It seems that some of Plato's students at the Academy took on this task.

The Spheres of Eudoxus

Plato had several notable students who did important work of their own and had a considerable reputation in ancient philosophy. The most important of these by a very wide margin was Aristotle, whose influence was greater than Plato's own. However, we get to Aristotle in the next chapter. Another important and original thinker who attended Plato's school was Eudoxus of Cnidus, who came to the Academy in 368. Eudoxus may have been more of an associate of Plato's than a pupil at the Academy. Eudoxus served as acting head of the Academy on at least one of Plato's periodic absences.

Eudoxus is known primarily as a mathematician. He created the theory of proportion that appeared later in Euclid and discovered various methods of measuring the area and volume of curved figures and solids. His importance here is that he came up with a scheme for explaining the motions of the planets.

Eudoxus' system was as follows: He began with the commonly held ancient view that the universe was spherical in shape, the Earth was an immobile sphere at the center of the universe, and the outer edge of the universe was a spherical shell which turned completely around on its axis every day and on which were imbedded the "fixed" stars. Between the sphere of the stars and the Earth was a transparent crystalline substance that was basically invisible while it filled all space. This much was not original to Eudoxus.

Chapter Five
The Task of the Philosopher

What Eudoxus contributed was the idea that the planets were situated on spherical shells concentric with the sphere of the fixed stars and the sphere of the Earth and which fit between them. Thus, there were layers of concentric shells like onion skins that built outward from the Earth to the stars. These layers, made of the transparent crystal, were invisible themselves, except for the planets that were carried on some of the layers. Now, the trick here is that while each of the crystalline shells was centered on the Earth, they all turned on different axes. If you have two concentric spheres with one inside the other, and the axis of the inner sphere is imbedded in the inner surface of the outer sphere, and you then set both spheres turning on their respective axes, when viewed from inside, any spot on the outer sphere appears to go round in a circle, while any spot on the inner sphere will have the combined motions of both spheres and look quite irregular.

This was the basic idea. The stars, on the outermost sphere, would appear to travel in regular circles, all moving together. The planets, on spherical shells nested inside each other, would have motions that combined all the motions of their own shell and all those further out to which they were attached. So much for the basic idea. The details became more difficult. Every planet had its own period—the time it took to come back to the same place in the sky, measured against the backdrop of stars. But all the planets travel round in a fairly tight band in the sky from which they never depart. (This is the band of the Zodiac.) To accommodate all the planets, each with its own rate of revolution, number of stops and duration of retrograde (i.e., backward) motion, it was necessary to add a lot of extra spherical shells. In total, Eudoxus claimed that the correct number was 27 shells. The Sun and the Moon, which were considered planets too, required three shells each to capture their motions. The planets Mercury, Venus, Mars, Jupiter, and Saturn required four shells each. And the fixed stars took the last shell.

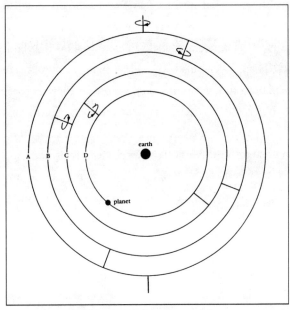

A Eudoxian four-sphere model for a single planet.

Each planetary system was independent of the others. For a typical planet, for example, Jupiter, the outermost shell produced the daily rotation. That is, it turned the same way as the sphere of the fixed stars. The next shell determined the annual motion of the planet as it moved through the Zodiac. This is the general "forward" motion of the planets through the stars. The third shell rotated the planet with respect to its position against the sun. The axis of the fourth shell was attached to the inside of the third shell at an angle which differed for each planet. On the equator of the fourth shell was the only thing visible in all of this, the planet itself. From its irregular but visible motions, all the other motions of the respective shells had to be inferred. Or, rather, in the spirit of saving the phenomena, if you postulated the existence of the invisible shells, their placement, and speed of rotation, you could account for the visible motion of the planets, which is the phenomena.

It did not seem to bother Eudoxus that he had no reason to explain why there should be such shells nor why they should be attached the way they were supposed to be nor why they should turn at all. Nor did the complexity of the system cause him to reject it as an improbable explanation. It did what it was supposed to do: it showed how the totally mysterious planetary motions could be reduced to understandable constant circular rotations.

At least that appears to be the intention. In practice, Eudoxus' model worked very poorly for some of the planets, especially Venus and Mars. But this could be viewed as a detail to be worked out with further tinkering.

For More Information

Alioto, Anthony M. *A History of Western Science,* 2nd ed. Englewood Cliffs, NJ: Prentice-Hall, 1987. Chapter 3.

Clagett, Marshal. *Greek Science in Antiquity.* New York: Collier, 1963.

Lindberg, David C. *The Beginnings of Western Science: The European Scientific Tradition in Philosophical, Religious, and Institutional Context, 600 B.C. to A.D. 1450.* Chicago: University of Chicago Press, 1992. Chapter 2.

Lloyd, G. E. R. *Early Greek Science: Thales to Aristotle.* London: Chatto & Windus, 1970.

O'Connor, J. J. and E. F. Robertson. "Plato." http://www-history.mcs.st-andrews.ac.uk/history/Mathematicians/Plato.html

Plato. *The Republic,* Benjamin Jowett, trans. New York: Modern Library.

Ronan, Colin A. *Science: Its History and Development Among the World's Cultures.* New York: Facts on File, 1985. Chapter 2.

A bust of Aristotle from the Kunsthistoriches Museum in Vienna.

Chapter Six

Aristotle and the World of Experience

Plato's most illustrious student at the Academy was Aristotle, a man who came to have greater influence than his master. Aristotle was born in 384 BCE in Stagirus, a Greek colony in Thrace, at the northern end of the Aegean Sea, near Macedonia. His father, Nichomachus, was a medical doctor from a long line of medical doctors, and his mother, Phaestis, was from a wealthy family in Chalcis on the island of Euboea. There are several notable facts related to this lineage.

The first is simply the name of the city where Aristotle was born, Stagirus. In later years when Aristotle was quoted so often in philosophical writings, efforts were made to find another way to refer to him other than to continue repeating his name. Frequently he was simply called "the philosopher," as though there were only one. Another nickname was "the Stagirite." It was expected that everyone would know that the Stagirite was Aristotle.

Second, his father's profession was one which normally continued from generation to generation. Medical knowledge was passed from father to son with the expectation that the son would simply carry on where the father left off. Aristotle certainly received some medical training from his father and would have gone on to succeed him, but his father died when Aristotle was only ten years old—before his father could have trained him adequately in medicine. Nevertheless, Aristotle had a strong appreciation for the process of life and did a great deal of important work in biological research. Later, Aristotle had a son whom he named Nichomachus, after his father, and he also named one of his most important works the *Nichomachean Ethics*.

As was common among physicians of the time, Nichomachus traveled from place to place, tending the sick wherever they were. Among his patients was Amyntas, the king of nearby Macedonia. Thus Nichomachus had good contacts in the court of Macedonia, and it is likely that young Aristotle also made contacts in the Macedonian court.

When Nichomachus died, and his mother seems to have also died soon after, Aristotle was raised by a guardian who may have been his uncle. His guardian, Proxenus, taught Aristotle other subjects to broaden his education. Then, when Aristotle was seventeen years old, Proxenus sent Aristotle to Athens to study at Plato's Academy.

Aristotle remained at the Academy for twenty years, until Plato's death. When Aristotle arrived at the Academy, Plato was in Syracuse, on one of his leaves of absence and the Academy was being run by Eudoxus. Aristotle was certainly Plato's most distinguished and accomplished student. Plato was said to have called him the "intelligence" of the Academy. Nevertheless, Aristotle and Plato had some very serious differences in their approaches to philosophy.

When Plato died in 347 BCE, Aristotle expected to be appointed to take over the Academy, but Plato had named his nephew Speusippus to the post instead. This may have been because of their philosophical differences or it may have been for political reasons since Aristotle's ties with Macedonia may have made him a liability. Macedonia's expansionist tendencies were viewed as a threat to Athens and rightly so since not long after Macedonia invaded Athens.

Speusippus was a minor intellect compared to Aristotle and moreover, their views differed fundamentally. Aristotle left, moved to the island of Assos, married the niece of Hermias, the local ruler, and they had a child. On Assos, Aristotle began some of his biological research. However, after three years, Assos was invaded by the Persians, Hermias was executed, and Aristotle fled to nearby Lesbos. After about a year on Lesbos, Aristotle accepted an invitation to move to Macedonia and join the court of Philip II, the son and heir of the king that Aristotle's father had served. It is usually reported that Aristotle was specifically invited back to Macedonia to be the personal tutor of Philip's thirteen-year-old son Alexander, who would later become the world leader, Alexander the Great. In any case, Aristotle did go to Macedonia and remained there for five years, most likely being Alexander's personal tutor as well as carrying on his own research.

When Philip died, Alexander assumed the throne, and Aristotle left the court. He returned to Athens and set up a rival school to the Academy which became known as the Lyceum (because it was adjacent to the temple of Apollo Lykaios).

Chapter Six
Aristotle and the World of Experience

The subjects pursued at the Lyceum were broader than those at Plato's Academy, including much more emphasis on the direct study of nature, especially biological topics. Aristotle arranged the curriculum in two parts: detailed studies for serious students in the mornings and more popular lectures in the evenings for the general public. Ancient sources report that Aristotle wrote dialogues of much literary merit in the style of Plato, but these have not survived. What has survived and is considered to be the opus of Aristotle are about 30 closely written treatises covering a wide range of topics. These may have been Aristotle's morning lectures or notes for those lectures or notes taken by students at those lectures. In any case, that is what we have. These works are considerably more dense than Plato's dialogues and require very close reading. What is extraordinary is that the bulk of western philosophy has been built upon the foundation of these terse monographs.

Aristotle's influence on the development of science is huge. In a nutshell, Aristotle outlined the basic methodology for the study of nature, the formation of explanations, and the construction of coherent, logical theories into which all the parts fit to make a general scientific description of the world. That said, almost every conclusion drawn by Aristotle about the natural world was wrong. He set science on a path to discovery, but he did not have the information necessary to draw sound conclusions.

Both Aristotle's methodology and his conclusions are important because, on the one hand, much that was productive came out of his systematic approach and, on the other hand, his incorrect conclusions were so convincing that were adopted as manifest truth for nearly two thousand years.

Aristotle's stay at the Lyceum was not nearly as long as Plato's at the Academy. Aristotle founded the Lyceum in 335. Twelve years later, in 323, Alexander the Great, who had been busily conquering the world and making a Greek empire, died of a sudden illness on one of his campaigns. Immediately, chaos descended on the Greek world. In Athens, anti-Macedonian feelings rose again, and Aristotle, who was associated with Macedonian interests, began to be persecuted. His enemies managed to get a trumped-up charge of impiety leveled against him. Aristotle decided to flee Athens rather than, as he is supposed to have said, give the Athenians another opportunity of sinning against philosophy—the first one being the charges against Socrates, which had resulted in his exe-

Raphael's famous fresco, *The School of Athens,* depicts most of the famous philosophers of antiquity assembled together in a hall. In the centre, framed by the archway, stand Plato and Aristotle.

cution. Aristotle retreated to Chalcis on Euboea, to the property of his mother's family, and within a year died there of a stomach disorder at the age of 62.

Empiricism

Plato's route to knowledge involved concentrating on the abstract forms accessible only to the mind and avoiding being distracted by the imperfect representations that could be discerned by the senses. The physical world was, as it were, merely a springboard to the world of ideas—the upper part of the Divided Line. Once the philosopher found the way "out of the cave" and up to true knowledge, there was nothing left to be learned from the world of sense experience. The reason that philosophers had to pay attention to the visible world after they had comprehended the intelligible world was to show other less fortunate people how the true state of affairs can explain the confusing aspects of the world of appearances—i.e., how it can save the phenomena.

Aristotle took issue with this viewpoint. For Aristotle, the real world *was* the world around, the world that one apprehends with the senses. The business of philosophy was to understand that world, not to get lost in some world of the imagination. To Aristotle,

Chapter Six
Aristotle and the World of Experience

Plato was in constant danger of making up a world that did not exist. Aristotle believed that true knowledge of nature came from close examination of the physical world.

Aristotle really established systematic observation of nature. He and his students at the Lyceum spent years collecting specimens of nature (especially animal and plant species), dissecting them, studying them, classifying them and trying to understand what they saw. It was essential for Aristotle that any theoretical explanation of nature fit with what was observed. This was as essential for him as it is now for modern science.

This difference between the viewpoints of Plato and Aristotle is fundamental. It is in a sense *the* essential difference of the two basic ways of understanding the world. For Plato, the senses mislead and don't reveal essential structures. For Aristotle, the senses are our only window on the world and we must make sure that our conclusions are based upon observation.

Detail from Raphael's *School of Athens* showing Plato on the left and Aristotle on the right.

This difference between the two men was clearly recognized in the Renaissance. The famous fresco by Raphael in the Vatican called *The School of Athens* features Plato and Aristotle in the center walking and conversing. Plato, characteristically, is pointing up, away from the Earth, indicating that the important matters are to be found in the eternal forms (or, if you prefer, in the heavens), while Aristotle is pointing downward indicating that the focus of attention should be on the world around. Characteristically, the Platonic viewpoint is best illustrated with an example from mathematics, while the Aristotelian view is best shown with an example from living things.

For Plato the ideal example is a geometrical object, such as a triangle, square, cube, sphere, etc. Any picture we draw or model we construct of any of these will be imperfect; that is, it will not fit the mathematical definition of the object. Therefore, we use the physical object only as an aid to bring the real entity, the mathematical form, to mind. When we analyze the properties of, say, a triangle, we do so with reference to the idealized form of a triangle, not with some three-sided drawn figure.

For Aristotle, his view can be illustrated with an animal, for example, a goat. Aristotle examines as many goats as he can and from his observation, draws conclusions about the essential features of goats. Those features, shared by all goats, define the form of a goat. But the form is merely a classification of the living things into a common category. Only the living goats are real. The form of goat is a useful concept for thinking about goats, nothing else.

The form of goat, or anything else, is an induction based on empirical evidence. That is, it forms general statements after examining particulars. This is the foundation of knowledge. It is a process known to all of us and is a natural part of learning about the world that every child has engaged in. Aristotle's method is often termed a common-sense approach because it is grounded in everyday observations.

Logic

But Aristotle does not stop there. If he satisfied himself with merely cataloguing and classifying all experience, he might be able to describe the world around him in some detail, but he would not have been able to discover truths that did not present themselves manifestly. To get beyond what presents itself to the senses, Aristotle relies upon logic. Aristotle was not the inventor of logic. That began with the pre-Socratics. In fact, Aristotle did not even use the term logic. The word logic was coined by his former colleague Xenocrates at the Academy. Aristotle used the word "analytics."

Logic, or "analytics," is a way of proceeding from what is known to learn something new that was previously unknown. The key tool is the *syllogism*. A typical syllogism has three parts: (1) a major premise, which is a statement expressing some general truth based upon induction from observation; (2) a minor premise, which is usually some particular observation; and (3) the conclusion, which is logically implied by the combination of the major and minor premises. The classic example, provided by Aristotle is as follows:

> Major premise: All men are mortal.
> Minor premise: Socrates is a man.
> Conclusion: Socrates is mortal.

The example is a trivial one, but it illustrates the basic principle. If you know that something is true about all the members of a certain group, and you know that a particular

Chapter Six
Aristotle and the World of Experience

individual is a member of that group, then you know something about that individual. The conclusion that Socrates is mortal is not based on any observation of the man himself, but ascribes to Socrates a characteristic that he must have because he is a man.

So far so good, and with the addition to this of well-established principles of logical inference that are independent of the content to which they are applied, one can build a body of knowledge consisting of logically valid conclusions that are true so long as the premises on which they are based are true. Since Aristotle intended that he would begin only with observations and inductions that were certain or self-evident, he could reasonably expect that he would be able to build an edifice of certain knowledge with these methods.

Alas, the methodology is sound, but his conclusions about the natural world were largely untrue. That is because the major premises that contained his inductions from experience were often wrong. Sense perception is deceiving, as Plato noted, and Aristotle was often misled by his observations into unwarranted generalizations. Only in the matter of biology where much careful observation is necessary have Aristotle's conclusions remained of much value.

The Four Causes

Aristotle held that to understand anything you have to understand its "causes." By "causes" he meant all the factors that went into making it what it is. There were, in Aristotle's analysis, four causes of any thing. Bear in mind that he used the word "cause" in a wider sense than that which we would use today.

The first cause is the "Material" cause. This is simply the material of which something is made. For example, wood is the material cause of a chair or a tree, flesh and blood is the material cause of an animal.

The second cause is the "Formal" cause. This is the shape or form into which the material is made. One can think of the form of a chair, tree, etc. Clearly a number of pieces of wood will only be a chair if they are assembled in a certain way.

Part I
Ancient Beginnings

The third cause is the "Efficient" cause. This is where we would use the word cause in our present sense. The efficient cause of something was what made it into its form. For a chair, it could be a carpenter; for a statue, a sculptor. For an animal the answer was not so obvious, but clearly Aristotle must have required that there be an answer before he could claim to "understand" the animal.

The fourth cause he calls the "Final" cause. This was the purpose for which the thing existed. For a manufactured object, these causes are easy to identify: the final cause of a chair is to sit on; of a sculpture is to look upon as a work of art; of a shoe, to protect the feet, and so on. For natural objects, the answers are not apparent, but Aristotle set out to find them using his method.

To get a sense of Aristotle's reasoning and the writing style of the works that have come down to us, I have reproduced below an excerpt from his *Physics*, Book II, Section 3, in which he discusses the four causes.

> Now that we have established these distinctions, we must proceed to consider causes, their character and number. Knowledge is the object of our inquiry, and men do not think they know a thing till they have grasped the "why" of it (which is to grasp its primary cause). So clearly we too must do this as regards both coming to be and passing away and every kind of physical change in order that, knowing their principles, we may try to refer these to principles each of our problems.
>
> In one sense, then, (1) that out of which a thing comes to be and which persists, is called "cause", e.g. the bronze of the statue, the silver of the bowl, and the genera of which the bronze and the silver are species.
>
> In another sense (2) the form or the archetype, i.e. the statement of the essence, and its genera, are called "causes" (e.g. of the octave the relation of 2:1, and generally number), and the parts in the definition.
>
> Again (3) the primary source of the change or coming to rest; e.g. the man who gave advice is a cause, the father is cause of the child, and generally what makes of what is made and what causes change of what is changed.
>
> Again (4) in the sense of end or "that for the sake of which" a thing is done, e.g. health is the cause of walking about. ("Why is he walking about?" we say. "To be healthy", and, having said that, we think we have assigned the cause.) The same is true also of all the intermediate steps which are brought about through the action of something else as means towards the end, e.g. reduction of flesh, purging, drugs, or surgical

instruments are means towards health. All these things are "for the sake of" the end, though they differ from one another in that some are activities, others instruments.

This then perhaps exhausts the number of ways in which the term "cause" is used.

As the word has several senses, it follows that there are several causes of the same thing (not merely in virtue of a concomitant attribute), e.g. both the art of the sculptor and the bronze are causes of the statue. These are causes of the statue qua statue, not in virtue of anything else that it may be--only not in the same way, the one being the material cause, the other the cause whence the motion comes. Some things cause each other reciprocally, e.g. hard work causes fitness and vice versa, but again not in the same way, but the one as end, the other as the origin of change. Further the same thing is the cause of contrary results. For that which by its presence brings about one result is sometimes blamed for bringing about the contrary by its absence. Thus we ascribe the wreck of a ship to the absence of the pilot whose presence was the cause of its safety.

All the causes now mentioned fall into four familiar divisions. The letters are the causes of syllables, the material of artificial products, fire, &c., of bodies, the parts of the whole, and the premises of the conclusion, in the sense of "that from which". Of these pairs the one set are causes in the sense of substratum, e.g. the parts, the other set in the sense of essence--the whole and the combination and the form. But the seed and the doctor and the adviser, and generally the maker, are all sources whence the change or stationariness originates, while the others are causes in the sense of the end or the good of the rest; for "that for the sake of which" means what is best and the end of the things that lead up to it. (Whether we say the "good itself" or the "apparent good" makes no difference.)

Such then is the number and nature of the kinds of cause.

Aristotle's Universe

The Sublunar World

Nothing illustrates Aristotle's methodology and how it led him to false conclusions more than his analysis of the general layout of the cosmos. Consider first the world around him. Aristotle noted that everything seemed to have a finite life span. Animals and plants came into being, lived, and then died. Works of human manufacture likewise came into being when they were made, existed for a while, and, it seemed inevitable, would disintegrate or be destroyed some day. This was also true of motion. Push an object and it may move, but it will come to a stop. Throw an object up in the air, and it will return to earth. Everything can be viewed as a finite event. Having satisfied himself that this was a general principle of earthly existence, Aristotle concluded, with a bit of syllogistic reasoning, that seemingly permanent features of the earth, rocks and mountains, for example, also participated in this process of coming to be and passing away (generation and corruption).

Aristotle had a more complicated version of this view wherein each of the four elements manifested the underlying qualities of hot or cold and wet or dry. Nevertheless, these resolved themselves into the four elements. Observation led Aristotle to conclude that these elements always sought to occupy their "natural" place. Fire is clearly always leaping upward. Earth (meaning virtually anything solid and heavy) would always fall to the lowest possible point. Water would also fall through the air, but not sink lower than pure earth (a rock, for example). Air would seek to rise above earth and water, but not as vigorously as would fire. Hence, Aristotle concluded, there were natural places for these elements. Earth, he contended, would seek to be as close to the center of the universe as possible. Surrounding the earth would be a layer of water. Above the water would be a band of air, and beyond that was fire. These are the natural places of the four elements. If you took any one of the elements out of its natural place and released it, it would quickly move back to where it belonged. Like most ancient thinkers, Aristotle thought that it was inconceivable that empty space should exist (cf., Parmenides' view of the impossibility of the existence of nothing). Therefore, these elements merely changed places with each other. A hundred years or so before Aristotle's time, the debate about the fundamental material of the world that had started with Thales was resolved as a sort of compromise by the philosopher Empedocles, who identified not one, but four basic elements from which everything was made. These were earth, air, fire, and water. Water had been proposed as the basic element by Thales and air by Anaximenes. Heraclitos said that what was fundamental was change and some saw fire as the element that embodies change. That left only the most obvious candidate, earth. Empedocles took all four and said that everything that existed in the world was produced by mixing these together in different combinations. Empedocles' view seemed most satisfactory to the ancient mind, and this was generally accepted as the correct view by most philosophers, including Aristotle.

Chapter Six
Aristotle and the World of Experience

Things moving to their natural places he called "natural motion." Anything that caused a body to move otherwise was "forced motion" (or "violent motion"). Like everything else in the world around him, Aristotle saw forced motions as having a finite span: they begin, continue for a bit, and then end. What causes a forced motion? The examples that presented themselves to Aristotle were pushes and pulls. One object collides with another and either pushes it along or grabs it and pulls it. The motion is caused by the thing pushing or pulling. When an object was no longer being pushed, it stopped its forced motion, and, if it were no longer in its "natural place," it would then assume its natural motion in order to get there. Since an object in its natural place would have no reason to move from it, the natural state of any object is to be motionless. Motion is therefore an unnatural state and must be caused by an agent (of pushing or pulling).

This analysis will suffice for explaining what happens when carts get pulled down city streets by draught animals, or when a pitcher of milk is raised, tipped, and the milk allowed to pour out—i.e., assume its natural motion downward. But what happens to an object thrown through the air? Suppose, say, an arrow is shot or a spear hurled. Aristotle's analysis would suggest that as soon as the arrow left contact with the bow or the spear was out of the warrior's hand it would immediately fall to the ground, that is, the moment it was no longer being pushed, it would assume its natural motion. Clearly this is not what happens, so Aristotle had to find a way to explain this.

Here is where logic completes what empirical classification could not provide. Aristotle reasoned that the arrow or spear, etc., would only continue to move horizontally if something were pushing it. Hence, he reasoned, there must be something that pushes the object along through the air. Continuing the train of logic, since the arrow, spear, etc., is traveling through the air, it must be pushing air out of the way. This air must go somewhere. Moreover, if the object is moving forward through space, the place that it just was a moment before would have nothing in it when the object left it. "Aha," says Aristotle. There is no such thing as empty space because that would mean that "nothing" exists. (Like Parmenides, he concluded that this was an impossibility.) Therefore, what must happen is that the air which was being pushed out of the way by the advancing arrow or spear or any projectile must travel immediately around to the back of the object and fill that otherwise empty space. When it fills that space, it gives a further push to the object, and that is what makes it continue to move forward.

This argument has come to be called the principle of *antiperistasis*. It is one of the weakest, most contrived of Aristotle's physical explanations. In fact, this argument is so preposterous that it became a focal point for Aristotelian scholars in later centuries who were trying to make better sense of Aristotle's physics. Aristotle seemed to have a compulsion to provide an answer to every question. If his direct evidence did not provide the answer, he would use logic to deduce one, no matter how contrived the argument might become.

Now, returning to the precept that all things and all events have a finite duration—a beginning, middle, and end, is this true? You could note the process of birth, life, and death and the beginning and end of motions for the things around you, but there was also what was happening in the sky. The sun came up every day. The seasons followed each other predictably. The stars appeared to turn inexorably. Astronomical records from Babylonia going back over a thousand years attested to the fact that things in the heavens went on and on and on. How was Aristotle to account for this?

His solution to this is to declare that there are really two distinct pieces to the cosmos: heaven and earth, and they each had different rules. Aristotle puts the boundary between these two worlds at the moon. Below the moon was the "sublunar" world where things move in straight lines and have beginnings and ends. Beyond the moon was the "superlunar" world where the stars and planets move in never ending circles. To make the break between these worlds complete, Aristotle says that the familiar four elements, earth, air, fire, and water, exist only in the sublunar world. In the superlunar world there was a different element. He calls this new element simply the "fifth element." In Latin, that translates as the "quintessence"—a name that has come to mean the ultimate or perfect type of something. (Many scholars have also translated the fifth element as the "aither.")

Chapter Six
Aristotle and the World of Experience

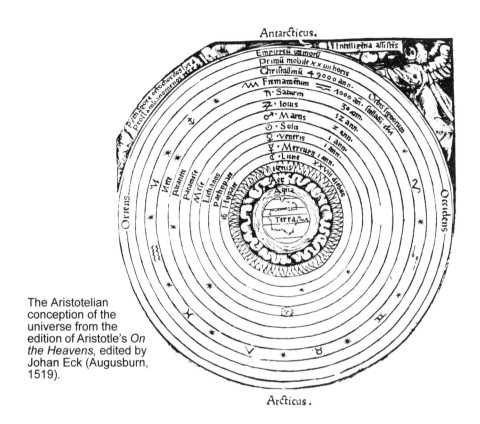

The Aristotelian conception of the universe from the edition of Aristotle's *On the Heavens,* edited by Johan Eck (Augusburn, 1519).

The Superlunar World

The quintessence is crystalline and invisible. That is, it is a material much like glass. It fills space completely, but you can see through it as though nothing is there. It was necessary of course for Aristotle to postulate the existence of the quintessence because he found the existence of a vacuum to be logically impossible. The stars and planets moved continuously. They did not move to a certain natural place and then stop, so there must be a different notion of natural motion in the heavens. How can something be moving forever, but also be in its natural place all the time? Its motion must somehow be motionless. Here comes another element of forced logic: If the heavenly bodies move in perfect circles, they travel on a path that does not change. The circle is fixed even though the object on it is moving. (This was not a new idea with Aristotle. It was almost universally adopted by Greek philosophers of his era.)

But Aristotle was not content to leave the heavens as some sort of mystery governed by arcane mathematical principles. He had to make physical sense out of it all. The general belief of the time was that the universe was spherical in shape, that at the outer edge of the universe was a spherical shell that carried the stars around with them, and that the planets were located between the stars and the earth. Moreover, the complex patterns of the motions of the planets had been given an explanation by Eudoxus.

Aristotle was prepared to accept the general arrangement of the planets and the spheres within spheres proposed by Eudoxus, but he had one serious problem with this scheme. Recall that for Aristotle, motion needed to be caused. It didn't just happen. What was making the spheres in the heavens rotate, taking the planets around with them? Eudoxus had proposed 27 spherical shells in order to carry round the stars and the planets (including the sun and moon). But each set of spherical shells belonging to one planet was independent of those belonging to another planet. What made any of them move? Aristotle did not find this satisfactory.

The solution that struck Aristotle as logically necessary was that at the outer edge of the universe there was something that caused all the motion within. That was the "unmoved mover." It caused the sphere carrying the stars to turn round every 24 hours. Connected to that was a sphere that turned on a somewhat different axis that was pushed around by a sort of friction contact with the sphere of the stars. And so on. Basically Aristotle adopted the general scheme of Eudoxus for explaining planetary motions, but in order to provide for a direct contact between the spheres so that one could push another around, he added several more intermediate spherical shells with different axes of rotation. He figured that the number required was 55 in all.

This complicated and convoluted picture of the universe was where Aristotle's empirical system aided by logic took him. At every stage of the way he had arguments to support his conclusions. With few exceptions, his arguments were logically valid. By present day standards, we would say his conclusions are nonetheless actually false. But they were immensely popular.

Chapter Six
Aristotle and the World of Experience

It's really extraordinary how Aristotle's work was received. His systematic approach of reasoning from observed facts provided a sound footing for building a coherent and realistic understanding of nature. His more scientific works (e.g., *The History of Animals*) is a model of careful observation and steady accumulation of facts (rather like the style of Darwin's works). Science, as we now understand it, would be nowhere without a foundation of work like this. Similarly, his logical analysis and powerful arguments provided a model of systematic theory building that can pull together the mass of observations into a cogent and reasonable understanding of nature.

Unfortunately, that is not enough. The task that Aristotle undertook was much too great for one man or one school of researchers to accomplish. It took only a few premature generalizations at the early stages to completely wreck the final conclusions. Aristotle's methodology was sound, but most of his conclusions were wrong.

Aristotle was such a powerful thinker that his works dominated philosophy for a very long time. But it's a lot of very hard work to slog through what we have of Aristotle's works and see just what the man was saying. It's far easier to let him do the thinking and just take his results.

People like answers. Aristotle provided answers to almost everything. Over the 1800 years or so after Aristotle's life, his work gained an almost scriptural authority. His wrong conclusions gained a dogmatic status. So, to the fresh thinkers of the Renaissance, Aristotle often appeared to be a wrong-headed armchair philosopher instead of the founder of the empirical study of nature.

For More Information

Alioto, Anthony M. *A History of Western Science,* 2nd ed. Englewood Cliffs, NJ: Prentice-Hall, 1987. Chapter 4.
Aristotle. *Basic Works of Aristotle*, ed. by Richard McKeon. New York: Random House, 1941.
_____. *The Complete Works of Aristotle*, ed. by Jonathan Barnes. Princeton: Princeton University Press, 1984.
Barnes, Jonathan. *The Cambridge Companion to Aristotle.* Cambridge: Cambridge University Press, 1995.
Cohen, I. Bernard. *The Birth of a New Physics,* Rev. ed. New York: Norton, 1985. Chapter 2.

Part I
Ancient Beginnings

Clagett, Marshal. *Greek Science in Antiquity.* New York: Collier, 1963.

Internet Encyclopedia of Philosophy. "Aristotle." http://www.utm.edu/research/iep/a/aristotl.htm

Lindberg, David C. *The Beginnings of Western Science: The European Scientific Tradition in Philosophical, Religious, and Institutional Context, 600 B.C. to A.D. 1450.* Chicago: University of Chicago Press, 1992. Chapter 3.

Lloyd, G. E. R. *Early Greek Science: Thales to Aristotle.* London: Chatto & Windus, 1970.

Lynch, John Patrick. *Aristotle's School: A Study of a Greek Educational Institution.* Berkeley and Los Angeles: University of California Press, 1972.

O'Connor, J. J. and E. F. Robertson. "Aristotle." http://www-history.mcs.st-andrews.ac.uk/history/Mathematicians/Aristotle.html

Ronan, Colin A. *Science: Its History and Development Among the World's Cultures.* New York: Facts on File, 1985. Chapter 2.

Chapter Six
Aristotle and the World of Experience

93

The Temple of Hephaistos, Athens.

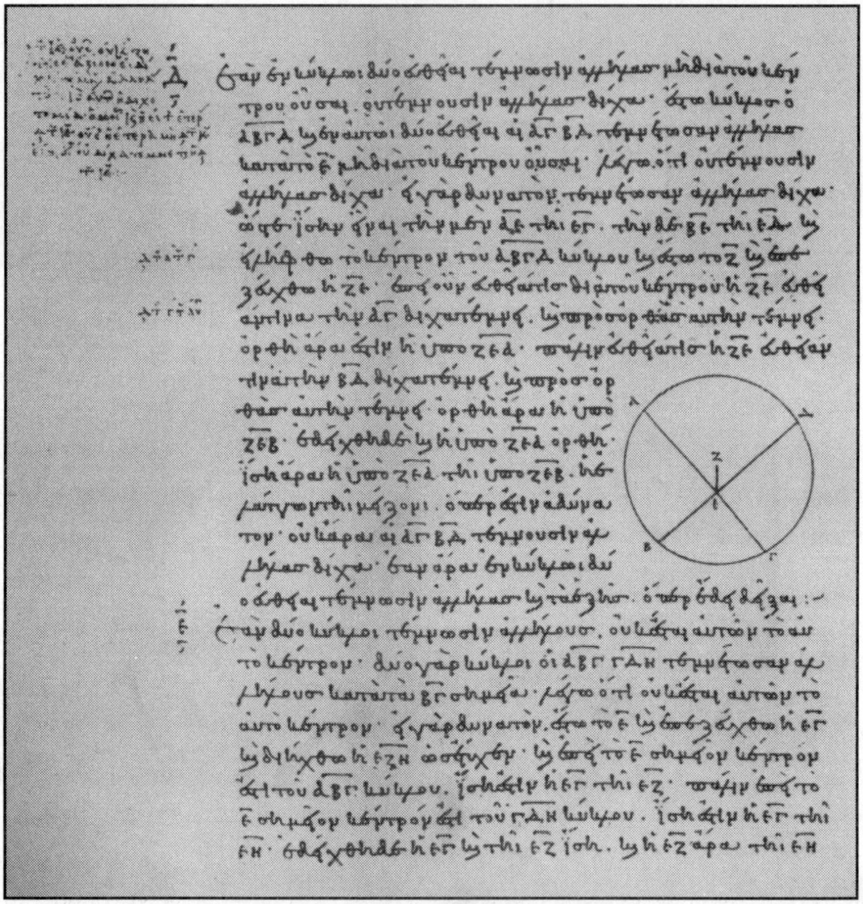

A page from a Greek text of Euclid's *Elements*.

Chapter Seven

Axioms and Proofs

Recall that for Plato, logic and mathematics had a special place in the hierarchy of what was real and important. It fit on the Divided Line above the major division, but on the lower part of that upper main section. It was the gateway to real knowledge and understanding. It also provided the entrance requirement for admission to the Academy. "Let no one who does not know geometry enter here" was inscribed above the entrance. To know mathematics was for Plato to know the rudiments of clear thinking. However, what made mathematics a lesser pursuit to other philosophical investigations was that mathematics was a one-way process of reasoning. One began with certain definitions and axioms which were to be taken for granted and then proceeded to show what logically followed from those via careful reasoning. A wonderful exercise, but Plato believed you could ultimately learn to dispense with the axioms and discover the true foundation of knowledge and being by reasoning backwards to the original Forms.

Aristotle, on the other hand, distrusted mathematics because it easily got away from tangible experience into a world of its own. But Aristotle's reasoning process had much in common with the mathematical approach. Aristotle built sure knowledge, in his view, by beginning with inductive generalizations about the world—his major premises—which he believed to be well-established inductions based upon ample experience, and therefore reliable. Then, to learn something he did not know directly in this way, he would apply syllogistic reasoning, using the rules of logic to derive results that he would not otherwise be able to reach. This was very much the same process as that which Plato described as mathematics. Recall the passage from the *Republic* where Socrates explains to Glaucon the nature of the mathematics portion of the Divided Line by outlining how geometry is pursued.

> You are aware that students of geometry, arithmetic, and the kindred sciences assume the odd and the even and the figures and three kinds of angles and the like in their several branches of science; these are their hypotheses, which they and everybody are supposed to know, and

> therefore they do not deign to give any account of them either to themselves or others; but they begin with them, and go on until they arrive at last, and in a consistent manner, at their conclusions? [For the whole quotation, see Chapter 4.]

Though Aristotle distrusted mathematics overall, his method of building knowledge—empirical inductions first, followed by logical deduction—is remarkably similar to Plato's description of mathematical studies. Considering the enormous influence Aristotle's thinking came to have for so long, this endorsement of reasoning carefully from the known to the unknown cannot but have given systematic mathematical studies a boost.

Plato's Academy was the site of much mathematical research and the academic home of several of the great mathematicians of ancient times. Eudoxus of Cnidos, for example, who "saved" the problem of the planets with his system of nested spherical shells, also produced some important works in pure mathematics while at the Academy. It is likely, though not certain, that one of the scholars working at the Academy in subsequent years was Euclid, author of *The Elements*.

Euclid

Little is known about the life of Euclid. A reasoned guesstimate of his date of birth is 325 BCE, but where it occurred is subject to much dispute. It is generally believed that he died in Alexandria in Egypt, and the estimated date of death is 265 BCE. Part of the confusion about his personal details is that Euclid was a fairly common name in that time and there are several prominent Euclids with whom he may have become confused. However, vague as this may be, it is pretty well established that he spent his productive years in Alexandria.

The Hellenistic Period

In his short life, Alexander the Great expanded Greek influence far beyond the traditional Greek-speaking city-states and colonies that constituted what might be called "Greece" at the time. Beginning from Macedonia, he conquered the rest of Greece, Mesopotamia, Palestine, Syria, Persia, and even made it into India. To the south, he conquered Egypt, where a city, Alexandria, was founded to honor him near the mouth of the Nile. Alexander had planned to establish a grand empire to be based in Babylon in Mesopota-

Chapter Seven
Axioms and Proofs

mia, but while there he contracted a fever and died at the age of thirty-three. Alexander's empire had been held together by the considerable force of his personality. When he died suddenly, the empire collapsed and was divided up by his generals. Alexander's death marks the traditional date of the end of the Hellenic ("Greek") period, when Greek culture was concentrated in the Greek-speaking city-states and had a compact and coherent tradition of cultural works of art, literature, and philosophy that formed the basis of the education of all learned people within its sphere.

But Alexander's conquests had spread Greek influence far and wide, putting it into direct contact with all the rich complexity of the Near East and India. Thus, began a new era of Greek influence, called the Hellenistic ("Greek-ish") period when the culture of Greece was spread to these other countries and mingled with it. The emphasis in Greek philosophy changed from creating all-encompassing systems to focusing on narrower problems.

Ptolemy Soter (Ptolemy the Preserver), a relative of Alexander the Great, was his general in Egypt and upon the division of Alexander's empire, became Egypt's new ruler, beginning a period of Greek pharaohs. Ptolemy established a great center of learning in the newly established city of Alexandria: the Museum. The Museum at Alexandria was so called because it was a temple to the Muses—the nine daughters of Zeus who were the patrons of culture. What the Museum was in fact was a research university devoted to the study of scientific problems. The centerpiece of the Museum was the Library, which housed the greatest collection of books in antiquity, estimated at about 600,000 papyrus rolls.

It was not long before the Museum at Alexandria became the center of all serious work in scientific subjects. Scholars came from all around Europe, North Africa, and the Middle East to study there. Ptolemy Soter sought the best scholars to lead the research efforts there. According to some sources, he appointed Euclid to head up the division devoted to mathematical studies.

There are several works attributed to Euclid; some of them may have actually been written by others. They include several works of pure geometry, a text on geometrical optics (perspective), another on mathematical astronomy, a book on logical fallacies, and a book on music theory. Today Euclid is remembered for one and only one work, his monumental *Elements*.

The Elements of Euclid

The ancient Greeks had many extraordinarily creative and important mathematicians among them. But if Euclid were to be judged solely from *The Elements*, he would not be included among them. *The Elements* is a derivative work. It is an assembly of much of the mathematics known at the time. It is a textbook for students. But it is also the most influential mathematical work of all time.

How the book was put together, how long it took to compile it, when it first appeared, whether it was used as a text in the Museum—we know none of these things. The usual date associated with the book is 300 BCE. That this is such a nice round figure, midway through the estimated life span for Euclid, suggests that it is really only given to situate the book in the history of mathematics: before this, after that, etc.

What makes *The Elements* such an extraordinary, definitive, and seminal work is its organization. The entire work is one coherent logical argument that builds from stated assumptions and carefully establishes one proposition after another until virtually everything that was known in mathematics in Euclid's time is shown to fit together in a single interlocking mass of mathematics.

The Elements is divided into thirteen sections, called "books." Book I begins just as Plato described mathematical study in *The Republic*: Mathematicians, Plato said, begin with assumptions which are not themselves discussed, and proceed on in a consistent manner to their conclusions. Euclid does the same; that is, he does not discuss his assumptions. But importantly, he does list them. In this regard, the structure of *The Elements* resembles Aristotle's arguments about the natural world. Aristotle would begin with stated assumptions, such as that the Earth is at rest in the center of the universe, and then go on to show what logically followed from those assumptions. Aristotle's stated assumptions were general statements based on inductions from observations. They were assumptions in that they were taken to be true without demonstration then and there, but the idea was that they were clearly and apparently true. These were the premises in his syllogisms. If

Chapter Seven
Axioms and Proofs

the premises were true, the conclusion must be true because the argument is logically valid.

The Definitions

It is much the same in Euclid. Euclid begins his treatise with a list of definitions. They are all of well-known mathematical terms, such as point, line, circle, and so on. Other mathematical works might have omitted this step, but Euclid makes sure everyone is using the terms in the same way and attributing nothing to them other than what is stated in the definition. For example, Euclid defines a right angle as what is produced when two straight lines cross each other making the angles on adjacent sides equal to each other.

These are some of the definitions in Book I:

1. A *point* is that which has no part.

2. A *line* is breadthless length.

3. The extremities of a line are points.

4. A *straight line* is a line which lies evenly with the points on itself.

5. A *surface* is that which has length and breadth only. ...

10. When a straight line set up on a straight line makes the adjacent angles equal to one another, each of the equal angles is *right*, and the straight line standing on the other is called a *perpendicular* to that on which it stands. ...

15. A *circle* is a plane figure contained by one line such that all the straight lines falling upon it from one point among those lying within the figure are equal to one another;

16. And the point is called the *centre* of the circle....

19. *Rectilineal figures* are those which are contained by straight lines, *trilateral* figures being those contained by three, *quadrilateral* those contained by four, and *multilateral* those contained by more than four straight lines.

20. Of trilateral figures, an *equilateral triangle* is that which has its three sides equal, an *isosceles triangle* that which has two of its sides alone equal, and a *scalene triangle* that which has its three sides unequal.

21. Further, of trilateral figures, a *right-angled triangle* is that which has a right angle, an *obtuse-angled triangle* that which has an obtuse angle, and an *acute-angled triangle* that which has its three angles acute. ...

23. Parallel straight lines are straight lines which, being in the same plane and being produced indefinitely in both directions, do not meet one another in either direction.

The Postulates and Common Notions

To these definitions, Euclid then adds five postulates and five "common notions." Three of the postulates simply assert that it is possible to draw certain figures, and he will not have to show how.

1. To draw a straight line from any point to any point.
2. To produce a finite straight line continuously in a straight line.
3. To describe a circle with any centre and distance.

Notice that this is important for assuring that he can discuss the pure Platonic form of a figure, not some physical representation of it. The circle that he postulates he can draw is a figure that matches the definition he gives of a circle. None of his discussion will have to deal with the imperfections of any actual drawing. The fourth postulate asserts:

4. That all right angles are equal to one another.

This is a feature of right angles that he wants to use but which was not absolutely implied in the definition.

Then there is the famous fifth postulate concerning parallel lines:

5. That, if a straight line falling on two straight lines make the interior angles on the same side less than two right angles, the two straight lines, if produced indefinitely, meet on that side on which are the angles less than the two right angles.

According to Euclid's Fifth Postulate, since the sum of angles *a* and *b* is less than two right angles, or 180 degrees, lines *de* and *fg* will, when extended, meet at point *c*.

Chapter Seven
Axioms and Proofs

It's called the parallel postulate, but notice that what it gives is a criterion for *not* being parallel. The usual modern version of this postulate turns this about and says that if the interior angles on the same side do equal two right angles, then the lines are parallel, or alternately, that one and only one line can be drawn through a point parallel to a given line (in the same plane).

This postulate caused a great deal of controversy in later years as mathematicians pored over the work looking for minor errors or improvements that could be made. This postulate seemed to state a self-evident fact. Surely, it was thought, it need not be stated as an assumption. It could be shown to be logically implied by the other definitions and postulates.

What this shows is that Euclid's *Elements* were seen as a system of maximum results from minimum assumptions. If something did not have to be assumed, it should not be assumed.

For centuries, mathematicians tried to prove the fifth postulate from the rest of the treatise. It did not work. Eventually it was realized that you cannot prove it from the rest of Euclid's *Elements*, and if you instead assumed something contrary (such as that there are no such thing as parallel lines), then you get an entirely different kind of geometry.

Finally, at the end of the beginning, as it were, there are the Common Notions:

1. Things which are equal to the same thing are also equal to one another.

2. If equals be added to equals, the wholes are equal.

3. If equals be subtracted from equals, the remainders are equal.

4. Things which coincide with one another are equal to one another.

5. The whole is greater than the part.

What's common about the Common Notions is that they are really very general principles that are not specific to geometry. These are virtually logical precepts rather than mathematical assumptions. Nevertheless, Euclid specified them because he planned to use them to draw inferences.

Altogether, the Definitions, Postulates, and Common Notions provide a complete list of what must be taken as true to begin with. These are the *axioms*.

The Axiomatic System

With this somewhat tedious and labored beginning, Euclid is ready to produce some mathematical results. From here on, nothing else is assumed. (To be more precise, some later scholars have quibbled that Euclid sneaks in a few other assumptions that really should have been laid out and uses some technical terms that he did not define, but these are very minor.) The idea now is that nothing else has to be assumed in order to prove a whole lot of mathematical theorems. That means that *if the arguments that follow are valid*–that is, they violate no rules of logic, *then the conclusions of the arguments must be true, provided that the assumptions are true*. In its entirely, Euclid's works, beginning with the stated axioms and going on to the demonstrated propositions, is a model example of an *axiomatic system*.

What really makes this so remarkable is that this is precisely what Aristotle had hoped for in his own systematic thought: begin with statements that can be reasonably taken as true—the axioms—and deduce results that must also be true. Where Aristotle went wrong himself was in taking as axioms generalizations that were unwarranted. An axiomatic system is only as good as the assumptions it makes at the beginning. If these are wrong in any detail, the deduced results will be completely unreliable.

Aristotle had a sound methodology for producing a body of complex knowledge, but he had few hard facts to begin with. Euclid was much more successful because he confined himself to very small beginnings. Euclid's axioms were assumptions and by the standards of present-day mathematics, they would be considered arbitrary assumptions, but clearly as far as the ancient experience was concerned, they could be taken as self-evident truths. What Euclid began from were simple statements about simple geometric objects that no one could or would deny. Therefore, what he was able to build from that was, by all accounts, a body of truth.

Chapter Seven
Axioms and Proofs

The Propositions

To see the true virtue of putting all the assumptions aside before starting on the mathematical demonstrations, it is useful to look at a few of Euclid's demonstrations. The key here is that everything builds sequentially. At the very beginning, Euclid can only call upon the axioms themselves to justify any deduction. After he has established one proposition, he may add that to the axioms as something he can take to be true. Every proposition he proves is another statement he can take to be true and use to find another result.

To start at the beginning, here is Proposition 1 of Book I. It is a simple geometrical construction. In his postulates, Euclid took for granted that he can draw a straight line between two points and he can draw a circle of any given radius. Proposition 1 shows that those tools also allow him to make an equilateral triangle:

> 1. On a given finite straight line to construct an equilateral triangle.
>
> Let *AB* be the given finite straight line.
>
> Thus it is required to construct an equilateral triangle on the straight line *AB*.

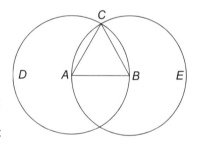

> With centre *A* and distance *AB* let the circle *BCD* be described. [NOTE: This is allowed because it is Postulate 3.]
>
> Again, with centre *B* and distance *BA* let the circle *ACE* be described [Postulate 3 again.]
>
> And from the point *C*, in which the circles cut one another, to the points *A*, *B* let the straight lines *CA*, *CB* be joined. [This is allowed by Postulate 1.]
>
> Now, since the point *A* is the centre of the circle *CDB*, *AC* is equal to *AB*. [This follows from Definition 15, the definition of a circle.]
>
> Again, since the point *B* is the centre of the circle *CAE*, *BC* is equal to *BA*. [Definition 15 again.]
>
> But *CA* was also proved equal to *AB*; therefore each of the straight lines *CA*, *CB* is equal to *AB*.
>
> And things which are equal to the same thing are also equal to one another. [This is Common Notion 1.]

>Therefore *CA* is also equal to *CB*.
>
>Therefore the three straight lines *CA, AB, BC* are equal to one another.
>
>Therefore the triangle *ABC* is equilateral. [Definition 20.]
>
>And it has been constructed on the given finite straight line *AB*.
>
>Being what it was required to do.

The proposition is extremely simple, but there is not a step in the proof that cannot be referred back to a stated assumption. And the entire book proceeds in this same fashion. Now Euclid can use equilateral triangles in other proofs, because he has demonstrated that he can construct one using only the abilities that he assumed from the beginning in the axioms.

Proposition I.47

Each different book of *The Elements* has a different set of related propositions to be proved. Book I contains many of the familiar theorems of plane geometry. Other books consider solid geometry, theory of proportions, and number theory. These are basically all the main branches of ancient Greek mathematical thought. Book I has a dramatic culmination in that the most famous of all theorems of plane geometry, the Pythagorean theorem. Euclid's proof of this theorem is ingenious and surprising. Since this theorem is both famous and important, it is the perfect capstone for his first book.

I have referred to the theorem as the "Pythagorean theorem" because that is how it is popularly known today, giving the credit for its discovery as a general theorem to Pythagoras of Samos who lived about 250 years before Euclid. Pythagoras was reputed to have had a proof of the theorem, but no one now knows what that was. Euclid's proof seems to be his own. But for many of the thousands of students who learned the rudiments of mathematics from Euclid, this theorem was better known simply by its number in Euclid's treatise, I.47, the 47th theorem in Book 1.

Proposition I.47 comes at the end of Book I (actually there is one more after it) because to get to it, he has to establish a number of intermediate results, which he does in earlier propositions. In particular, Euclid requires the results of propositions I.4, I.14, and I.41 as steps in his proof of I.47. Briefly, these propositions show the following: Proposition I.4 is the familiar side-angle-side theorem of plane geometry that states that if two triangles have two sides of one triangle equal to two sides of the other triangle plus the angle between the sides that are equal in each triangle is the same, then the two triangles are congruent, i.e., they are essentially the same triangle. The wording in Euclid (in Thomas Heath's translation) is:

> If two triangles have the two sides equal to two sides respectively, and have the angles contained by the equal straight lines equal, they will also have the base equal to the base, the triangle will be equal to the triangle, and the remaining angles will be equal to the remaining angles respectively, namely those which the equal sides subtend.

Proposition I.14 asserts that two right angles next to each other make a straight line. Recall that Definition 10 stated that the angles formed on either side of a line perpendicular to another were right angles. Euclid wants to assert the converse, that if you put two right angles together, each having the same point as a vertex and sharing one side, then the other two sides of the angles make a straight line.

> If with any straight line, and at a point on it, two straight lines not lying on the same side make the adjacent angles equal to two right angles, the two straight lines will be in a straight line with one another.

Proposition I.41 asserts that the area of a triangle is one-half the area of a parallelogram having the same base and height.

> If a parallelogram have the same base with a triangle and be in the same parallels, the parallelogram is double of the triangle.

In addition to these, Euclid needs to be able to construct figures meeting certain specifications. He established that he can do this in propositions I.31 and I.46. I.31 asserts that given a point, a line may be constructed parallel to another line. I.46 asserts that given a straight line, a square can be constructed.

Now, using these propositions and any of the definitions, postulates, and common notions that he needs, Euclid shows that in a right triangle, squares built on each of the sides of the triangle that form the right angle added together are equal in area to the square built on the side across from the right angle.

Part I
Ancient Beginnings

For the proof, Euclid constructs squares on each of the sides of the right triangle and then divides the square across from the right angle into two rectangles by drawing a line from the vertex of the right angle parallel to the sides of that square. From there on, the proof shows that each of the *other* two squares is equal in area to one of the rectangles which together make the larger square. In the most interesting and ingenious part of the proof, Euclid draws lines connecting vertices of the squares opposite them, thus creating triangles which have bases and heights equal, respectively, to those of the rectangles which form the larger square and those of the other two squares. Euclid then shows that pairs of these (new) triangles are congruent to each other (and therefore have equal area).

Since squares and rectangles are also parallelograms, the area of each triangle is exactly half of the area of one of the squares or rectangles. Therefore each smaller square is equal in area to one of the rectangles, therefore the sum of the areas of the smaller squares is equal to the area of the larger square.

Here it is as expressed by Euclid.

I.47 In right-angled triangles the square on the side subtending the right angle is equal to the squares on the sides containing the right angle.

Let *ABC* be a right-angled triangle having the angle *BAC* right;

I say that the square on *BC* is equal to the squares on *BA*, *AC*.

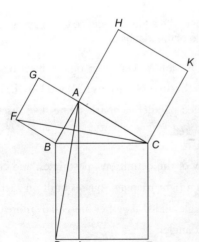

For let there be described on *BC* the square *BDEC*, and on *BA*, *AC* the squares *GB*, *HC*; [Proposition I.46] through *A* let *AL* be drawn parallel to either *BD* or *CE*, and let *AD*, *FC* be joined.

Then, since each of the angles *BAC*, *BAG* is right, it follows that with a straight line *BA*, and at the point *A* on it, the two straight lines *AC*, *AG* not lying on the same side make the adjacent angles equal to two right angles; therefore *CA* is in a straight line with *AG*. [Proposition I.14]

For the same reason *BA* is also in a straight line with *AH*.

Chapter Seven | 107
Axioms and Proofs

And, since the angle *DBC* is equal to the angle *FBA*: for each is right: let the angle *ABC* be added to each; therefore the whole angle *DBA* is equal to the whole angle *FBC*. [Common Notion 2]

And, since *DB* is equal to *BC*, and *FB* to *BA*, the two sides *AB, BD* are equal to the two sides *FB, BC* respectively, and the angle *ABD* is equal to the angle *FBC*; therefore the base *AD* is equal to the base *FC*, and the triangle *ABD* is equal to the triangle *FBC*. [Proposition I.4]

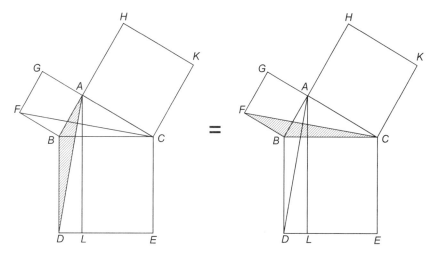

Now the parallelogram *BL* is double of the triangle *ABD*, for they have the same base *BD* and are in the same parallels *BD, AL*. [Proposition I.41]

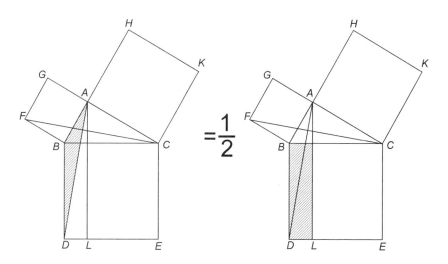

And the square GB is double of the triangle FBC, for they again have the same base FB and are in the same parallels FB, GC. [Proposition I.41 again]

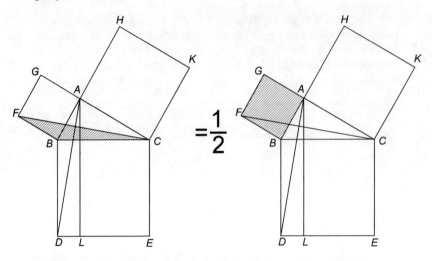

[But the doubles of equals are equal to one another.] Therefore the parallelogram BL is also equal to the square GB.

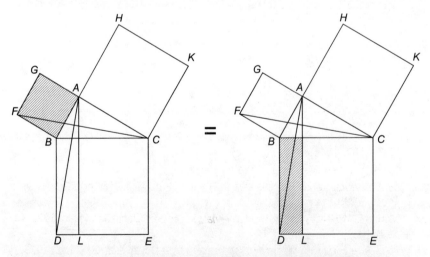

Similarly, if AE, BK be joined, the parallelogram CL can also be proved equal to the square HC; therefore the whole square BDEC is equal to the two squares GB, HC. [Common Notion 2]

And the square *BDEC* is described on *BC*, and the squares *GB*, *HC* on *BA*, *AC*.

Therefore the square on the side *BC* is equal to the squares on the sides *BA*, *AC*.

Therefore etc.

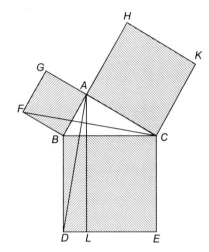

This is what mathematicians call elegant. Each step is simple enough, but the result is not obvious until the end. We don't know what proof Pythagoras found, but he was said to have sacrificed a hundred oxen to honor the occasion. As for Euclid's proof, a long line of famous mathematicians and scientists have said that they realized that they had to pursue a career in mathematics or science when they first read and appreciated the clarity and beauty of Euclid's *Elements*. Several of those specifically mention this proof.

For More Information

Alioto, Anthony M. *A History of Western Science*, 2nd ed. Englewood Cliffs, NJ: Prentice-Hall, 1987. Chapter 5.

Clagett, Marshal. *Greek Science in Antiquity.* New York: Collier, 1963.

Euclid. *The Elements*, trans. Thomas Heath. 3 vols. Cambridge: Cambridge University Press, 1908.

Heath, Thomas L. *A History of Greek Mathematics.* Oxford: Clarendon Press, 1921. Chapter 11.

Knorr, Wilbur. *The Evolution of the Euclidean Elements.* Dordrecht: D. Reidel, 1975.

Lloyd, G. E. R. *Greek Science after Aristotle.* London: Chatto & Windus, 1973.

Neugebauer, Otto. *The Exact Sciences in Antiquity,* 2nd ed. New York: Dover, 1969.

O'Connor, J. J., and E. F. Robertson. "Euclid of Alexandria" http://www-history.mcs.st-andrews.ac.uk/history/References/Euclid.html

Ronan, Colin A. *Science: Its History and Development Among the World's Cultures.* New York: Facts on File, 1985. Chapter 2.

A woodcut depicting the Ptolemaic Universe.

Chapter Eight

Saving the Heavens

Euclid's treatise served a number of purposes. It was a secure foundation for increasing mathematical knowledge. It was a ready toolbox of techniques and formulations to help solve calculation problems of all sorts. And, it was a model of how to build a complex system of knowledge. It served these purposes for a very long time. Of more immediate significance was the milieu in which it was composed, the research-oriented Museum of Alexandria. The Museum was quickly becoming the center of all serious scientific research in the Mediterranean area. Having Euclid's *Elements* as a text for the training of mathematicians must surely have given a great advantage to those who studied there.

Whether due to Euclid or to other factors, Alexandria was where answers were found to many of the questions about nature that were asked back in the Hellenic age. Or, to put it in Platonic terms, where many of the phenomena of nature were "saved." Saving phenomena was showing how they could be explained as natural results of unchanging (mathematical) principles (see Chapter 5 for more details). Euclid provided the analysis that helped account for these connections.

Eratosthenes and the Size of the Earth

As an example of how Euclid could be used to solve seemingly intractable problems, consider the question of how large the Earth is. The only reasonably reliable information anyone had came from reports by travelers to distant places who could estimate how far away some place was that they had visited—the Orient, for example, or south into Africa, and these estimates were notably unreliable, especially those involving long treks over irregular terrain where distances actual distances would be difficult to distinguish from the distance on the ground as it was traveled.

But in the century after Euclid, another researcher at the Museum in Alexandria found an ingenious way to calculate the size of the Earth using one fairly reliable traveler's report and the size of a shadow on the ground. The key for putting these together was a proposition from Book I of Euclid's *Elements,* Proposition I.29:

> A straight line falling on parallel straight lines makes the alternate angles equal to one another, the exterior angle equal to the interior and opposite angle, and the interior angles on the same side equal to two right angles.

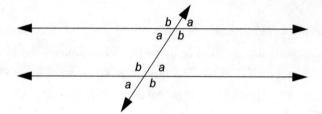

All angles marked *a* are equal, as are all marked *b*.

The clever person who figured out how to use this proposition was Eratosthenes, an extraordinarily productive scholar of Hellenistic times. He was born around 276 BCE in the Greek colony of Cyrene in North Africa, which is now the town of Shahhat, Libya. One of Eratosthenes' teachers in Cyrene was the poet and scholar Callimachus, who was the second person appointed as the Librarian of the vast collection at the Museum at Alexandria. As a young man Eratosthenes traveled to Athens and studied at Plato's Academy, where he showed much promise. On the death of Callimachus, Eratosthenes was appointed his successor as the Librarian at the Museum. Eratosthenes worked in many different subjects. He did important work in astronomy, in mathematics, in geography (in fact he coined the word "geography" and drew maps of the world), and he wrote both poetry and history. His prolific accomplishments in many fields earned him the nickname "Beta"—the second letter in the Greek alphabet—with the implication that he was not the best in the world at anything, but he was the second best at a great many things.

Eratosthenes

According to legend, Eratosthenes had heard that in Syene in Lower Egypt (now the city of Aswan), there was a deep well where, at noon on the summer solstice (June 21 on our calendar) the sun would shine directly down into the well so that a person looking down would see its reflection. Thus, the sun was directly overhead at that time, a most unusual event. On investigating, Eratosthenes also learned of a fairly accurate estimate of the dis-

Chapter Eight
Saving the Heavens

113

tance from Syene to Alexandria, because there was a well-traveled camel route over the desert between the two cities for trade. He realized that he could use these two pieces of information, plus one measurement that he could easily make in Alexandria, apply Euclid's Proposition I.29, and come up with a calculated figure for the size of the Earth.

He reasoned as follows:

1. The Sun is large and very far away, so any light coming from it to the Earth can be regarded as traveling in parallel lines.

2. The Earth is a perfect sphere. (This was the almost universal assumption among ancient scientists in Eratosthenes' time.)

3. A vertical shaft (like the well in Syene) and a *gnomon* (a stick placed upright perpendicular to the ground) can both be in principle extended downwards and will pass directly through the center of the Earth.

4. Alexandria is directly north of Syene, or close enough for these purposes.

The one other piece of information he needed was the angle of the sun when it was at its zenith (i.e., as high as it will get) on the summer solstice in Alexandria. This was a simple enough measurement to make. He placed a gnomon in the ground at Alexandria on the day of the solstice and watched it as noon approached. At noon (when the sun was at its zenith), the shadow of the gnomon would be its shortest and pointed directly north. He then measured the length of the shadow and from that determined the angle at which the sun met the top of the gnomon.[1] Now he was ready to apply Proposition I.29. The angle that he had measured in Alexandria could be seen as an interior angle produced by a line crossing two parallel lines. The parallel lines were the sun's rays striking the Earth in Syene and Alexandria. The line crossing those parallel lines was the line produced by the gnomon and extended all the way to the center of the Earth, where it would cross the extension of the sun's rays from Syene.

---

1. He could do this by connecting the top of the gnomon to the far edge of the shadow with a straight-edge and then measuring the angle between gnomon and his straight-edge, or he could construct a right triangle with the length of the gnomon and the length of the shadow as sides, then measure the angle.

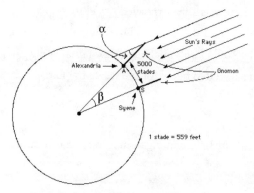
Eratosthenes' method of calculation.

Euclid's proposition showed that the angle measured from the center of the Earth between Syene and Alexandria was the same as the angle produced by the sun's rays off the gnomon in Alexandria. So, by measuring a small shadow in Alexandria at the right time, Eratosthenes determined the angular distance between the two cities as it would be "seen" from the center of the Earth. That angle turned out to be 7°12′, which, fortuitously, is exactly one-fiftieth of a circle.

Now for the camel route. The estimated land distance from Syene to Alexandria was 5000 stades. A stade was the standard long measure of distance in ancient Greece. Eratosthenes had determined that the angle between Syene and Alexandria was one-fiftieth of a circle, and the distance on the surface of the Earth was 5000 stades. Therefore, he reasoned, the circumference of the Earth must be 50 times 5000 stades, or 250,000 stades.

How close was he? Well, it depends on what the size of the stade is, and about that there is much disagreement among scholars. A stade is the length of the running track at an ancient Greek stadium. (A stade is where the word stadium comes from.) These varied. The stade at the Athens stadium was 185 m. Other Greek stadia were close to this figure. However, the term was also used in Egypt, where Eratosthenes was, and there the common figure was 157.5 m. Also a bit later, Eratosthenes did a more careful measurement and determined that the correct figure was 252,200 stades.

Then there is the question of the shape of the Earth. Eratosthenes, as mentioned, assumed the Earth to be a perfect sphere, but in fact it has a bit of a bulge around the middle (as a result of rotation), and therefore its circumference around the equator is greater than its circumference measured on a great circle going through the poles. We therefore have at least two measures of the stade, two calculations by Eratosthenes of the number of stade required, and two modern measures of the circumference.

Chapter Eight
Saving the Heavens

115

The four possible Eratosthenes' calculations and the two modern figures are:

Eratosthenes' figures

Stade Length	Number of Stades	Circumference in km.
157.5 m	250,000	39,375
157.5 m	252,000	39,690
185 m	250,000	46,250
185 m	252, 000	46,620

Modern figures, measured by orbiting spacecraft

Polar Circumference	39,942
Equatorial Circumference	40,074

Even taking the very worst interpretation of Eratosthenes' figures, he was off by less than 17%. However, if you accept that the Egyptian stade was the likely measure he used, work with his revised number of stades, and compare it to the polar circumference (since after all, he was calculating the circumference in a north-south direction), then you get an error of less than 1%.

More than anything, what Eratosthenes' calculation showed was the power of some very simple mathematical relationships to answer seemingly impossible questions. Philosophers could reason forever about how nature must be, because of their preconceived views, but this could and would all be swept away by mathematical analysis applied to direct measurements.

The Astronomical Problems with New Tools to Solve Them

The problem of explaining the planets' orbits was attacked by Eudoxus in the fourth century BCE, and his basic solution was adopted with refinements by Aristotle where it became the standard view. But while the Eudoxus-Aristotle scheme of concentric spherical shells fit with other Aristotelian precepts, it did not do a very good job of tracking the movements of the planets. It just did not have the mathematical sophistication to model the complex combinations of movements that appeared to be the true paths of the heavenly bodies. This all changed in the Hellenistic period. Euclid's work brought together the existing mathematical knowledge of the time in an organized format. Other mathematicians gathered in Alexandria and began solving many problems, both pure mathematical ones and problems that applied mathematics to questions about the physical world, as Eratosthenes had done.

In about 150 BCE, a man named Hipparchus developed a number of mathematical tools to help in his astronomical studies. One of them, the Table of Chords, is a circle-based equivalent to trigonometry. With it he could calculate relative distances among the stars with a new accuracy. Among the discoveries he made was that there is a very slow process in the heavens that causes annual events such as equinoxes and solstices to move ever so slightly earlier every year. This process, called *precession* (or "precession of the equinoxes") has a cycle of 26,000 years before everything comes back to the same place. Hipparchus could give no reason to explain precession, but what is extraordinary is that he was able to calculate it. In modern terms, it is caused by a tiny wobble of the Earth's axis of daily rotation. Since the ancients took the view that the Earth was still, this would have made no sense to them.

Actually, not everyone was convinced that the Earth was still and in the center of the universe. Aristarchos of Samos, who lived about the time of Eratosthenes, had the notion that the sun was in the center of the world, and the Earth revolved around it annually while rotating daily on its axis. This has all the general features of the system of Copernicus, who lived 1700 years later. But Aristarchos' idea was too far off from the general view to get much of a hearing, so the Earth-centered view remained.

Chapter Eight
Saving the Heavens

Claudius Ptolemy

In around 150 CE, three hundred years after Hipparchus, another scholar at the Museum in Alexandria, Claudius Ptolemy, used the mathematical tools developed by Hipparchus and the wealth of other mathematics developed at the Museum to attack the problem of accounting for the heavens once again—this time with much greater success. What Plato had wanted philosophers to do to "save the phenomena," Ptolemy succeeded in doing, using the vast resources of astronomical observations available to him at the Museum and the more sophisticated mathematics of the late Hellenistic period.

Claudius Ptolemy.

Not much is known about the life of Ptolemy. His family name is the same as the dynasty of Greek pharaohs in the Hellenistic period, but there is no indication that he was related to them. Probably it signifies that he was from a wealthy Greek family that had settled in Egypt some generations before his birth. His given name, Claudius, is Roman, which may have meant that he had been granted Roman citizenship as an honor. His birth and death dates are not known, but he refers to observations he made in the years 127-141 CE, and the date generally assigned to his major work is 150 CE.

Ptolemy wrote several works which have survived, among them are works on astrology, optics, and music theory. More important was his *Geography*, a systematic work containing maps of many parts of the world (in this sense it is an early example of an atlas), as well as details on how to construct maps, taking into account the curvature of the earth. This work, lost for a time during the Dark Ages, was of considerable importance when rediscovered during the Renaissance, when interest in world exploration created a demand for the best maps available before starting a voyage, and knowledge of the principles of perspective made people realize what Ptolemy was getting at with his map drawing procedures. Columbus was said to have sailed to the New World with a set of (recreated) Ptolemy's maps of the known world.

His major work, and the main reason he is remembered today, was his system of astronomy, which Ptolemy called simply *The Mathematical Composition (Mathematical Syntaxis)*. After Ptolemy's time, his work began to be referred to not just as a *Mathemati-*

cal Composition, but more grandly as *The Greatest Composition*. Ptolemy's work is among those many scientific works (including Euclid's *Elements*) that were lost to Europe after the fall of Rome. These works were, however, retained in the Arab world where many of the great libraries were located. Arab scholars simply transliterated the title, taking the Greek word for "greatest," *megiste*, and putting the Arabic word for "the" before it to make *Al Megiste*. When ultimately this work was translated into Latin for the use of European scholars, the Arabic title was the one transliterated, so the work came to be known in Europe as *The Almagest*, which of course literally means "The the Greatest (Composition)." And this is the title which has stuck.

The Almagest followed the lead of Euclid in organization, beginning with statements of the assumptions that will be made without rigorous proof, and then producing, one after another, mathematically derived theorems that follow from them. These are arranged in "books" of different topics (13 books in fact, just like Euclid) that develop the means to plot, track, measure, and predict the movements of the heavenly bodies. Though the work has the general axiomatic structure of Euclid's works, it fails to have the thoroughgoing logical validity that was the hallmark of Euclid because it is not, despite its title, about the world of mathematics per se, but about the world of astronomy. Astronomy is a study of part of the visible world, not just a theoretical subject; therefore, it is necessary that the conclusions drawn in the work match up with observations of astronomical events. Aristotle's model of arguing logically from empirical data must be followed much more closely by Ptolemy than by Euclid, who was in effect developing part of Plato's "intelligible world" that was not accessible to the senses. Therefore, Ptolemy shared with Aristotle the serious problem that he had to begin with good empirical data and had to make sound inductions from that data to get major premises to begin with.

These are severe strictures. Ptolemy's work suffered from some of the same problems that plagued Aristotle's arguments about the physical world: he began with commonsense generalizations, for which he had a certain amount of supporting evidence, but which nevertheless were incorrect. On that foundation he built his elaborate and mathematically valid system, which did an excellent job in the context of the general worldview he began with. Because his original worldview was wrong, his system was ultimately wrong, but that does not mean that it did not have tremendous value. What Ptolemy's *Almagest* showed was that a complex system of the size and scope of Euclid's *Elements*

Chapter Eight
Saving the Heavens

was able to do the job that Eudoxus' and Aristotle's far simpler schemes could not do: save the phenomena of the heavens, i.e., provide a mathematical system that could account for and predict the mysterious movements in the heavens. And so long as the basic error of the major premises was not discovered, Ptolemy's work seemed to be a major triumph of natural philosophy to explain the world.[2]

The Almagest

Ptolemy's system of the heavens is an elaborate model involving combinations of circular motions in an extremely complex manner. By modern standards, that very complexity is a signal that something must be wrong. But to the ancient mind, complexity alone was not a detriment; to the contrary, if that is what it took to map the heavens, so be it. The incredible complexity is a consequence of the "reasonable" assumptions that Ptolemy makes about the shape and layout of the universe. This is where the book begins. The following is an excerpt from Book I where Ptolemy explains the rationale for his worldview, and in particular makes arguments to support the following: (1) that the heavens move spherically, (2) that the Earth is spherical, (3) that the Earth is in the middle of the heavens, (4) that the Earth has the ratio of a point to the heavens—i.e., that the Earth is so small in relation to the heavens that it can be considered to be a point for the purpose of measurements, and (5) that the Earth is immobile. These are all arguments

The opening page of a medieval edition of Ptolemy's *Almagest*.

---

2. The incorrect premises were not the only flaws in Ptolemy. In addition to the general observations that led Ptolemy to get his worldview wrong, he necessarily relied upon a great many direct observations of star and planetary positions to which he had to fit his data. Most of those observations are very good, considering when they were made and the primitive technology available to take accurate measurements. But later scholars detected something else. Ptolemy's models work out just a bit too well. Some of the data he reported have systematic errors that fit his explanations. Ptolemy has been accused of fudging his data. This is certainly not the only major figure in the history of science to be accused of data tampering, but it may have been the first significant one. For details see Robert R. Newton, *The Crime of Claudius Ptolemy* (Baltimore: Johns Hopkins University Press, 1977).

that had been put forth again and again by Greek philosophers from the pre-Socratic philosophers through Aristotle and later. Ptolemy repeats them here to give some justification for making them the axioms on which he will build his system. In themselves they serve very well as a summary of the general ancient view of the cosmos.

That the Heavens Move Spherically

It is probable that the first notions of these things came to the ancients from some such observation as this. For they kept seeing the sun and moon and other stars always moving from rising to setting in parallel circles, beginning to move upward from below as if out of the earth itself, rising little by little to the top, and then coming around again and going down in the same way until at last they would disappear as if falling into the earth. And then again they would see them, after remaining some time invisible, rising and setting as if from another beginning; and they saw that the times and also the places of rising and setting generally corresponded in an ordered and regular way.

But most of all the observed circular orbit of those stars which are always visible, and their revolution about one and the same center, led them to this spherical notion. For necessarily this point became the pole of the heavenly sphere; and the stars nearer to it were those that spun around in smaller circles, and those farther away made greater circles in their revolutions in proportion to the distance, until a sufficient distance brought one to the disappearing stars. And then they saw that those near the always-visible stars disappeared for a short time, and those farther away for a longer time proportionately. And for these reasons alone it was sufficient for them to assume this notion as a principle, and forthwith to think through also the other things consequent upon these same appearances, in accordance with the development of the science. For absolutely all the appearances contradict the other opinions.

If, for example, one should assume the movement of the stars to be in a straight line to infinity, as some have opined, how could it be explained that each star will be observed daily moving from the same starting point? For how could the stars turn back while rushing on to infinity? Or how could they turn back without appearing to do so? Or how is it they do not disappear with their size gradually diminishing, but on the contrary seem larger when they are about to disappear, being covered little by little as if cut off by the earth's surface? But certainly to suppose that they light up from the earth and then again go out in it would appear most absurd. For if anyone should agree that such an order in their magnitudes and number, and again in the distances, places, and times is accomplished in this way at random and by chance, and that one whole part of the earth has an incandescent nature and another a nature capable of extinguishing, or rather that the same part lights the stars up for some people and puts them out for others, and that the same stars

Chapter Eight
Saving the Heavens

happen to appear to some people either lit up or put out and to others not yet so—even if anyone, I say, should accept all such absurdities, what could we say about the always-visible stars which neither rise not set? Or why don't the stars which light up and go out rise and set for every part of the earth, and why aren't those which are not affected in this way always above the earth for every part of the earth? For in this hypothesis the same stars will not always light up and go out for some people, and never for others. But it is evident to everyone that the same stars rise and set for some parts, and do neither of these things for others.

In a word, whatever figure other than the spherical be assumed for the movement of the heavens, there must be unequal linear distances from the earth to parts of the heavens, wherever or however the earth be situated, so that the magnitudes and angular distances of the stars with respect to each other would appear unequal to the same people within each revolution, now larger now smaller. But this is not observed to happen. For it is not a shorter linear distance which makes them appear larger at the horizon, but the steaming up of the moisture surrounding the earth between them and our eyes, just as things put under water appear larger the farther down they are placed.

The following considerations also lead to the spherical notion: the fact that instruments for measuring time cannot agree with any hypothesis save the spherical one; that, since the movement of the heavenly bodies ought to be the least impeded and most facile, the circle among plane figures offers the easiest path of motion, and the sphere among solids; likewise that, since of different figures having equal perimeters those having the more angles are the greater, the circle is the greatest of plane figures and the sphere of solid figures, and the heavens are greater than any other body.

Moreover, certain physical considerations lead to such a conjecture. For example, the fact that of all bodies the ether [i.e., the quintessence, Aristotle's fifth element] has the finest and most homogeneous parts; but the surfaces of homogeneous parts must have homogeneous parts, and only the circle is such among plane figures and the sphere among solids. And since the ether is not plane but solid, it can only be spherical. Likewise the fact that nature has built all earthly and corruptible bodies wholly out of rounded figures but with heterogeneous parts, and all divine bodies in the ether out of spherical figures with homogeneous parts, since if they were plane or disc-like they would not appear circular to all those who see them from different parts of the earth at the same time. Therefore it would seem reasonable that the ether surrounding them and of a like nature be also spherical, and that because of the homogeneity of its parts it moves circularly and regularly.

That Also the Earth, Taken as a Whole, is Sensibly Spherical

Now, that also the earth taken as a whole is sensibly spherical, we could most likely think out in this way. For again it is possible to see that the sun and moon and the other stars do not rise and set at the same time for every observer on the earth, but always earlier for those living towards the orient and later for those living towards the occident. For we find that the phenomena of eclipses taking place at the same time, especially those of the moon, are not recorded at the same hours for everyone—that is, relatively to equal intervals of time from noon; but we always find later hours recorded for observers towards the orient than for those towards the occident. And since the differences in the hours is found to be proportional to the distances between the places, one would reasonably suppose the surface of the earth spherical, with the result that the general uniformity of curvature would assure every part's covering those following it proportionately. But this would not happen if the figure were any other, as can be seen from the following considerations.

For, if it were concave, the rising stars would appear first to people towards the occident; and if it were flat, the stars would rise and set for all people together and at the same time; and if it were a pyramid, a cube, or any other polygonal figure, they would again appear at the same time for all observers on the same straight line. But none of these things appears to happen. It is further clear that it could not be cylindrical with the curved surface turned to the risings and settings and the plane bases to the poles of the universe, which some think more plausible. For then never would any of the stars be always visible to any of the inhabitants of the curved surface, but either all the stars would both rise and set for observers or the same stars for an equal distance from either of the poles would always be invisible to all observers Yet the more we advance towards the north pole, the more the southern stars are hidden and the northern stars appear. So it is clear that here the curvature of the earth covering parts uniformly in oblique directions proves its spherical form on every side. Again, whenever we sail towards mountains or any high places from whatever angle and in whatever direction, we see their bulk little by little increasing as if they were arising from the sea, whereas before they seemed submerged because of the curvature of the water's surface.

That the Earth is in the Middle of the Heavens

Now with this done, if one should next take up the question of the earth's position, the observed appearances with respect to it could only be understood if we put it in the middle of the heavens as the center of the sphere. If this were not so, then the earth would either have to be off the axis but equidistant from the poles, or on the axis but farther advanced

Chapter Eight
Saving the Heavens

towards one of the poles, or neither on the axis not equidistant from the poles.

The following considerations are opposed to the first of these three positions—namely, that if the earth were conceived as placed off the axis either above or below in respect to certain parts of the earth, those parts, in the right sphere, would never have any equinox since the section above the earth and the section below the earth would always be cut unequally by the horizon. Again, if the sphere were inclined with respect to these parts, either they would have no equinox or else the equinox would not take place midway between the summer and winter solstices. The distances would be unequal because the equator which is the greatest of those parallel circles described about the poles would not be cut in half by the horizon; but one of the circles parallel to it, either to the north or to the south, would be so cut in half. It is absolutely agreed by all, however, that these distances are everywhere equal because the increase from the equinox to the longest day at the summer tropic are equal to the decreases to the least days at the winter tropic. And if the deviation for certain parts of the earth were supposed either towards the orient or the occident, it would result that for these parts neither the sizes and angular distances of the stars would appear equal and the same at the eastern and western horizons, nor would the time from rising to the meridian be equal to the time from the meridian to setting. But these things evidently are altogether contrary to the appearances.

As to the second position where the earth would be on the axis but farther advanced towards one of the poles, one could again object that, if this were so, the plane of the horizon in each latitude would always cut into uneven parts the sections of the heavens below the earth and above, different with respect to each other and to themselves for each different deviation. And the horizon could cut into two even parts only in the right sphere. But in the case of the inclined sphere with the nearer pole ever visible, the horizon would always make the part above the earth less and the part below the earth greater with the result that also the great circle though the center of the signs of the zodiac would be cut unequally by the plane of the horizon. But this has never been seen, for six of the twelve parts are always and everywhere visible above the earth, and the other six invisible; and again when all these last six are all at once visible, the others are at the same time invisible. And so—from the fact that the same semicircles are cut off entirely, now above the earth, now below—it is evident that the sections of the zodiac are cut in half by the horizon.

And, in general, if the earth did not have its position under the equator but lay either to the north or south nearer one of the poles, the result would be that during the equinoxes, the shadows of the gnomons at sunrise would never perceptibly be on a straight line with those at sunset in planes parallel to the horizon. But the contrary is everywhere seen to occur. And it is immediately clear that it is not possible to advance the third position since each of the obstacles to the first two would be present here also.

In brief, all the observed order of the increases and decreases of day and night would be thrown into utter confusion if the earth were not in the middle. And there would be added the fact that the eclipses of the moon could not take place for all parts of the heavens by a diametrical opposition to the sun, for the earth would often not be interposed between them in their diametrical oppositions, but at distances less than a semicircle.

That the Earth Has the Ratio of a Point to the Heavens

Now, that the earth has sensibly the ratio of a point to its distance from the sphere of the so-called fixed stars gets great support from the fact that in all parts of the earth the sizes and angular distances of the stars at the same times appear everywhere equal and alike, for the observations of the same stars in the different latitudes are not found to differ in the least.

Moreover, this must be added: that sundials placed in any part of the earth and the centers of armillary spheres can play the role of the earth's true center for the sightings and the rotations of the shadows, as much in conformity with the hypotheses of the appearances as if they were at the true midpoint of the earth.

And the earth is clearly a point also from this fact: that everywhere the planes drawn through the eye, which we call horizons, always exactly cut in half the whole sphere of the heavens. And this would not happen if the magnitude of the earth with respect to its distance from the heavens were perceptible; but only the plane drawn through the point at the earth's center would exactly cut the sphere in half, and those drawn through any other part of the earth's surface would make the sections below the earth greater than those above.

That the Earth Does Not in Any Way Move Locally

By the same arguments as the preceding it can be shown that the earth can neither move in any one of the aforesaid oblique directions, nor ever change at all from its place at the center. For the same things would result as if it had another position than at the center. And so it also seems to me superfluous to look for the causes of the motion to the center when it is once for all clear from the very appearances that the earth is in the middle of the world and all weights move towards it. And the easiest and only way to understand this is to see that, once the earth has been proved spherical considered as a whole and in the middle of the universe as we have said, then the tendencies and movements of heavy bodies (I mean their proper movements) [Note: in other words, what Aristotle called "natural motions" in the sublunar world] are everywhere and always at right angles to the tangent plane drawn

Chapter Eight
Saving the Heavens

through the falling body's point of contact with the earth's surface. For because of this it is clear that, if they were not stopped by the earth's surface, they too would go all the way to the center itself since the straight line drawn to the center of a sphere is always perpendicular to the plane tangent to the sphere's surface at the intersection of that line.

All those who think it paradoxical that so great a weight as the earth should not waver or move anywhere seem to me to go astray by making their judgment with an eye to their own affects and not to the property of the whole. For it would not still appear so extraordinary to them, I believe, if they stopped to think that the earth's magnitude compared to the whole body surrounding it is in the ratio of a point to it. For thus it seems possible for that which is relatively least to be supported and pressed against from all sides equally and at the same angle by that which is absolutely greatest and homogeneous. For there is no "above" and "below" in the universe with respect to the earth, just as none could be conceived of in a sphere. And of the compound bodies in the universe, to the extent of their proper and natural motion, the light and subtle ones are scattered in flames to the outside and to the circumference, and they seem to rush in the upward direction relative to each one because we too call "up" from above our heads to the enveloping surface of the universe; but the heavy and coarse bodies move to the middle and center and they seem to fall downwards because again we all call "down" the direction from our feet to the earth's center. And they properly subside about the middle under the everywhere-equal and like resistance and impact against each other. Therefore the solid body of the earth is reasonably considered as being the largest relative to those moving against it and as remaining unmoved in any direction by the force of the very small weights, and as it were absorbing their fall. And if it had some one common movement, the same as that of the other weights, it would clearly leave them all behind because of its much greater magnitude. And the animals and other weights would be left hanging in the air, and the earth would very quickly fall out of the heavens. Merely to conceive such things makes them appear ridiculous.

The next section is particularly interesting. There were, in ancient times, a few renegade thinkers who entertained the idea that perhaps the Earth was in motion and the heavens were still. This was so far removed from the general viewpoint of the time that it was never taken very seriously. But Ptolemy felt the need to explain why that view was untenable. Fourteen hundred years later, Copernicus took up the same issue and made the same points, but drew the conclusion that Ptolemy was the person in error.

> Now some people, although they have nothing to oppose to these arguments, agree on something, as they think, more plausible. And it seems to them there is nothing against their supposing, for instance the heavens immobile and the earth as turning on the same axis from west

to east very nearly one revolution a day; or that they both should move to some extent, but only on the same axis as we said, and conformably to the overtaking of the one by the other.

But it has escaped their notice that, indeed, as far as the appearances of the stars are concerned, nothing would perhaps keep things from being in accordance with this simpler conjecture, but that in the light of what happens around us in the air such a notion would seem altogether absurd. For in order for us to grant them what is unnatural in itself, that the lightest and subtlest bodies either do not move at all or no differently from those of contrary nature, while those less light and less subtle bodies in the air are clearly more rapid than all the more terrestrial ones; and to grant that the heaviest and most compact bodies have their proper swift and regular motion, while again these terrestrial bodies are certainly at times not easily moved by anything else—for us to grant these things, they would have to admit that the earth's turning is the swiftest of absolutely all the movements about it because of its making so great a revolution in a short time, so that all those things that were not at rest on the earth would seem to have a movement contrary to it, and never would a cloud be seen to move toward the east nor anything else that flew or was thrown into the air. For the earth would always outstrip them in its eastward motion, so that all other bodies would seem to be left behind and to move towards the west.

For if they should say that the air is also carried around with the earth in the same direction and at the same speed, none the less the bodies contained in it would always seem to be outstripped by the movement of both. Or if they should be carried around as if one with the air, neither the one nor the other would appear as outstripping, or being outstripped by, the other. But these bodies would always remain in the same relative position and there would be no movement or change either in the case of flying bodies or projectiles. And yet we shall clearly see all such things taking place as if their slowness or swiftness did not follow at all from the earth's movement.

Starting then from this view of the cosmos, its size, shape, what moves and what does not—this "common sense" view—, Ptolemy then proceeds to develop a model that is unbelievably complex and anything but common sense. All at once, Ptolemy seems to have crossed the line into the Pythagoras-Plato camp. From then on in the book, Ptolemy's analyses are mathematical mysteries that are difficult or even impossible to conceive as a physical reality and instead gain their power from their adherence to preconceived ideas about heavenly motions (e.g., they must be eternal, changeless, and circular), and their ability to account for the positions of the stars and planets. Having set out his assumptions, Ptolemy then goes on to develop the mathematical model that he

Chapter Eight
Saving the Heavens

needs to capture ("save") the observed planetary positions in a predictive system. You will note how different his assumptions are from Euclid's definitions, postulates, and common notions. Ptolemy's assumptions come with a detailed rationale and read much more like an Aristotelian analysis of the configuration of the cosmos. In fact, much of what has been quoted above from the *Almagest* actually is a rehash of Aristotelian philosophy and pretty well reflects commonly held views in Ptolemy's day (which is about 500 years after Aristotle). The conclusions are basically what would occur to one at the outset, and the arguments simply show that other arrangements would defy these views.

Ptolemy's Model in Brief

Like most astronomers before him, Ptolemy took the view that the stars were all fixed to the inner surface of a spherical shell that was as large as (or nearly as large as) the universe. This shell, the "celestial sphere" turned all the way around once a day, carrying with it all the "fixed" stars. Between the celestial sphere and the immobile earth at the center of the universe were the planets, which were Mercury, Venus, Mars, Jupiter, Saturn, as well as the Sun and Moon. This much was common to many other astronomers. But unlike the Eudoxus model as amended by Aristotle, Ptolemy did not provide any physical explanation of what made planets move, nor what held them in their positions. Instead he simply described, precisely, a combination of abstract circles upon circles that moved at certain rates and placed the planets in different positions at different times.

The basic model is called the *epicycle-deferent system*. Each planet is situated on the edge of a circle, called an *epicycle*, while the center of the epicycle is situated on the edge of a larger circle called the *deferent*. The center of the deferent is either at the center of the earth (i.e., the center of the universe) or somewhat displaced from the center. The epicycle turns around its center carrying the planet round as it does so, and the deferent turns, carrying the epicycle around. Meanwhile, the deferent itself is carried round with the daily rotation of the stars. The combination of motions makes the motions of the of the planet quite complex.

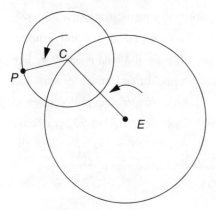

A simple epicycle-deferent combination. Planet *P* is carried around the Earth, *E*, by the revolutions of the epicycle, whose center is *C*, and the deferent centered on the Earth.

Every planet has its own epicycle-deferent system. For the Sun and the Moon (which were considered planets), one could either use an epicycle-deferent combination or just a deferent that is widely displaced from the center of the earth. For all the other planets, both an epicycle and deferent are required. Therefore, Ptolemy requires five epicycles and five deferents, plus two deferent-like circles without epicycles (called *eccentrics*). That's 12 circles in the sky. But are they really there? For Aristotle's model, the circular motions in the sky were merely the paths carved out by a rotating sphere which had a physical, material existence (made of the quintessence). But circles are pure geometrical abstractions made of curved lines having no breadth whatsoever. In other words they are pure form, in the Platonic sense. Ptolemy busies himself with providing mathematical formulae that calculate the changing positions of the heavenly bodies and leaves aside completely the question of what causes their motions and what holds them in place.

This is an interesting separation of *kinematic* (description of motions) from *dynamic* (accounting for forces and causes of motion). Aristotle was unable, or unwilling, to give much thought to motions alone without explaining their causes. Ptolemy's analyses address only what Aristotle called the "formal cause" and ignore the three other "causes" (material, efficient, and final) that Aristotle said were necessary to understand and explain anything. What Ptolemy does, in effect, is separate the mathematical calculation and deduction from the empirical and inductive analysis of nature. This is a distinction that became very much more important later during the Scientific Revolution of the Renaissance. Then science was seen to be most productive when it answered only the questions it could and left other issues to the philosophers.

By Aristotelian standards, Ptolemy's system made no sense. But because that did not prevent Ptolemy from continuing his work, he was able to do what earlier astronomers

Chapter Eight
Saving the Heavens

could not do: he created a system that did specify with reasonable accuracy where the heavenly bodies—stars and planets—were to be seen in the heavens in a reliable way.

The epicycle-deferent system was not Ptolemy's invention. About 300 years before, Hipparchus, who developed the table of chords and discovered the precession of the equinoxes, had set out the basic principles of an epicycle and deferent system and shown how it could track some of the movements of the planets. But Ptolemy took this device much farther. Ptolemy's system succeeded where others had not, because he elaborated the mathematical devices well beyond any simple design in order to make them fit the observations. Clearly, Ptolemy took his task to be one of finding the mathematical formulae to track the planetary positions, not to find the "true" meaning of the organization of the heavens.

This becomes clearer when some of the elaborations to make a better fit are taken into account. Alas, the pure epicycle-deferent combinations did not suffice to capture the planetary motions. For one thing, the planets seem to move more quickly through one part of their orbit than another. Since it was virtually an article of faith that everything in the heavens was eternal and unchanging, it was necessary that there be no changes in speed of a planet in its orbit. Aside from the philosophical requirement of changelessness, the mathematics of Euclid was just not capable of tracking an object that sped up and slowed down.

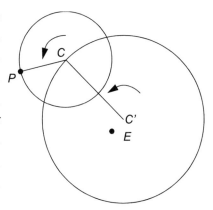

A more complex epicycle-deferent combination. Again, planet *P* is moved by the revolutions of both circles, but this time the center of the deferent, *C'*, does not correspond with the position of the Earth, *E*.

Ptolemy's first attempt to solve this was to move the center of the deferent away from the Earth. Then the motions could still be uniform, but from an observer on Earth, one-half of the journey appears to go faster than the other. This works quite well for the Sun and accounts for the fact that the time from the vernal to the autumnal equinox is six days longer than that from autumnal to vernal.

This solved some problems, but there were others where the position of the longer part of the orbit changed also. Hence, the next addition was to place the center of the eccentric deferent on another smaller deferent or even on an eccentric deferent. This is the system of *movable eccentrics*. The Moon, for example, required a large eccentric carried round on a much smaller deferent centered on the Earth.[3]

The Equant

But the added complexities did not stop there. The strangest of all was the device that was later called the *equant*. It was important that anything that moved did so at a constant rate, for both philosophical reasons and in order to be calculable with ancient mathematics. A circle was a figure that maintained a constant distance from a single point, its center, and if it turned, it must turn at a constant rate. One would naturally think that a circle would turn around its center, but Ptolemy took a different view. Suppose a point on a circle moved so that it always maintained a constant distance from its center (otherwise it would not be a circle), but moved at a constant angular rate around a point other than its center. Though the point moves at a constant angular rate, its linear speed going round the circumference of the circle would be always changing. The position could be calculated without reference to the changing linear speed and therefore was within the capability of ancient mathematics.

Here's how it was used. Consider an eccentric circle around the earth. In the diagram, the point P on the eccentric is a planet, the point C is the center of the eccentric, and the point E is the Earth.. Draw a line from E through C ending at a distance that is as far from C on one side as E is from C on the other side. Call that point Q, for *equant*

---

3. The typical deferent is a large circle on the edge of which is centered a small epicycle. Here what is described is the same arrangement, but the small circle is on the inside and the large circle around it. It would be consistent to call this an epicycle-deferent combination with the relative sizes reversed but it is more usual to call it an eccentric centered on a deferent.

Chapter Eight
Saving the Heavens

Let P, the planet, move around the eccentric, maintaining its distance from C but moving at an angular rate that is constant from Q. If this is observed from E, the Earth, P would appear to be speeding up and slowing down, but from Q it would appear to be moving steadily. It is difficult to find a physical explanation for any of the models used by Ptolemy, but this one is especially troublesome since the planet would only appear to turn at a constant rate from a point out in space which is neither the center of the universe (i.e., the center of the Earth) nor the center of its path of motion.

Ptolemy used all the models described in different combinations. Eccentrics, epicycle-deferent systems centered on the Earth, epicycle-deferent systems eccentric to the Earth, moving eccentrics, eccentrics with equant points, and eccentric epicycle-deferent systems with equant points. With all these possible combinations and with epicycles, deferents, and epicycles of many different sizes and speeds of rotation, he was able to fit his models well enough to the observed celestial phenomena that he effectively saved the phenomena of the heavens and solved the problem of the planets. His work was convincing enough that it was the standard work of astronomy and the standard from which the calendar was reckoned for the next 1400 years.

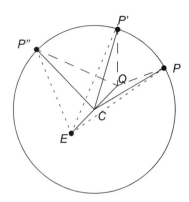

The use of an equant point allowed Ptolemy to account for irregularities in planets' motions. For example, consider a planet moving over a given period of time from point P to P' and then, in an equal period, from point P' to P''. For an observer on the Earth, E, comparing the two angles of apparent motion, PEP' and $P'EP''$, it appears that the planet accelerated, traveling a greater distance in the second time period than it did in the first. By placing the center of the planet's orbit away from the Earth, at point C, the anomaly can be diminished (angles PCP' and $P'CP''$ are closer to equal than angles PEP' and $P'EP''$) but it cannot be resolved. No point can be found which can be both the center of the planet's circular orbit and the point around which the planet revolves at a regular speed. But, if the center of the orbit is not the point around which the planet moves at a constant angular rate, the anomaly can be explained. A point Q can be found such that for any equal periods of time, angle PQP' equals angle $P'QP''$, thereby maintaining the regularity of the planet's motion.

◆

Part I
Ancient Beginnings

The ancient era drew to a close with the fall of Rome around 476 CE, and after that there was little original scholarship for a considerable time. The level of literacy went down in Europe, the fabric of society disintegrated, and many of the great works of philosophy and science were lost altogether or became rare. Scientific works, in particular, were not of great interest in Europe, though they survived in the great libraries of the Middle East, where, centuries later, they were discovered by the vibrant Islamic culture, translated into Arabic, and thereby saved from eternal oblivion.

Of the works that did survive, a small number seemed to encompass definitively all that was necessary to say in their respective subjects. Aristotle's works became the standard in philosophy (just as he was often referred to simply as "the Philosopher"), Euclid's *Elements* served for mathematics, Ptolemy's *Almagest* for astronomy. And in the life sciences, works by Aristotle and his follower Theophrastus were definitive catalogues of plants and animals, and the Greek-Roman physician Galen's *Physiology* was the standard treatise in medicine. By the later Middle Ages when these works had been translated into Latin and were available to European scholars, they had achieved a near scriptural status. They were the foundations upon which all other work was erected and also were effective barriers to original thought which must be destroyed before other viewpoints could take their place. This was the legacy of antiquity to the Scientific Revolution of the Renaissance.

Chapter Eight
Saving the Heavens

For More Information

Alioto, Anthony M. *A History of Western Science,* 2nd ed. Englewood Cliffs, NJ: Prentice-Hall, 1987. Chapter 5.

Clagett, Marshal. *Greek Science in Antiquity.* New York: Collier, 1963.

Cohen, I. Bernard. *The Birth of a New Physics,* Rev. ed. New York: Norton, 1985. Chapter 3.

Fraser, P. M. *Ptolemaic Alexandria,* 3 vols. Oxford: Clarendon Press, 1972.

Kuhn, Thomas S. *The Copernican Revolution: Planetary Astronomy in the Development of Western Thought.* Cambridge: Harvard University Press, 1957. Chapter 1.

Lindberg, David C. *The Beginnings of Western Science: The European Scientific Tradition in Philosophical, Religious, and Institutional Context, 600 B.C. to A.D. 1450.* Chicago: University of Chicago Press, 1992. Chapter 5.

Lloyd, G. E. R. *Greek Science after Aristotle.* London: Chatto & Windus, 1973. Chapter 8.

Ptolemy, Claudius. *The Almagest*, trans. by R. Catesby Taliaferro. In Robert Maynard Hutchins, ed. in chief, *Great Books of the Western World*. Chicago: Encyclopedia Britannica, 1952.

_____. *Ptolemy's Almagest*, ed. and trans. by G. J. Toomer. New York: Springer, 1984.

Ronan, Colin A. *Science: Its History and Development Among the World's Cultures.* New York: Facts on File, 1985. Chapter 2.

Wall, Byron Emerson. "Anatomy of a Precursor: The Historiography of Aristarchos of Samos" *Studies in the History and Philosophy fo Science* 6(1975): 201-228.

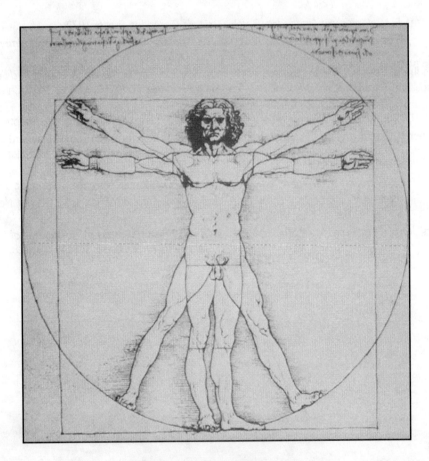

Leonardo da Vinci's sketch of man as the measure of all things.

PART TWO

Science Emerges

The death of Alexander the Great in 323 BCE marked the beginning of the decline of the political supremacy of Greece in the Mediterranean and Near East, though it was also the beginning of the Hellenistic period, during which some of the greatest Greek work on scientific topics was done. Greek scholars continued to have great influence on European and Near Eastern cultures because they became the teachers in the rising Roman Empire. The Roman Empire expanded all across Europe and well into the Near East in the millennium between 500 BCE and 500 CE before collapsing under its own weight, leading to the Dark Ages of Europe. Some important scientific work was done in the Roman Empire, though most of it was done by scholars who either were themselves Greek or were trained by Greeks. However, by and large, the Roman Empire was interested in other matters than the questions of science.

In 395 CE, the Roman Empire was partitioned into two parts, an Eastern Roman Empire and a Western Roman Empire. The Western Roman Empire proved too hard to govern from its capital, Rome, and succumbed to invading Germanic tribes from the north and west. By 500, nothing was left of it.

The Byzantine Empire

But the Eastern Roman Empire fared much better. It was based in the ancient Greek colony of Byzantium so that in the 17th century scholars came to refer to it as the Byzantine Empire. Byzantium had been the eastern headquarters of Constantine, who had become the first Christian Roman emperor in 324 and had tried to hold the entire Roman Empire together. Byzantium was renamed Constantinople in his honor. (It was renamed Istanbul in 1930.) The Byzantine Empire spanned Greece and Asia Minor—by and large the countries where Greek influence was greatest in the Hellenistic period. The language of scholarship in the Byzantine Empire quickly became Greek instead of Latin.

The geographical features of the Byzantine Empire made invasion much more difficult than in the Western Roman Empire. As a result, it thrived and remained independent for over a thousand years until finally conquered by the Ottoman Turks in 1453. In the Western Roman Empire, scholarly works were lost or destroyed, but the classical Greek and Latin works were preserved and studied in the Byzantine world.

Islam

To the east of the Byzantine Empire lay the Arabian Peninsula, a largely desert land populated by nomadic tribes. In the 7th century, Arabia was a fairly disorganized land where loyalties were to family clans within tribes, which bickered frequently; there were many religions practiced and many gods worshipped. Into this culture, around 570, was born Muhammad, who brought Islam to the Arabs, a new religion with a single god, interpreted by a single prophet, Muhammad, and ultimately with a single authoritative book to guide and unify the culture, the Koran. This cohesion gave Islam strength both as a religion and as a governing institution. In the 100 years from 630 to 730, Islam spread from the Arabian peninsula eastward through the Middle East and into India, and westward across the north of Africa and into southern Spain. It was an extraordinary transformation from disorganization to an efficient and powerful empire, capable of imposing its will wherever it went.

Unlike the Germanic tribes that invaded the Western Roman Empire and mindlessly destroyed whatever they found, when the Muslims conquered a land they tended to preserve and learn from the local culture. This was especially important for the development of science, because the lands conquered by the Muslims housed some of the best collections of ancient manuscripts from the Hellenistic period, including such works as Euclid's *Elements* and Ptolemy's *Almagest*. Muslim scholars translated many of the ancient Greek and Roman works into Arabic, especially the more scientific works. These Arabic copies made their way into libraries throughout the Islamic world, of which there were many. Also, the Arab scholars continued to study these topics and advanced them considerably, especially mathematics, optics, astronomy, and chemistry (i.e., alchemy).

Europe

Meanwhile, in the lands that had been the western part of the Roman Empire, the very opposite was happening. The Empire was fractured into small warring groups, the level of literacy declined, and respect for learning and scholarly pursuits was low. Intellectual leadership came from the Christian church, but even there corruption and laziness rose while curiosity and scholarship sank. In the 8th century, midway through the Dark Ages, a charismatic leader who became known as Charlemagne took for himself the title Holy Roman Emperor and set about to reform Europe and get back some of its power by concentrating on raising the level of education. He founded schools all over Europe in the churches and monasteries (where the only literate people could be found).

Charlemagne's schools originally had only the remnants of the classical literature to work with, along with what had been written and preserved in Europe since then. There was not much of a scientific nature among these. After his death some of the impetus for enlightened learning was lost, but many of the schools continued. By 1200, some of those schools had developed into important centers of learning where scholars gathered to study what was available to them. Then around this time, an extraordinary avalanche of scholarship descended upon these European schools, turning them into the first European universities.

The avalanche came from the Muslim world and then from the Byzantine Empire. What prompted this were partly the Crusades, military sorties into Muslim-held lands to "liberate" them for Christianity. Among the things the Crusaders discovered were the vast libraries in the Muslim world, stocked with scholarly works that were totally unknown to Europeans. Over a period of one hundred to two hundred years, Europe absorbed its lost past. The ancient works were translated into Latin from Arabic, which had itself been translated from Latin or Greek. Added to these were the Arabic works that had continued the tradition of scholarship.

Part II
Science Emerges

After a pause of about a thousand years, Europe was ready to continue its extraordinary enterprise of trying to understand the world rationally. Why the thousand-year hiatus? Why did this enterprise not flourish equally well in the Byzantine or Muslim worlds? We can only speculate here, but a likely reason would be that neither the Byzantine nor the Muslim cultures were conducive to open-ended inquiry into the nature of the universe; the kind of persistent questioning and discussion necessary for these activities was discouraged there more than it was in Europe.

The Scientific Revolution

In any case, what happened in Europe starting around 1500 was the extraordinary blossoming of inquiry, learning, and productive thought that we call the Renaissance. Out of that science, as we know it, came into existence.

The Scientific Revolution of 1500 to 1700 was the work of many people working on many different issues from many different viewpoints. Trying to understand what happened during that two-hundred-year period has become the life work of scores of historians of science, who have analyzed and written extensively on the people and the events that made science come to life during this period. It is a complex and fascinating area.

Nevertheless, there are a few key people and developments that figure prominently in almost everyone's analyses of the period. Here we shall look briefly at the lives and work of some of these people and thereby try to get a reasonable sense of what happened and what it meant. The people to be considered in this section are: Copernicus, Kepler, Galileo, Descartes, and Newton. Each one is extraordinary. Each one is a different story.

Part II | 139
Science Emerges

The cathedral in Florence with the vast dome designed by the "Renaissance Man" Fillipo Brunelleschi using the latest in mathematical and engineering techniques.

net, in quo terram cum orbe lunari tanquam epicyclo contineri diximus. Quinto loco Venus nono mense reducitur. Sextum denique locum Mercurius tenet, octuaginta dierum spacio circu currens. In medio uero omnium residet Sol. Quis enim in hoc

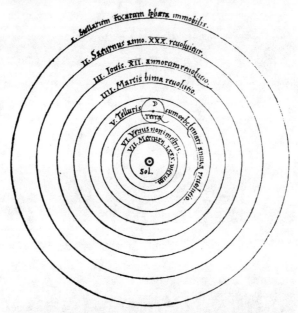

pulcherrimo templo lampadem hanc in alio uel meliori loco po neret, quàm unde totum simul possit illuminare? Siquidem non inepte quidam lucernam mundi, alij mentem, alij rectorem uocant. Trimegistus uisibilem Deum, Sophoclis Electra intuentē omnia. Ita profecto tanquam in solio regali Sol residens circum agentem gubernat Astrorum familiam. Tellus quoq; minime fraudatur lunari ministerio, sed ut Aristoteles de animalibus ait, maximā Luna cū terra cognationē habet. Concipit interea à Sole terra, & impregnatur annuo partu. Inuenimus igitur sub hac

A page from Copernicus' *De Revolutionibus*.

Chapter Nine

Copernicus Chooses Formal Elegance over Common Sense

The Julian Calendar

In ancient Rome, the calendar was a mess. An extra short month was inserted every few years to try to bring the calendar back in line with the seasons. When Julius Caesar came to power he was persuaded by an Egyptian astronomer named Sosigenes to introduce a new calendar that would stay in harmony with the seasons. This Caesar did in 45 BCE. It came to be called after him—the Julian calendar—and was far more accurate than anything that had gone before it. The Julian calendar comprised a year of 365 days, with an extra "leap day" inserted every four years. Ptolemy's *Almagest*, written about 200 years later, used the Julian calendar as the basis of all its date calculations.

Though the Julian calendar was more accurate than anything that preceded it, it was not perfect. The astronomical year, measured from vernal equinox to vernal equinox is not 365 days and 6 hours, which the Julian calendar implied, but 365 days, 5 hours, 48 minutes, and 48 seconds. This is a very small difference over a few years time, but in 400 years, the Julian calendar is ahead of the astronomical events by about 3 days. By 1500, the Julian calendar was running about 10 days ahead of where it should have been to match astronomical observations.

This might be viewed as a minor detail having little to do with everyday life. If the calendar was used primarily for agricultural purposes, deciding when to plant and when to harvest, ten days would make some, but not a vast difference. And the change was so slow that farmers would adjust their thinking to match it. But the calendar had a much more crucial role in religion. The Catholic Church divided the year into several different segments marked by special days, some determined by the calendar alone, such as Christmas, on December 25, some determined by astronomical events, such as Easter—the first Sunday after the first full moon, after the vernal equinox. There were different rituals to be followed in the different parts of the church year. It was therefore important to get it right. When the Catholic Church set up the Council of Trent in 1545 to deal with the Protestant threat, one of their tasks was to consider what was necessary to reform the calendar. They found that the research work of a Polish Canon of the Church named Nicholas Copernicus most useful in determining the problem with the old calendar. In 1582, Pope Gregory adopted a new calendar to replace the Julian calendar, which we still use and call the Gregorian calendar. To adjust for the inaccuracies of the Julian calendar, three leap days every four hundred years were eliminated.[1] Then to correct for the errors that had been introduced, when the Gregorian calendar was fixed in place on October 4, 1582, the following day was deemed to be October 15.

Nicholas Copernicus

People such as Nicholas Copernicus, who helped sort out the calendar problems, were characteristic of the period we call the Renaissance. These are the intellectuals who were so fascinated by the rich heritage of ancient culture and so confident of their own abilities to digest and master what there was to know that they studied everything they could find and became experts in many disciplines. The typical Renaissance Man was highly skilled in mathematics and technical knowledge and could apply this knowledge to issues outside of their usual domain. In the arts, the names that spring to mind include Leonardo da Vinci, Leon Battista Alberti, and Fillippo Brunelleschi. In the humanities, Erasmus, Giovanni Boccaccio, Petrarch; and in science and technology, Agricola (George Bauer), Vanocchio Biringuccio, and Paracelsus. These are some of the more famous people of the Renaissance who were instrumental in advancing one or more fields of endeavor and who were known to have wide interests and expertise. Copernicus fits well among these.

---

1. Like the Julian calendar, years divisible by four are leap years, having one extra day (February 29). But in the Gregorian calendar, century years, such as 1800, 1900, etc., are not leap years despite being divisible by four—except that those divisible by 400, such as 2000, *are* deemed to be leap years. This brings the Gregorian calendar almost in line with the astronomical observations.

Chapter Nine | 143
Copernicus Chooses Formal Elegance over Common Sense

Nicholas Copernicus was born in Torun, Prussia (now Poland), about 100 miles south of Gdansk in 1473. Copernicus' father died when Nicholas was ten, and he was raised by an uncle who became a bishop in the Catholic Church. Copernicus was sent to the University of Cracow, where he studied medicine, but while there discovered the ancient classics of mathematics and astronomy and became fascinated by them. Before he could complete his medical studies, his uncle summoned him home in order that he could take up a post as canon in the Cathedral at the Polish port town of Frauenberg.

Nicholas Copernicus.

A canon is an administrative position in the Catholic church, typically held for life. It has moderate bureaucratic duties and provides a living wage, with lots of free time to pursue other interests. Copernicus' uncle, as bishop, had the power to appoint people to the position of canon when there were vacancies, and he saw this as an opportunity to secure a comfortable living for his nephew.

Copernicus' uncle called his nephew home when the incumbent canon was gravely ill and not expected to live. Copernicus could immediately be appointed to succeed him. Actually, the arrangement was that when there were vacancies in the position of canon at a cathedral, the bishop appointed someone to fill the vacancy in even months and the pope appointed the successor in odd-numbered months. As it turned out, the dying man was sicker than the uncle had expected and died in an odd-numbered month. So the pope appointed the new canon, and Copernicus interrupted his studies for nothing.

Copernicus went back to his studies, but not back to the University of Cracow. This time he went to Italy, first to Bologna and then to Padua. Finall,y another canon died at the right time, and Copernicus did get appointed to the position of Canon at Frauenberg Cathedral, a position he held for the rest of his life. Nevertheless, he did not take up the position right away. Instead, he continued his studies for another eight years. Copernicus began his studies in medicine, but pursued many subjects, especially mathematics and astronomy, in his long university career. Ultimately, he took a doctorate in Canon Law, as befit the position he would take up. Strangely, he got his doctorate at the University of

Ferrara, where he submitted his thesis, but never attended. Even then, though having the position of canon in hand, he did not take up the position for another six years, during which he is said to have lived at his uncle's castle, serving as house physician. It was during this period at his uncle's castle that Copernicus first formed and wrote his ideas about astronomy.

When his uncle died in 1512, Copernicus finally took up residence and his duties at the Frauenberg Cathedral, where he remained the rest of his life. Officially, Copernicus functioned as a physician and church administrator, but had ample leisure time and devoted much of it to his interest in astronomy. He completed a draft of the work for which he is remembered, *On the Revolutions of the Heavenly Spheres*, in 1530, but locked it away, telling only a few people about it and making only a few corrections to his original draft. In 1543, when he was in poor health, he was finally persuaded to publish the book. He died in the same year.

The Copernican System

Copernicus took an interest in astronomy in part because it was perceived as a critical subject for understanding all that was most important. It was necessary to understand the order of the heavens to understand the mysteries of nature, and ultimately, to understand God. Moreover, the understanding provided by ancient astronomy was clearly in need of attention.

The Julian calendar was not correctly tracking the heavens. Even allowing for that, the positions of the heavenly bodies predicted by Ptolemy were no longer matching observations. Ptolemy's system had some inherent inaccuracies that accumulated over time, so that by the Renaissance the Ptolemaic account of the heavenly motions appeared to be rife with problems. Moreover, Ptolemy's purely mathematical representation did not sit well with a literal-minded intelligentsia that accepted Aristotle's general description of the heavens as filled with crystal spheres that carried the planets around.

The Aristotelian system of nested spheres, modeled on Eudoxus' plan (see Chapter 6), and the Ptolemaic abstract epicycle and deferent system are basically incompatible. Aristotle's spheres moved in complex patterns, but they were all concentric with the

Chapter Nine | 145
Copernicus Chooses Formal Elegance over Common Sense

Earth and nested inside each other, and they all turned around axes that passed through their centers. Ptolemy's deferents were often eccentric to the Earth, carried epicycles that turned around in space though connected to nothing at all, and when the equant point was evoked, the deferents turned on axes that were not their centers.

Medieval scholars were determined to reconcile these two cosmic systems. So, a compromise system was evolved wherein the Ptolemaic deferents were given solid form by being the paths drawn by a spot on an Aristotelian solid spherical shell as it rotated on its axis. The Ptolemaic epicycles were thought of as small rotating spheres between Aristotelian shells, like ball bearings. This reconciliation served to give a material being to the Ptolemaic paths, but it only worked for the simplest cases of deferents concentric with the earth that rotated at constant speeds.

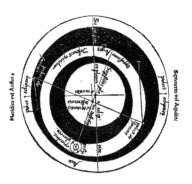

A diagram of the compromise Aristotelian-Ptolemaic system.

The Equant

Some of the devices that Ptolemy used to get his calculations to match observations simply seemed impossible to put in a physical form. The worst of these was the equant. (See Chapter 8.) The equant point was a point within the rotating deferent. The deferent rotated maintaining a constant distance from its geometric center—just as the circumference of a rotating sphere maintains a constant distance from its center. But the angular rotational speed is constant around the off-center equant. How could a physical system be conceived that would maintain rotational distance around one point and angular speed around another?

To Copernicus, this seemed to be a clear problem area in Ptolemy. Copernicus therefore undertook to investigate this problem at some length, with the idea that perhaps he could repair the Ptolemaic system and solve the problems that were leading to inaccurate calculations and maybe even show what was wrong with the calendar.

Make the Planets Go Round the Sun

In 1496, a summary version of *The Almagest* was published by Johann Müller, who took the Latin name Regiomontanus, in which he showed that a more complicated version of epicycles could do the work of the equant. Copernicus tried this and found that it made calculations nearly impossible. But the problem could be simplified by making the planets go around the Sun, as Regiomontanus had suggested.

This Copernicus tried and discovered that it did serve the purpose of the equant. Bear in mind here that when Copernicus is conceiving the planets going around the Sun, he was using the term planet as the ancients did. In other words, the bodies circling the Sun did not include the Earth, which was not considered a planet. Thus the Sun continued to go around the Earth, as in the Ptolemaic system. But here a problem arose. For Mercury and Venus, this system worked okay, but for Mars it raised other problems. The orbit of Mars around the Sun that fit this model had a radius of about one and one-half times that of the radius of the orbit of the Sun around the Earth. If all these orbits were caused by solid spheres in rotation, then the sphere carrying Mars would cut into the sphere carrying the Sun. This model seemed to introduce as many problems as it solved.

Now Make the Earth Go Around the Sun

After wrestling with this impasse for some considerable time, Copernicus took a step away from the common-sense viewpoint of the Aristotelians and toward the mysterious worldviews of the Neo-Platonists, with whom he had had contact in his university career. Just as Plato distrusted common-sense understanding that was based upon sense perceptions, the Neo-Platonists shirked the everyday world of experience and sought the meaning of life in mysteries that could be understood only by the mind.

In a move that embraced mathematical harmony and defied common sense altogether, Copernicus solved the problem of the intersecting orbits by claiming that the Sun stood still in the center of the cosmos while the Earth, the very definition of what was stable and motionless, was whirling about in the heavens, going around the sun once a year and turning on its axis every day.

For someone who had been so concerned that the Ptolemaic model have a literal, physical representation, it may seem to be a complete reversal for him to now embrace

Chapter Nine
Copernicus Chooses Formal Elegance over Common Sense

the Platonic priorities, where the elegance of a consistent mathematical system trumped the foundation of sense experience. But this was the Renaissance, a time in which the truly original thinkers were able to hold contradictory views in their minds and evaluate each of them. Another factor which may have helped was the contact that Copernicus had had with the Hermetic writings that were very popular at Bologna and Padua.[2] In the Hermetic tradition, the Sun was often described as the "giver of life" and of central importance in the world. It was perhaps easier then to also ascribe to the Sun a central position in the cosmos.

Once Copernicus cut himself loose from the bounds of a common-sense, stationary Earth, he began to see more and more clearly that many of the ancient problems and mysteries of the heavens were explainable as optical illusions. He then set himself to work to create a new system to replace Ptolemy's *Almagest* altogether.

However, realizing that his views would cause an uproar of protest and he would be ridiculed as a fool or persecuted for unorthodoxy, he kept his thoughts to himself until very late in life when a few of his trusted friends who knew of his work convinced him that he must publish it before he died.

On the Revolutions

For anyone familiar with Ptolemy's *Almagest*, the first thing one notices upon picking up *On the Revolutions of the Heavenly Spheres* is that Copernicus has modeled his work very closely on Ptolemy's. The layout of the book is similar, the style of argument is exactly parallel, the general assumptions about the universe are the same. What is different is the placement of the bodies that we would now call the solar system. Everything

---

2. The Hermetic writings were a dozen or so treatises that were supposed to have been written by an ancient Egyptian figure names Hermes Trismegistus (Hermes, the blessed three times), who is now believed to have never existed. The Hermetic writings were very popular in Italy in the 16th and 17th centuries. They purported to be written by a mystic living in Egypt at the time of Moses. They stressed the importance of the Sun as a symbol of God, and even suggested that the Sun was the center of the universe. Moreover, the writings implied that the world was full of secrets and riddles that could only be understood by the few who could see beyond the surface into the mathematical language of nature.

about *On the Revolutions* is ancient in tone. Except for the use that Copernicus makes of later observational data, his book could have been written in, say, the year 200.

For Ptolemy, the Earth is in the center and the Sun, Moon, Mercury, Venus, Mars, Jupiter, and Saturn are all planets, meaning that they are heavenly bodies (i.e., stars) that move around. For Copernicus, the Sun is in the center, the Earth is now a planet, and it and the other planets all circle the Sun, except that the Moon is now no longer a planet, but a satellite of the Earth. Otherwise, the universe is the same. The shape of the universe is spherical; the stars are all affixed to a spherical shell at the outer edge; everything else is between the stars and the body in the middle, which does not move—except for Copernicus that body is the sun. The fixed stars no longer rotate around once a day. That effect is caused by the Earth spinning on its axis once daily.

Ptolemy laid out in his Book I the general assumptions about the configuration of the universe, on which he would build his system. Copernicus does just the same. In fact, except for the differences about the Earth, the points and the arguments are similar and in the same order. Compare this with the sections quoted from Ptolemy in Chapter 8. Copernicus opens with the following arguments: (1) The Universe is Spherical, (2) The Earth too is Spherical, (3) How Earth Forms A Single Sphere With Water, (4) The Motion Of The Heavenly Bodies Is Uniform, Eternal, And Circular Or Compounded Of Circular Motions, (5) Does Circular Motion Suit The Earth? What Is Its Position?, (6) The Immensity Of The Heavens Compared To The Size Of The Earth, (7) Why The Ancients Thought That The Earth Remained At Rest In The Middle Of The Universe As Its Center, (8) The Inadequacy Of The Previous Arguments And A Refutation Of Them, And (9) Can Several Motions Be Attributed To The Earth? The Center Of The Universe. From here he goes on into more technical arguments, just as Ptolemy did. But note that points (1), (2) and (6) are the same as Ptolemy's and point (7) is Ptolemy's argument about the Earth's immobility, repeated by Copernicus in order to be refuted in point (8).

Points (4), (5), (8) and (9) contain the crux of Copernicus' non-mathematical argument, justifying putting the Earth in motion. They read as follows:

The Motion Of The Heavenly Bodies Is Uniform, Eternal, And Circular Or Compounded Of Circular Motions

Chapter Nine | 149
Copernicus Chooses Formal Elegance over Common Sense

I shall now recall to mind that the motion of the heavenly bodies is circular, since the motion appropriate to a sphere is rotation in a circle. By this very act the sphere expresses its form as the simplest body, wherein neither beginning nor end can be found, nor can the one be distinguished from the other, while the sphere itself traverses the same points to return upon itself.

In connection with the numerous [celestial] spheres, however, there are many Motions. The most conspicuous of all is the daily rotation, which the Greeks call νυχθημερινος, that is, the interval of a day and a night. The entire universe, with the exception of the earth, is conceived as whirling from east to west in this rotation. It is recognized as the common measure of all motions, since we even compute time itself chiefly by the number of days.

Secondly, we see other revolutions as advancing in the opposite direction, that is, from west to east; I refer to those of the sun, moon, and five planets. The sun thus regulates the year for us, and the moon the month, which are also very familiar Periods of time. In like manner each of the other five planets completes its own orbit.

Yet [these motions] differ in many ways [from the daily rotation or first motion]. In the first place, they do not swing around the same poles as the first motion, but run obliquely through the zodiac. Secondly, these bodies are not seen moving uniformly in their orbits, since the sun and moon are observed to be sometimes slow, at other times faster in their course. Moreover, we see the other five planets also retrograde at times, and stationary at either end [of the regression]. And whereas the sun always advances along its own direct path, they wander in various ways, straying sometimes to the south and sometimes to the north; that is why they are called "planets" [wanderers]. Furthermore, they are at times nearer to the earth, when they are said to be in perigee; at other times they are farther away, when they are said to be in apogee.

It stands to reason, therefore, that their uniform motions appear nonuniform to us. The cause may be either that their circles have poles different [from the earth's] or that the earth is not at the center of the circles on which they revolve. To us who watch the course of these planets from the earth, it happens that our eye does not keep the same distance from every part of their orbits, but on account of their varying distances these bodies seem larger when nearer than when farther away (as has been proved in optics). Likewise, in equal arcs of their orbits their motions will appear unequal in equal times on account of the observer's varying distance. Hence I deem it above all necessary that we should carefully scrutinize the relation of the earth to the heavens lest, in our desire to examine the loftiest objects, we remain ignorant of things nearest to us, and by the same error attribute to the celestial bodies what belongs to the earth. We must acknowledge, nevertheless, that their motions are circular or compounded of several circles, because these nonuniformities recur regularly according to a constant law. This could not happen unless the motions were circular, since only the circle can bring back the past. Thus, for example, by a composite motion of circles the sun restores to us the inequality of days and nights as well as the four seasons of the year. Several motions are discerned herein, because a simple heavenly body cannot be moved by a single sphere nonuniformly. For this nonuniformity would have to be caused either by an inconstancy, whether imposed from without or generated from within, in the moving force or by an alteration in the revolving body. From either alternative, however, the intellect shrinks. It is improper to conceive any such defect in objects constituted in the best order.

Does Circular Motion Suit The Earth? What Is Its Position?

Now that the earth too has been shown to have the form of a sphere, we must in my opinion see whether also in this case the form entails the motion, and what place in the universe is occupied by the earth. Without the answers to these questions it is impossible to find the correct explanation of what is seen in the heavens. To be sure, there is general agreement among the authorities that the earth is at rest in the middle of the universe. They hold the contrary view to be inconceivable or downright silly. Nevertheless, if we examine the matter more carefully, we shall see that this problem has not yet been solved, and is therefore by no means to be disregarded.

Every observed change of place is caused by a motion of either the observed object or the observer or, of course, by an unequal displacement of each. For when things move with equal speed in the same direction, the motion is not perceived, as between the observed object and the observer, I mean it is the earth, however, from which the celestial ballet is beheld in its repeated performances before our eyes. Therefore, if any motion is ascribed to the earth, in all things outside it the same motion will appear, but in the opposite direction, as though they were moving past it. Such in particular is the daily rotation, since it seems to involve the entire universe except the earth and what is around it. However, if you grant that the heavens have no part in this motion but that the earth rotates from west to east, upon earnest consideration you will find that this is the actual situation concerning the apparent rising and setting of the sun, moon, stars and planets. Moreover since the heavens, which enclose and provide the setting for everything, constitute the space common to all things, it is not at first blush clear why motion should not be attributed rather to the enclosed than to the enclosing, to the thing located in space rather than to the framework of space. This opinion was indeed maintained by Heraclides and Ecphantus, the Pythagoreans, and by Hicetas of Syracuse, according to Cicero. They rotated the earth in the middle of the universe, for they ascribed the setting of the stars to the earth's interposition, and their rising to its withdrawal.

Chapter Nine
Copernicus Chooses Formal Elegance over Common Sense

If we assume its daily rotation, another and no less important question follows concerning the earth's position. To be sure, heretofore there has been virtually unanimous acceptance of the belief that the middle of the universe is the earth. Anyone who denies that the earth occupies the middle or center of the universe may nevertheless assert that its distance (therefrom] is insignificant in comparison with [the distance of] the sphere of the fixed stars, but perceptible and noteworthy in relation to the spheres of the sun and the other planets. He may deem this to be the reason why their motions appear nonuniform, as conforming to a center other than the center of the earth. Perhaps he can [thereby] produce a not inept explanation of the apparent nonuniform motion. For the fact that the same planets are observed nearer to the earth and farther away necessarily proves that the center of the earth is not the center of their circles. It is less clear whether the approach and withdrawal are executed by the earth or the planets.

It will occasion no surprise if, in addition to the daily rotation, some other motion is assigned to the earth. That the earth rotates, that it also travels with several motions, and that it is one of the heavenly bodies are said to have been the opinions of Philolaus the Pythagorean. He was no ordinary astronomer, inasmuch as Plato did not delay going to Italy for the sake of visiting him, as Plato's biographers report.

But many have thought it possible to prove by geometrical reasoning that the earth is in. the middle of the universe; that being like a point in relation to the immense heavens, it serves as their center; and that it is motionless because, when the universe moves, the center remains unmoved, and the things nearest to the center are carried most slowly...

The Inadequacy Of The Previous Arguments [that the Earth is at rest in the center of the universe] And A Refutation Of Them

For these and similar reasons forsooth the ancients insist that the earth remains at rest in the middle of the universe, and that this is its status beyond any doubt. Yet if anyone believes that the earth rotates, surely he will hold that its motion is natural, not violent. But what is in accordance with nature produces effects contrary to those resulting from violence, since things to which force or violence is applied must disintegrate and cannot long endure. On the other hand, that which is brought into existence by nature is well-ordered and preserved in its best state. Ptolemy has no cause, then, to fear that the earth and everything earthly will be disrupted by a rotation created through natureís handiwork, which is quite different from what art or human intelligence can accomplish.

But why does he not feel this apprehension even more for the universe, whose motion must be the swifter, the bigger the heavens are than the earth? Or have the heavens become immense because the indescribable violence of their motion drives them away from the center? Would they also fall apart if they came to a halt? Were this reasoning sound, surely the size of the heavens would likewise grow to infinity. For the higher they are driven by the power of their motion, the faster that motion will be, since the circumference of which it must make the circuit in the period of twenty-four hours is constantly expanding; and, in turn, as the velocity of the motion mounts, the vastness of the heavens is enlarged. In this way the speed will increase the size, and the size the speed, to infinity. Yet according to the familiar axiom of physics that the infinite cannot be traversed or moved in any way, the heavens will therefore necessarily remain stationary.

But beyond the heavens there is said to be no body, no space, no void, abso- lutely nothing, so that there is nowhere the heavens can go. In that case it is really astonishing if something can be held in check by nothing. If the heavens are infinite, however, and finite at their inner concavity only, there will perhaps be more reason to believe that beyond the heavens there is nothing. For, every single thing, no matter what size it attains, will be inside them, but the heavens will abide motionless. For, the chief contention by which it is sought to prove that the universe is finite is its motion. Let us therefore leave the question whether the universe is finite or infinite to be discussed by the natural philosophers.

We regard it as a certainty that the earth, enclosed between poles, is bounded by a spherical surface. Why then do we still hesitate to grant it the motion appropriate by nature to its form rather than attribute a movement to the entire universe, whose limit is unknown and unknowable? Why should we not admit, with regard to the daily rotation, that the appearance is in the heavens and the reality in the earth? This situation closely resembles what Vergil's Aeneas says: *Forth from the harbor we sail, and the land and the cities slip backward* [Aeneid, III, 72].

For when a ship is floating calmly along, the sailors see its motion mirrored in everything outside, while on the other hand they suppose that they are stationary, together with everything on board. In the same way, the motion of the earth can unquestionably produce the impression that the entire universe is rotating.

Then what about the clouds and the other things that hang in the air in any manner whatsoever, or the bodies that fall down, and conversely those that rise aloft? We would only say that not merely the earth and the watery element joined with it have this motion, but also no small part of the air and whatever is linked in the same way to the earth. The reason may be either that the nearby air, mingling with earthy or watery matter, conforms to the same nature as the earth, or that the air's motion, acquired from the earth by proximity, shares without resistance in its unceasing rotation. No less astonishingly, on the other hand, is the celestial movement declared to be accompanied by the uppermost belt of air. This is indicated by those bodies that appear suddenly, I mean, those that the Greeks called "comets" and "bearded stars". Like the other heavenly bodies, they rise and set. They are thought to be generated in

Chapter Nine
Copernicus Chooses Formal Elegance over Common Sense

that region. That part of the air, we can maintain, is unaffected by the earth's motion on account of its great distance from the earth. The air closest to the earth will accordingly seem to be still. And so will the things suspended in it, unless they are tossed to and fro, as indeed they are, by the wind or some other disturbance. For what else is the wind in the air but the wave in the sea?

We must in fact avow that the motion of falling and rising bodies in the framework of the universe is twofold, being in every case a compound of straight and circular. For, things that sink of their own weight, being predominantly earthy, undoubtedly retain the same nature as the whole of which they are parts. Nor is the explanation different in the case of those things, which, being fiery, are driven forcibly upward. For also fire here on the earth feeds mainly on earthy matter, and flame is defined as nothing but blazing smoke. Now it is a property of fire to expand what it enters. It does this with such great force that it cannot be prevented in any way by any device from bursting through restraints and completing its work. But the motion of expansion is directed from the center to the circumference. Therefore, if any part of the earth is set afire, it is carried from the middle upwards. Hence the statement that the motion of a simple body is simple holds true in particular for circular motion, as long as the simple body abides in its natural place and with its whole. For when it is in place, it has none but circular motion, which remains wholly within itself like a body at rest. Rectilinear motion, however, affects things which leave their natural place or are thrust out of it or quit it in any manner whatsoever. Yet nothing is so incompatible with the orderly arrangement of the universe and the design of the totality as something out of place. Therefore rectilinear motion occurs only to things that are not in proper condition and are not in complete accord with their nature, when they are separated from their whole and forsake its unity.

Furthermore, bodies that are carried upward and downward, even when deprived of circular motion, do not execute a simple, constant, and uniform motion. For they cannot be governed by their lightness or by the impetus of their weight. Whatever falls moves slowly at first but increases its speed as it drops. On the other hand, we see this earthly fire (for we behold no other), after it has been lifted up high, slacken all at once, thereby revealing the reason to be the violence applied to the earthy matter. Circular motion, however, always rolls along uniformly, since it has an unfailing cause. But rectilinear motion has a cause that quickly stops functioning. For when rectilinear motion brings bodies to their own place, they cease to be heavy or light, and their motion ends. Hence, since circular motion belongs to wholes, but parts have rectilinear motion in addition, we can say that "circular" subsists with "rectilinear" as "being alive" with "being sick". Surely Aristotle's division of simple motion into three types, away from the middle, toward the middle, and around the middle, will be construed merely as a logical exercise. In like manner we distinguish line, point, and surface, even though one cannot exist without another, and none of them without body.

As a quality, moreover, immobility is deemed nobler and more divine than change and instability, which are therefore better suited to the earth than to the universe. Besides, it would seem quite absurd to attribute motion to the framework of space or that which encloses the whole of space, and not, more appropriately, to that which is enclosed and occupies some space, namely, the earth. Last of all, the planets obviously approach closer to the earth and recede farther from it. Then the motion of a single body around the middle, which is thought to be the center of the earth, will be both away from the middle and also toward it. Motion around the middle, consequently, must be interpreted in a more general way, the sufficient condition being that each such motion encircle its own center. You see, then, that all these arguments make it more likely that the earth moves than that it is at rest. This is especially true of the daily rotation, as particularly appropriate to the earth. This is enough, in my opinion, about the first part of the question.

Can Several Motions Be Attributed To The Earth? The Center Of The Universe

Accordingly, since nothing prevents the earth from moving, I suggest that we should now consider also whether several motions suit it, so that it can be regarded as one of the planets. For, it is not the center of all the revolutions. This is indicated by the planets, apparent nonuniform motion and their varying distances from the earth. These phenomena cannot be explained by circles concentric with the earth. Therefore, since there are many centers, it will not be by accident that the further question arises whether the center of the universe is identical with the center of terrestrial gravity or with some other point. For my part I believe that gravity is nothing but a certain natural desire, which the divine providence of the Creator of all things has implanted in parts, to gather as a unity and a whole by combining in the form of a globe. This impulse is present, we may suppose, also in the sun, the moon, and the other brilliant planets, so that through its operation they remain in that spherical shape which they display. Nevertheless, they swing round their circuits in divers ways. If, then, the earth too moves in other ways, for example, about a center, its additional motions must likewise be reflected in many bodies outside it. Among these motions we find the yearly revolution. For if this is transformed from a solar to a terrestrial movement, with the sun acknowledged to be at rest, the risings and settings which bring the zodiacal signs and fixed stars into view morning and evening will appear in the same way. The stations of the planets, moreover, as well as their retrogradations and [resumptions of] forward motion will be recognized as being, not movements of the planets, but a motion of the earth, which the planets borrow for their own appearances. Lastly, it will be realized that the sun occupies the middle of the universe. All these facts are disclosed to us by the principle governing the order in

Chapter Nine
Copernicus Chooses Formal Elegance over Common Sense

which the planets follow one another, and by the harmony of the entire universe, if only we look at the matter, as the saying goes, with both eyes.

The Three Motions of the Earth

Other than fitting the parameters of his model to fit the observed data as best he can, the burden of *On the Revolutions* is to show how all the well-known phenomena of astronomy are accounted for by considering a moving Earth and a change of perspective. Copernicus actually proposes that the Earth moves in three different ways, each of which explains, or, as Plato might have said, "saves" different phenomena:

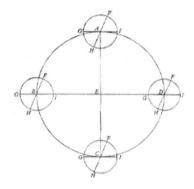

The Triple Motion of the Earth as illustrated in *On the Revolutions*.

The First Motion: The Earth rotating on its own axis.

1. The Earth rotates daily on its own north-south axis. This replaces the motions of the celestial sphere, which carried the fixed stars and also dragged around the Sun, Moon, and planets. Copernicus argues (as quoted above) that it is much simpler to explain all this by supposing that the Earth revolves west to east than to suppose the entire universe revolves east to west.

2. The Earth revolves in a circle around the sun, taking a year to complete one revolution. This explains that pesky issue, the problem of the planets. All that business of planets seeming to move slowly through the stars in a west-to-east motion, then stopping, then going backwards, then stopping again before resuming the west-to-east migration, all of that which led to Eudoxus' interlocking spheres and Ptolemy's epicycles Copernicus

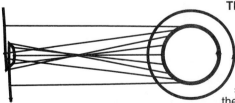

The Appearance of Retrograde Motion is explained by Copernicus as the result of the different rates of planetary motion. If the inner of the concentric circles represents the Earth's orbit and the outer Jupiter's, and lines are drawn between points on the two circles that show the progress of the two planets in their orbits over a given period of time, then by extending these lines, so as to get a rough sense of where Jupiter would appear in relation to the stars for an observer on the Earth, one can see how the planet appears at times to be moving in one direction in the sky while at other times the same planet appears to be moving in the opposite direction.

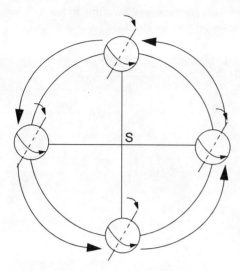

The First, Second, and Third Motions: As the Earth revolves around the Sun its axis rotates in the opposite direction at an almost identical rate. This third motion, the annual rotation of the Earth in the plane of its orbit, is what keeps the north pole pointed at the same star in the sky.

shows is just an optical illusion caused by the combination of motions around the sun at different speeds of the Earth and the other planets.

3. The third motion postulated by Copernicus concerns the tilt of the Earth's axis. The plane in which the Earth circles the Sun is called the Ecliptic.

(The ancients thought of it as the plane in which the Sun circles the Earth, but it comes to the same thing.) If the axis on which the Earth rotated were perpendicular to that plane, we would always have the same amount of daylight and nighttime everyday. But as we know, the amount of daylight varies during the year, with more daylight in the summer and less daylight in the winter. The amount of daylight is indeed the cause of the seasons. The reason that we have varying amounts of daylight is because the axis on which the Earth rotates is angled to the Ecliptic at about 23 ½ degrees. On the summer solstice, the north pole points away from the sun at that angle and on the winter solstice, the north pole points toward the sun at the same angle. We no longer think of the universe as a closed sphere, so we would view the north-south poles as basically pointing always to the same point in the sky. But because Copernicus was thinking of a spherical universe, he related all positions to the center, i.e., the Sun. Therefore to him the axis of the Earth running through the poles does move, taking one year to complete a rotation between pointing toward and pointing away from the sun. This then is the motion that, according to Copernicus, provides the seasons. Actually it is just a little bit more complicated than that. Remember that back in 150 BCE Hipparchus noticed that there was a very slow cycle in the heavens that made the equinoxes occur earlier in each successive year. This phenomenon, *precession*, had a period of 26,000 years before everything was back in the same place in the heavens. Copernicus took care of this phenomena with his third motion of the Earth by making the period of revolution of the third motion very

Chapter Nine
Copernicus Chooses Formal Elegance over Common Sense

slightly different from the annual revolution of the Earth round the Sun. At present, we view this as a slight wobble of the Earth's axis.

Mathematical Elegance versus Common Sense

The ancient philosophers, particularly Plato and Aristotle, gave the world two distinct routes to knowledge and understanding. For Plato, as for the Pythagoreans, understanding and truth lay in a hidden structure. The formal relations of things to each other provided the keys to knowledge. Those formal relations could be best expressed in mathematical terms. Therefore, the test of truth lay in whether those mathematical relations revealed a harmony, an elegance of symmetry. For Aristotle, on the other hand, this road led too easily to flights of fancy. Truth and understanding were built on careful observation of the world and proceeded in careful steps to put those observations together in a logical, consistent system that made the most sense. One did not begin by turning everything upside down or ascribing meaning to unverifiable tricks.

How did the thinking of Copernicus fit with these two contrasting approaches? If you look at Book I of *On the Revolutions*, Copernicus looks very Aristotelian in some of his arguments. He sets out his starting assumptions just as Ptolemy did, as generalizations and implications of everyday observations. Every assertion is buttressed by supporting evidence, e.g., that the Earth must be spherical because different stars are seen in different latitudes, distant objects can be seen at sea from the top of the mast of a ship, but not from the deck, and so on. These are nearly identical to the examples cited by Ptolemy and are themselves mostly taken from Aristotle's own writings. But once he starts on his refutations of the ancient views, the tone changes to that of showing how appearances can be deceptive, that the same phenomena can be produced in different ways, and how looking at it his way is more harmonious. In short, for Aristotle, it would be very difficult to get away from the everyday observation that the Earth was motionless and the heavens moved. All it would take to disprove this notion were the sort of objections cited by Ptolemy about clouds rushing to the west and heavy objects not falling straight down to show that this was nonsense.

Consider the argument, quoted above, that Copernicus offers for the Earth's daily rotation rather than the rotation of the rest of the universe.

> [I]t would seem quite absurd to attribute motion to the framework of space or that which encloses the whole of space, and not, more appropriately, to that which is enclosed and occupies some space, namely, the earth.

Or consider his comments, also quoted above, on why it is more fitting to attribute motions to the Earth and stability to the Sun and thereby more easily account for the problem of the planets. These arguments end with the invocation of an overriding orderliness:

> All these facts are disclosed to us by the principle governing the order in which the planets follow one another, and by the harmony of the entire universe, if only we look at the matter, as the saying goes, with both eyes.

Overall, Copernicus argues that the phenomena of the heavens can be accounted for either by the Ptolemaic system or by his system, but his is better because it is more elegant. This is a very Platonic or Pythagorean view. The ultimate test of the theory will be that it "saves the phenomena," that is, that his calculated positions of the stars and planets do correspond to what is observed, and that the mathematical demonstration is compact and compelling.

How did he do? One way to judge is to see how he disposes of cases where his theory implies certain phenomena that are not observed. Consider these three: (1) the motion of the Earth, (2) the phases of Venus, and (3) stellar parallax.

For the motion of the Earth, his arguments, quoted above, have a very labored quality to them. Rotation, he says is natural to a sphere, so the Earth is naturally in motion. On the other hand, the Sun, which he also claims to be a sphere, he says is fixed at the center of the universe. He wants to believe that the Earth is in motion, so he must find an explanation of why motion is not detected. The clouds and air do not rush to the west, he claims because they must be carried around with the Earth, and anyway, if rotation is natural to a sphere, then everything that is part of that sphere would have natural motion along with the rotation. This is an Ad Hoc argument; i.e., one made up to fit the conclusion.

The phases of Venus present an unexpected problem. In the Ptolemaic system, Venus is always closer to the Earth than the Sun and travels on an epicycle that cycles from one side of the Sun to the other, making it sometimes east of the Sun, when it is the "evening star," and sometimes west of the Sun, when it is the "morning star." If we saw Venus because it reflected light from the Sun, it would always be glancing off Venus and

Chapter Nine
Copernicus Chooses Formal Elegance over Common Sense

we would catch a piece of it. But in the Copernican system, Venus had an orbit *around* the Sun, closer to it than the orbit of the Earth. Sometimes Venus would be between us and the Sun and sometimes it would be on the other side of the Sun. There would therefore be a large difference in what we should see. When it was between the Sun and ourselves, light would glance off it, as in the Ptolemaic system, but when Venus was on the far side of the Sun, we would be looking straight at the fully lit side of the planet. Hence Venus should show these different angles of light. It should have phases, like the Moon does. In fact, if it did show phases, that would be a point in favor of Copernicus .

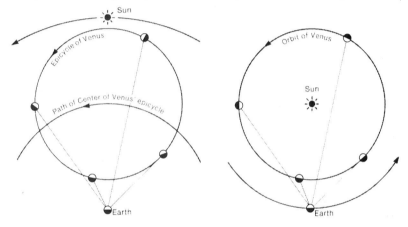

The Phases of Venus according to Ptolemy's system (left) and Copernicus' Heliocentric system (right). In the Ptolemaic system, Venus will vary between showing a sliver of reflected light (when near the Earth) and half-full (when near the sun). This is far less than according to Copernicus, whose system predicts that Venus could appear nearly full when it and the Earth are on opposite sides of the sun.

Alas, to the naked eye, Venus appears the same all the time, a bright "wandering star." This is a case where the Copernican system would seem to be contradicted by the facts and where Copernicus fails to save the phenomena. Copernicus coped with this problem by coming up with another Ad Hoc hypothesis. The reason that we see no phases, says Copernicus, is because Venus is not lit by the reflection of the Sun. Instead it has its own light, like a star. This is a very unsatisfactory answer. Not only is it just invented to make the problem go away, it does not fit the whole scheme of the difference between stars and planets. Copernicus, in effect, asks us to believe in his system not because it corresponds to sense experience, but because it is beautiful.

Of course, Venus actually does show phases and Copernicus would have been vindicated eventually, but it takes a telescope to see them and the telescope had not yet been invented.

The third problem is interesting in another way. It is also a question of geometry and this one is even more purely mathematical because it hinges on the measurement of an angle. Consider this: in Ptolemy's world, the Earth is at the direct center of the universe and the stars are fixed to a sphere that turns around that center. An astronomer watching the stars will always perceive them as being the same distance apart. Take for example, two stars some distance apart in the sky, say 30°. No matter where the stars are in the visible sky, near the horizon, directly overhead, etc., they should appear to be the same 30° apart. Likewise, it really would not matter what season of the year it was since seasons have to do with the motions of the Sun only.

But to Copernicus it should matter. The Earth is not in the center of the universe. The Earth therefore gets closer to the stars at one time of year and farther away at another. Even a few months should suffice to bring the Earth closer or farther away from the stars. And the point is, the angle that is seen between the stars varies according to how close they are. The closer you are to two separated objects, the wider they seem apart. The farther you move away the closer they appear. This is the phenomenon of *stellar parallax*, based upon a simple principle of linear perspective, which was a mathematical subject that was all the rage in Copernicus' day.

But once again, the observations let Copernicus down. There was no stellar parallax to be seen. And once again, Copernicus was ready with an Ad Hoc explanation to tidy up the problem: You can't detect any stellar parallax because the entire orbit of the Earth around the Sun is so tiny compared with the size of the celestial sphere that it can be considered a single point for the purpose of geometric measurements. This must have seemed an outrageous dodge to his contemporaries, since at the time one generally held idea about the size of the universe was that it was about as far from the Earth to Saturn, the farthest (known) planet, as it was from Saturn to the stars. The irony here is that this time, Copernicus was right. The universe is far larger than anyone had imagined: far larger than Copernicus had imagined, too. His desperate Ad Hoc stopgap turned out to be a correct guess.

Chapter Nine
Copernicus Chooses Formal Elegance over Common Sense

Ultimately, when backed into a corner, Copernicus wriggled his way out of inconsistencies by dreaming up stopgap answers. It seems clear that his intellect had led him in the right direction and problems of correspondence with empirical evidence were in comparison minor. This is very Platonic. Like Plato and Pythagoras before him, ultimately Copernicus settles for speaking only to the initiated who can follow his arguments. At one point in Book I, Copernicus admits this:

> Though these views (of mine) are difficult, counter to expectation, and certainly uncommon, yet in the sequel we shall, God willing, make them abundantly clear at least to mathematicians.

The Reception of On the Revolutions

The common-sense view of the world was dominant in Copernicus' time, as it had been for over a millennium. It was not very likely that ideas as strange as his and as dependent as his were on abstruse mathematical calculations would gain many adherents. Despite that, his system was used to prepare a new set of astronomical tables, the *Prutenic Tables*, published in 1551, which were superior to other existing tables. It was these tables that were used as the basis of the Gregorian calendar instituted in 1582.

In general, Copernicus' work managed to be used for its improved calculations without being taken seriously as a new cosmology. This was made all the easier by the man who in the end was Copernicus' editor and who saw *On the Revolutions of the Heavenly Spheres* through the publication process. This was the Lutheran minister, Andreas Osiander, who took it upon himself to add an unsigned preface to the work, which therefore appeared to be by Copernicus himself. The preface completely undermined the notion that Copernicus was serious. Osiander left the impression with the reader that all Copernicus' talk about the Sun being in the center and the Earth in motion was just a convenient fiction to help structure some mathematics that worked. The critical comment in the preface was:

> [I]t is the duty of an astronomer to compose the history of the celestial motions through careful and expert study. Then he must conceive and devise the causes of these motions or hypotheses about them. Since he cannot in any way attain to the true causes, he will adopt whatever suppositions enable the motions to be computed correctly from the principles of geometry for the future as well as for the past. The present author has performed both these duties excellently. For these

hypotheses need not be true nor even probable. On the contrary, if they provide a calculus consistent with the observations, that alone is enough.

On the Revolutions therefore caused very little stir when it first came out. A few people did take it literally and wrote off Copernicus as a nut, or to quote Martin Luther, "The fool wants to overturn the whole art of astronomy." Also, a very small number of mathematicians and astronomers took it literally and were themselves convinced, or at least intrigued, by it. A larger number followed the lead of Osiander and accepted the work as some sort of weird mental gymnastics that helped astronomers figure out where the planets would be at given times.

For a book that ultimately changed the fundamental nature of our understanding of the world and our place in it, *On the Revolutions of the Heavenly Spheres* got off to a very slow start.

For More Information

Alioto, Anthony M. *A History of Western Science,* 2nd ed. Englewood Cliffs, NJ: Prentice-Hall, 1987. Chapters 12-13.

Cohen, I. Bernard. *The Birth of a New Physics,* Rev. ed. New York: Norton, 1985. Chapter 3.

_____. *Revolution in Science.* Cambridge: Harvard University Press, 1985. Chapter 7.

Copernicus, Nicolaus. *On the Revolutions of the Heavenly Spheres,* trans. by Charles Glen Wallis. In Robert Maynard Hutchins, ed. in chief, *Great Books of the Western World.* Chicago: Encyclopedia Britannica, 1952.

_____. *Nicholas Copernicus On the Revolutions,* trans. and commentary by Edward Rosen. Baltimore and London: Johns Hopkins University Press. Available on line at http://www.dartmouth.edu/~matc/readers/renaissance.astro/1.0.Copernicus.html.

Hall, A. Rupert. *The Revolution in Science: 1500-1750.* London: Longman, 1983.

Kearney, Hugh. *Science and Change: 1500-1700.* New York: McGraw-Hill, 1971.

Kuhn, Thomas S. *The Copernican Revolution: Planetary Astronomy in the Development of Western Thought.* Cambridge: Harvard University Press, 1957.

MacLachlan, James. *Children of Prometheus: A History of Science and Technology,* 2nd ed. Toronto: Wall & Emerson, Inc., 2002. Chapter 9.

Ronan, Colin A. *Science: Its History and Development Among the World's Cultures.* New York: Facts on File, 1985. Chapter 7.

Spielberg, Nathan, and Bryon D. Anderson. *Seven Ideas that Shook the Universe,* 2nd ed. New York: Wiley, 1995. Chapter 2.

Chapter Nine | 163
Copernicus Chooses Formal Elegance over Common Sense

Wall, Byron E. "What the Copernican Revolution is All About," *The Nature of Science: Classical and Contemporary Readings.* Toronto: Wall & Emerson, Inc., 1990. Pages 45-54.

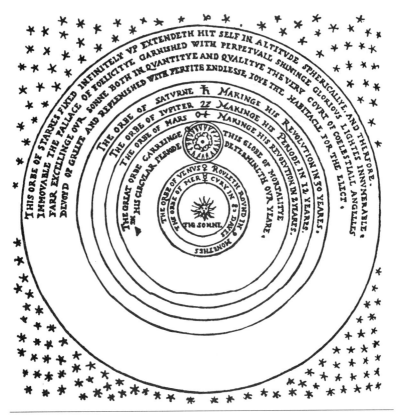

The Copernican system illustrated by Thomas Digges in 1576.

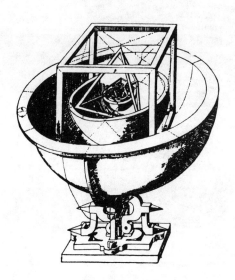

The model pictured above is a representation of Kepler's conception of the cosmos. In the center is the Sun and around it are built successive concentric spheres between which the planets orbit. The spheres' radii are in such a proportion to one another that regular solids can be fitted exactly between them. That is, each regular solid, when circumscribed around a particular sphere, will also be inscribed inside of the next larger sphere, so that every regular solid touches two spheres, one on the inside and the other on the outside.

Chapter Ten

Kepler's Celestial Harmony

One of those very few people who did read Copernicus' *On the Revolutions of the Heavenly Spheres*, understood its mathematical subtleties, and was convinced by it, was the mathematician, astrologer, and astronomer, Johannes Kepler. Kepler was the perfect reader for Copernicus' work. He was a Pythagorean at heart; mathematical relationships were more important to him than almost anything. He could put aside conventional common-sense notions whenever there was a mysterious hidden meaning in the offing. And he had the patience, perseverance, and just plain obsession to devote years of his life to calculating and recalculating and mulling over astronomical theories. If Copernicus' *On the Revolutions* can be said to have had its major impact by influencing only a handful of important people, Kepler would certainly have to be one of them.

Kepler was born in 1571 in Weil, Swabia (now southwest Germany) to a family notorious for its alleged connections with the Devil. His mother had been raised by an aunt who was burned at the stake as a witch. In later years, Kepler had to return home to help keep his mother, then seventy years old, from the same fate. Kepler had a fascination for, and it appears a faith in, astrology all his life.[1] He was a sickly child and always suffered from poor health. Curiously, for someone who devoted his life to understanding the heavens, he had very poor vision.

He attended the University of Tübingen, a Lutheran institution, to study theology and philosophy, but there became interested in mathematics and astronomy. Kepler's astronomy professor, Michael Mästlin, taught the Ptolemaic system, but was said to espouse Copernicus privately. Kepler became enamored of the Copernican view and defended it in public, an action that virtually assured that he would not be offered a faculty position

---

1. That is, he believed in the general principles of astrology—that the movements of the planets influenced life on Earth. He did however complain that the subject was in a very sorry state and was full of nonsense.

at Tübingen on graduation. But Kepler did get appointed professor of mathematics and astronomy at the University of Graz (now in Austria) and went there in 1594. One of the duties of the position was to cast horoscopes and make predictions based upon astrology. Though having misgivings himself, he complied with the demands of his job and predicted a hard winter and also a Turkish invasion. Both predictions turned out to be correct and Kepler was accordingly treated with new respect.

The Cosmographical Mystery

While teaching mathematics at Graz, Kepler realized something that set him off on a bizarre quest to explain how some of the mysteries of the heavens are explained ("saved" in the Platonic sense) by some features of Euclidean geometry. The particular mysteries that Kepler thought he could explain had to do with the placement of the planets in relation to the Sun. Kepler had fully adopted the Copernican viewpoint by then.

Why were there precisely six planets (Mercury, Venus, Earth, Mars, Jupiter, and Saturn) and why were they spaced as they were? In Euclid, or more precisely, in ancient mathematics, he thought he found the answer. To explain this requires a discussion of regular figures and solids.

The Platonic Solids

In Euclidean geometry, a two-dimensional linear figure is called "regular" if all its sides are the same length and all the angles are the same. An equilateral triangle is the simplest regular figure. Next is the square, then the regular pentagon, hexagon, and so on. One way to make a regular figure is to take a circle and divide the circumference by the desired number of sides of the figure, making arcs of equal length. Then connect the end points of the arcs by straight lines. This can be done for any number of sides desired: 3, 4, 5, ..., 17, ..., 108, and so on. There is no particular limit to the number of different kinds of regular figures.

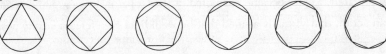

Regular figures inscribed in circles.

Chapter Ten
Kepler's Celestial Harmony

To fit a regular figure into a circle this way, so that every vertex of the figure just touches the edge of the circle is called *inscribing* the figure in a circle. You can also *circumscribe* the regular figure around a circle, so that each side of the figure just touches and is tangent to the edge of the circle within.

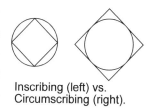

Inscribing (left) vs. Circumscribing (right).

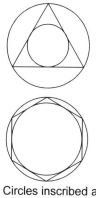

Circles inscribed and circumscribed around a triangle and a hexagon.

It is easy enough to see that each particular regular figure determines the relative ratios of the inscribed and circumscribed circles. The equilateral triangle, for example, spaces them far apart, while regular figures with more sides, for example a regular hexagon will make the two circles closer to the same size. Therefore, given one circle of a specified size and one regular figure to circumscribe around it, the size of the circle that will inscribe that figure is already determined.

And the same features apply in three dimensions, where regular solids can be fit exactly inside a sphere (inscribed), with each vertex touching the inner surface of the sphere, or fit exactly around a sphere (circumscribed), with each plane surface of the solid just touching the outer edge of the sphere within it. The definition of a regular solid is a closed three-dimensional convex object, each face of which is a regular figure.

But adding the third dimension changes matters considerably. It turns out that the extra dimension introduces a constraint that limits the number of possible regular solids. That number turns out to be exactly five, no more, no less. They are: the tetrahedron (4 sides), the cube (6 sides), the octahedron (8 sides), the dodecahedron (12 sides), and the icosahedron (20 sides). This surprising mathematical fact was known in ancient times and was considered to have some mysterious portent. Plato wrote about them in his dialogue *The Timaeus* and because of that they are often referred to as the Platonic solids. In *The Timaeus* he expressed the view that these different shapes determined the constituent elements, earth, air, fire, and water—with the fifth solid said to represent the whole cosmos. This was a very Pythagorean assertion that the tangible matter of the world was in reality an abstract mathematical object underneath.

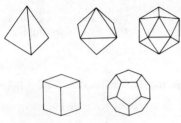

The five Platonic solids.

The mathematical proof that there are five and only five regular solids is rather tricky and builds upon a great deal of prior mathematical results. It was given a very special position by Euclid in his *Elements* because it comes at the very end of Book 13, the last book of the treatise, and is in effect a sort of capstone for the whole work.

Use the Five Solids to Space the Six Planets

What got Kepler's special attention were the circumscribing and inscribing properties of the solids. Only these five solid figures could be inscribed and circumscribed in spheres. You could therefore nest them all inside each other: from the outside going inward, *sphere 1*, solid 1, *sphere 2*, solid 2, and so on, with a final 6th sphere inscribed in solid 5. The five solids therefore determined the relative placement of six spheres. Kepler shared the common view that the planetary orbits were circular and those circles were actually the circumferences of great invisible spheres with a common center.

Kepler's inspiration was that the "spheres" of the six planets were separated from each other by sandwiching between each pair of spheres one of the regular solids. If he could only figure out the particular order of the solids, from inside to outside, he would know the mathematical reason (the "formal cause") of the seemingly erratic placement of the planets around the sun.

In 1596, Kepler was confident enough that he had worked out the answers that he published his theory in a book entitled *The Cosmographical Mystery (Cosmographicum Mysterium)*. Viewed from the outside, the sphere of Saturn circumscribed the cube; inscribed in the cube was the sphere of Jupiter; inside Jupiter was the tetrahedron; inside that was the sphere of Mars; then the dodecahedron, Earth, the icosahedron, Venus, the octahedron, and finally the sphere of Mercury.

Chapter Ten
Kepler's Celestial Harmony

This work is typical of all of Kepler's work, which is itself so characteristic of the Platonic/Pythagorean viewpoint shared by Copernicus of the supremacy of mathematical elegance over empirical evidence. Kepler's calculations worked, but only approximately. It really didn't work for Jupiter. But this did not cause him to throw out a beautiful theory. Instead he concluded that the theory had to be right and the observational data was corrupt. What he needed therefore, to be more convincing, was better data.

Tycho Brahe

Kepler was right about the observational data. It was a hodgepodge of astronomical records going back to the Babylonians, collected in a variety of ways by a variety of observers of different levels of skill and care. There were no standards that were agreed upon and enough different reference points and units of measure to guarantee that there was a fair discrepancy in the data for any planet. The telescope had yet to be invented and the sighting tools in use were of poor quality and unreliable. It was not just sour grapes to say that the data was in error.

This thought had occurred to more people than just Kepler. In fact, the quality of astronomical data was the complete preoccupation of another person living at the time. This was Tycho Brahe.

Tycho Brahe using a great quadrant to make very precise observations without the aid of a telescope.

Tycho Brahe was born in 1546 in the town of Skåne, then in Denmark, but now part of Sweden. His parents were wealthy and aristocratic and had several other children. He was raised, nevertheless by an uncle with the agreement of his natural parents because the uncle and his wife were childless and wanted a child to bring up. At the age of thirteen, Tycho was sent off to the Lutheran University of Copenhagen to study, where he developed an interest in mathematics and astronomy. This did not please the uncle, who wanted him to study law, as befit his social standing. Hence, Tycho was shipped off to Leipzig to commence legal studies. Tycho continued to study mathematics and astronomy in secret until his uncle died in 1565. Around this time, the young Tycho, who throughout his life had an acerbic personality and a terrible temper, got into a duel with another student over who was the better mathematician and lost a piece of his nose. For the rest of his life he wore a metal plate over his nose, which can be seen in all existing portraits.

Also during this period, Tycho came to the realization that even the best astronomical tables were seriously defective. In 1564, for example, he had observed a near conjunction of Jupiter and Saturn in the sky and found that the tables listed them as much farther apart at the time. Then, in 1572, Tycho observed a star in the sky that had simply not been there before. Since the stars were supposed to be fixed forever, never changing their position except by rotating all together, this was further proof to him that the observational records were totally inadequate. (The star he saw is now called a "nova"–literally a "new star"—and is caused by a distant, dim star exploding at the end of its life, giving off a huge burst of light that lasts for several months before disappearing forever.) Tycho made measurements of the light from the new star and compared it with observations by other astronomers across Europe. On the basis of these comparisons, he determined that it must in fact be very distant, that is, in Aristotelian terms, it must be in the superlunar, not the sublunar world. This was crucial because according to Aristotle, anything that had a beginning and an end had to be in the sublunar world—below the moon—and anything above the moon had to remain the same.

The Observatory of Uraniborg.

When Tycho wrote up his observations and calculations, he came to the attention of the scientific community, and also of the King of Denmark, Frederick II, who invited Tycho to set up an observatory at the king's expense and make those observations that he deemed necessary to put astronomy on a sound footing.

Tycho was installed on the island of Hven in the Danish Sound where he built an observatory, Uraniborg, dedicated to accurate measurements. He dug the observatory into the ground to minimize disturbance from the winds; he built huge metal instruments that could be calibrated to an unheard of accuracy, and he built in redundancy

Chapter Ten
Kepler's Celestial Harmony

(also unheard of) in order to be able to check one observation against another and guard against systematic error. He stayed at Uraniborg for twenty years, making the best observations of stars and planets ever made before telescopes. He might have stayed longer, but his sour personality had made him many enemies, including the tenants on the island of Hven after whom Tycho was supposed to be looking. When King Frederick died, royal funding for the observatory dried up and Tycho found himself unwelcome.

He managed to find another patron for his work; this time it was the Holy Roman Emperor, Rudolph II, who ruled from Prague. Tycho moved to Prague where he was given the title of Imperial Mathematician to the Holy Roman Emperor.

Tycho had realized how inadequate the Aristotle/Ptolemy worldview was and he also realized that there was much to be said for the better results obtained from the Copernican system. Nevertheless, he could not accept the idea of a moving Earth, both on common-sense and religious grounds. Luther himself had attacked Copernicus for contradicting the Bible, and Tycho remained a devout Lutheran.

So Tycho adopted the compromise position that Copernicus seems to have held at an earlier stage: the planets circle the Sun, but the Sun goes round the Earth, carrying the planets, while the Earth remains still. This is still called the Tychonic system.

Tycho wished to show how his new improved observations would clarify the issue and prove that his view was the correct one. Unfortunately, the mathematical work involved in doing so was horrendous. Tycho needed an assistant who could devote himself to doing the calculations. Tycho wrote to the astronomer Mästlin at the University of Tübigen and asked him if he knew of anyone who might be appropriate whom he could nominate to be his assistant. Mästlin was not only Kepler's former teacher, he also was the person who helped Kepler get the *Cosmographical Mystery* published in Tübigen. Mästlin was in a position to recommend Kepler without qualification. However, independently, Kepler, trying to get access to better astronomical data had sent a copy of his book to Tycho. Most likely, Tycho was not impressed by the geometrical symmetry argued for by Kepler, but he would have been very impressed by Kepler's abilities to make detailed calculations. Tycho invited Kepler to come to Prague.

Kepler in Prague

If Kepler had any misgivings about resigning his position as professor at Graz and joining the ill-tempered Tycho in a subservient role just to get to his data, he had another incentive at home in Graz. In 1598, Archduke Ferdinand, new ruler of the Hapsburg Empire, closed down the University at Graz in a move to rid Austria of Protestants. Kepler had the choice of converting to Catholicism or leaving the country. In 1600, he left for Prague to seek work with Tycho.

Their collaboration was, as it turns out, very short. The very next year Tycho became ill at a banquet and died shortly afterwards. Kepler had to scramble to get access to Tycho's data, which his family had wanted to sell immediately. After a struggle and some ill will, Kepler prevailed, and was appointed Tycho's successor as Imperial Mathematician to the Holy Roman Emperor.

Kepler, then, had the data he needed to sift through, calculating and recalculating in a thoroughly mind-numbing process, looking for the secrets that would explain the cosmos. It is difficult for us in the present age to realize, first, how utterly daunting were the tasks that Kepler threw himself into, and, second, what could possibly have motivated him to devote more than thirty years of his life to this limited and tedious work.

First, on the difficulties he faced: there were, of course, no mechanical aids to calculation whatsoever. The only possible device that Kepler may have had was the abacus. This would only have been of help in problems of addition and subtraction, which were not a significant part of Kepler's work. Moreover, he may not have trusted any calculation where he did not have a written record to recheck. The early calculating devices that could multiply and divide and handle trigonometric functions were just beginning to be developed in the last years of Kepler's life. Logarithms were invented in 1614, which might have been of some assistance to Kepler, but he had done most of his important work by then. (He actually wrote an important treatise on the logarithm in 1624.) An early version of the slide rule based on logarithms appeared in 1622, but even if Kepler had been able to obtain and use one, it would not have had the accuracy he needed.

Then, there is the mathematics itself. Basically what Kepler needed was the Calculus, a branch of mathematics that would not be created until after his death. Actually Ke-

Chapter Ten
Kepler's Celestial Harmony

pler did work out some of the principles of the Calculus in order to obtain measures of the area of irregular figures, but his system was extremely tedious.[2] Second, what was he looking for that was so important? Like many astronomers from antiquity to the Renaissance, Kepler viewed the heavens as not just what happened to be above us, but as a key to understanding the vital issues of existence. Because celestial events reoccurred in a predictable fashion, they partook of the eternal and the changeless. What was changeless was the foundation on which all else was built. Moreover, if there was a pattern to be discovered in the celestial world, that pattern would be the structure of the world. It would be the most important thing to know.

In *The Cosmographical Mystery*, it was the number of planets and the placement of their orbits that Kepler sought to explain. There were other features of the heavens (more specifically of the solar system) that also seemed arbitrary. Kepler sought answers to these as well: (1) What is the orbit of each planet? Individual circles did not seem to work and so far epicycle-deferent systems had not been found that matched observations satisfactorily. (2) How do the planets move through their orbits? Ptolemy realized that planets seem to complete one-half of their orbit in less time than the other half. To account for this, he had introduced the eccentric orbits and then the equant point. But that was thinking of the planets as orbiting the Earth. What is the necessary structure if they orbit the Sun? (3) The planets that are closer to the Sun have shorter periods of revolution than those farther out. Why is this? What is the precise formulation that accounts for the fact that the "year" of, say, Saturn, is thirty times as long as that of the Earth? These are the kinds of questions that occupied Kepler all his life, and indeed, he found answers to all of them, which is why he is important for the history of science. Consider the *Cosmographical Mystery*, for example. Kepler thought he had found an explanation as to why there were six planets and why they were spaced as they were. If his analysis were true then he would have nailed down some piece of eternity for examination. He would have

---

2. It is said that he got the idea for this new calculating system when attending his second wedding festivities in 1613 and noticing that the wine merchants measured the amount of wine remaining in each of several differently shaped wine barrels using the same measuring stick thrust diagonally through the bung hole in each barrel. He questioned whether this method could possibly be accurate, but to check it he needed to find a way to calculate the volume of solids of revolution. When he put his mind to this, he developed a method of calculating infinitesimals based in part on work by Archimedes. He published his analysis in a book, *New Steriometry of Wine Barrels*, in 1615.

some understanding of the mind of God. What could be more important? Because Kepler was a committed Pythagorean/Platonist, the answers to these important questions would all be analyses of formal structure. To find a mathematical formulation that specified the arrangement of the heavens unambiguously was worth all the effort he could give it.

Kepler's Laws

In a typical introductory astronomy or physics textbook, Kepler will be listed as the author of three laws. Kepler's "laws" amount to three sentences distilled from his life work and given as crucial steps in our understanding of planetary astronomy and in the dynamics of classical mechanics. These laws are answers to the three questions listed above. They are, to be sure, answers of the sort that satisfied Kepler. They were enough to motivate Kepler to do all the work to get them. These analyses of formal structure are not, however, statements of cause and effect, just precise descriptions of what happens in a compact form applicable to all the planets.

The first two laws appeared in a work which Kepler published in 1609 entitled *The New Astronomy (Astronomia Nova)*. This was the first important work of Kepler's based on Tycho Brahe's data. Tycho had put Kepler to work on figuring out the orbit of Mars, which Tycho had found very troublesome. The orbit of Mars deviated widely from the near circular orbits of the other planets, and Tycho could not fit it easily into his system. It turns out that it was the process of working out this very orbit that helped Kepler find out that contrary to all assumptions about heavenly motions made in ancient times and continued into the Renaissance, planetary orbits were not circles or combinations of circles.

The frontispiece of *Astronomia Nova*.

Conic Sections

Kepler was an accomplished mathematician, well versed in all the classical mathematical studies, such as Euclid and Ptolemy, and what was available from the Arab scholars in the Middle Ages. One of the works that would have been known to him was a study done by Apollonius of Perga, one of the Greek scholars in the Hellenistic Age, living in the century after Euclid and ultimately working at the Museum in Alexandria. He is said to have been the originator of the con-

Chapter Ten | 175
Kepler's Celestial Harmony

A **Parabola** is formed by cutting the cone with a plane parallel to one side.

An **Ellipse** is the intersection line of the cone and a cutting plane which is not parallel to one side and only meets the cone one side of its vertex.

A **Hyperbola** is created by cutting the cone by a plane which intersects the cone on both sides of its vertex. Note that the hyperbola is actually composed of two separate curves.

cept of epicycles and deferents later used by Hipparchus and then by Ptolemy. Apollonius wrote several books but the only one that has survived is his *On Conics*. A conic, short for "conic section" is the curve produced when a plane is cut through a cone.[3] Depending on the respective placement of the cone and plane, the figure produced by their intersection will be curves of different shapes: a circle, an ellipse, a parabola, and a hyperbola. Apollonius named the new curves and described their properties, though in his time they remained as objects of abstract mathematical interest only.

Kepler therefore knew about the ellipse and knew of its particular mathematical properties. It could even qualify as an appropriate heavenly pathway because, as Apollonius showed, it could be generated by an epicycle-deferent combination. What's more, the path of Mars that Kepler had calculated appeared to have the general shape of an ellipse. At some point, Kepler stopped trying to calculate combinations of circles and tried ellipses.

---

3. For purposes of mathematics, a "cone" is the shape you get by rotating a line around an axis not necessarily perpendicular to the line. Thus a cone has two pieces, a top and a bottom. Think of two ice cream cones placed point to point directly over each other. But strictly speaking, the cone would go on forever in both directions.

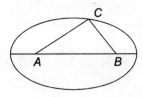

An ellipse can be defined as a curve that includes every point that is at a constant distance from two points, called the *foci*, taken as a sum. So, if A and B are the foci and C is any point on the ellipse, AC + BC is equal to some constant, say, K. You can draw such an ellipse by fixing the ends of a loose string (of length K) to two fixed points (A and B), for example by tying the string to posts at A and B. Then pull the string tight with a pencil and draw a closed path that goes all the way around the points A and B. The figure drawn will be an ellipse.

While fiddling with trying to fit Mars' orbit to an ellipse, Kepler's thinking was also being influenced by a speculative book published in 1600 by the personal physician to Queen Elizabeth of England, William Gilbert, called *On the Magnet* (*De Magnete*). Gilbert had written that the Earth itself behaved like a giant lodestone and this was the reason that compass needles always pointed north. The magnetic attraction, said Gilbert, is also the reason for the Earth's daily rotation. Kepler reasoned that if the Earth was a magnet, why not the Sun too? Then, maybe the reason that the planets went round the Sun was some sort of attraction to the Sun as the planets went round it. This was further strengthened when he discovered that if the orbit of Mars was an ellipse, the Sun would be situated at one of those foci.

Moreover, Mars seemed to travel faster along its elliptical orbit when it was nearer the Sun than when it was farther from it. Perhaps, reasoned Kepler, the Sun "animated" the planet, making it move more quickly as it neared the Sun and slowed it down as it receded. These were speculations, and Kepler treated them as such. Nothing would be worth remarking on until he had found a mathematical formula to express this.

Kepler's First and Second Law

It all came together quickly in Kepler's mind, though there were years of calculations necessary to find the general principle and demonstrate that it worked. When he had done so, he published *The New Astronomy*. In the book are his grand results, first from working with Mars, then working with all the other planets to see that the same principle applied to them all. These are what we call Kepler's first and second laws.

Chapter Ten | 177
Kepler's Celestial Harmony

The first law says that the orbits of all planets are ellipses, and in each ellipse, the Sun is at one focus. His study of Mars led him to this conclusion, but then he checked it against data for all the other planets, including the Earth, and found that they all worked. The orbits of the other planets (with the exception of Mercury) are more circular than the orbit of Mars, but even a perfect circle is an ellipse (where the two foci occupy the same point), so the general principle holds. All that bother about deferents, epicycles, eccentrics, and so on that occupied the attention of so many astronomers was unnecessary. Even Copernicus had not succeeded in finding the simple orbits and had to resort to epicycles. The orbits were not circles, but they were another simple and well defined geometrical figure. Kepler had succeeded in defining the orbit of each planet and showed how they all followed a common pattern. Just this law alone simplifies so much about astronomy. A single principle will define the orbit of all the planets, not a completely different set of rules for each one.

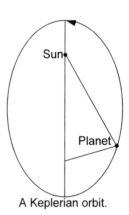
A Keplerian orbit.

The second law says that a line drawn from a planet to the Sun sweeps out equal areas in equal times. The planets indeed do move more quickly when near the Sun, but until Kepler could capture this in a formula he did not have much new to report. To understand this law, imagine a planet at part of its orbit that is farther the Sun, say, at point A, then check its position again three months later and call that position point B. Draw a line from A to the Sun and another line from B to the sun. Then those two lines, together with the curved line that is the path of the planet from A to B, makes a triangle with a curved base. Calculate the area of that triangle.[4] Now take a point on the

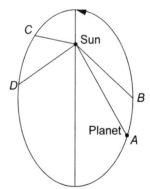
Kepler's second law illustrated. The rounded triangle formed by A, B, and the sun is equal to the rounded triangle formed by C, D, and the sun.

---

4. This is not so easy because of the curved base, but Kepler had worked out a way to approximate the area by using many ordinary triangles to fill the space as much as possible. This was similar to the system he worked out later for estimating the volume of wine barrels. (See footnote on page 167.)

orbit some time later, for example, when the planet is near to the Sun. Call that point C. Go ahead three months and locate the position of the planet and label that point D. Do the same as before, making the triangle with the curved edge and calculate its area. The second law states that if the time elapsed is the same, the area of the triangles "swept out" will also be the same.

Compare the formulation of the second law to the speculative idea that the Sun energizes the planets and makes them run faster. The latter is the kind of vague argumentation that was endemic in ancient philosophical explanations of nature. It also appeared in precise works such as Ptloemy's *Almagest* and even Copernicus' *On the Revolutions* as justifications for their assumed worldviews. Kepler entertained these ideas. He even wrote about them in his book and they may have led him to his findings. What he deemed worthy of publishing was not the speculation, but the mathematical result that the planets sweep out equal areas in equal times. This result can be completely separated from his speculations about causes and considered by itself, without any justification as to why it should be so. Were this not so, we would not remember Kepler today.

The Third Law

The next several years were hard for Kepler. First his young son died, then his wife died too, leaving him as the sole parent of three young children. Then his employer, the Emperor Rudolf abdicated due to poor health and his successor was not so tolerant of Protestants. Kepler had to leave Prague and move to Linz in Austria. Soon after, he married again and his life began to settle down, but he remained in a precarious position, in ill health, and frequently was owed money that he could not collect.

Still, Kepler kept a dream alive, that of returning to his theory of the arrangement of the planets using the five Platonic solids that he published in *The Cosmographical Mystery* and now, with the better data at his disposal, finally prove that it was correct. In 1619, he was ready with his sequel, *The Harmony of the World* (*Harmonices Mundi*). In this work, Kepler continued with his conception of the planetary orbits being spaced apart by the five regular solids, but with his new understanding of the shape and size of each orbit, he was able to refine his model.

Chapter Ten
Kepler's Celestial Harmony

Kepler's third law is found in this book, but the first surprising thing about it is that it is not what the book is about. *The Harmony of the World* was already in press when Kepler discovered the relationship that we call his third law. He had to make last-minute revisions to get it into the book at all. What the book is really about is also surprising, but first to the law.

The third law is rather more complicated to state than either of the first two laws. What it states is a precise relationship between the period of revolution of a planet and its distance from the Sun. The law can be expressed in a number of equivalent ways, several of which will be stated here. To begin, the most colorful statement of the law is Kepler's own from *The Harmony of the World*, which also tells a bit about how he thought of it.

> For after finding the true intervals of the spheres by the observations of Tycho Brahe and continuous labor and much time, at last, at last the right ratio of the periodic times to the spheres "though it was late, looked to the unskilled man, yet looked to him, and, after much time, came," and, if you want the exact time, was conceived mentally on the 8th of March in this year One Thousand Six Hundred and Eighteen but unfelicitously submitted to calculation and rejected as false, finally summoned back on the 15th of May, with a fresh assault undertaken, outfought the darkness of my mind by the great proof afforded by my labor of seventeen years on Brahe's observations and meditation upon it uniting in one concord, in such fashion that I first believed I was dreaming and was presupposing the object of my search among the principles. But it is absolutely certain and exact that the ratio which exists between the periodic times of any two planets is precisely the ratio of the 3/2th power of the mean distances, i.e., of the spheres themselves; provided, however, that the arithmetic mean between both diameters of the elliptic orbit be slightly less than the longer diameter. And so if any one take the period, say, of the Earth, which is one year, and the period of Saturn, which is thirty years, and extract the cube roots of this ratio and then square the ensuing ratio by squaring the cube roots, he will have as his numerical products the most just ratio of the distances of the Earth and Saturn from the sun.[5]

To put this in somewhat simpler language: Take any two planets, A and B. Call their periods of revolution T_A and T_B. Compute the ratio of the squares of those periods: T_A^2/T_B^2. That ratio will be the same as the ratio of the cubes of the distances, D, of those planets from the Sun. That is,

---

5. *The Harmonies of the World*, Book 5, Chapter 3. Translated by Charles Glenn Wallis.

$$T_A^2/T_B^2 = D_A^3/D_B^3$$

To be more exact, since the planets' orbits are not circles, their distance from the Sun varies. Kepler uses the average distance from the Sun.

Another way to view this is, suppose you have a good figure for the distance of the Earth from the Sun. You know the length of the Earth year, and you have measured the length of the period of revolution of some planet, such as Saturn. You can now calculate the distance of Saturn from the Sun using Kepler's third law.

Still another way to state the law is to say that the ratio of the cube of the distance to the Sun of a planet to the square of the period of revolution is a constant for all planets. That is,

$$D^3/T^2 = K$$

where K is the same for all planets. This is not the way that Kepler expressed the law, but in a sense it is the best way because it identifies that there is some constant ratio between distance and time of revolution. Kepler did not get further into this relationship than the statement of his law. That was enough. The task of showing why this ratio is a constant was performed by Newton, who, by the way, is the person who identified these results as Kepler's Laws.

The Harmony of the World

To return to the book *The Harmony of the World*, what it was intended to be was a continuation of *The Cosmographical Mystery*, that is, it was going to explain more of the secret and marvelously elegant plan of the universe that fit the vision of hidden forms that controlled all. As the title of the book suggests, the hidden structure that Kepler saw this time was a musical one. Like Pythagoras, Kepler saw musical intervals as fixed ratios of numbers: 2:1 was the octave, 3:2 the perfect fifth, 4:3 the perfect fourth, and so on. There were also plenty of ratios to be found in the positions of the planets.

In *The Harmony of the World*, Kepler took on music theory itself. He was going to show that the intervals that the ear hears as consonant are produced by the planets as they go through their orbits. The argument is an extremely convoluted one and does more to show how Kepler could get fixated on arcane mathematical relationships than it

Chapter Ten
Kepler's Celestial Harmony

illuminates anything about either astronomy or music. So this is a brief synopsis. As Kepler's first and second laws state, the planets move in elliptical orbits and move more quickly in those orbits when they are near the sun. The point where a planet is closest to the Sun is called the *perihelion*; where it is farthest is called the *aphelion*. As implied by the second law, the planet will have greater angular speed at the perihelion than at the aphelion. Those two different angular speeds can be compared to each other as a ratio. A ratio determines a musical interval.

Each planet therefore carves out a musical interval as it goes through its orbit. Since those intervals are fixed in the heavenly bodies, they must have special significance. In particular, Kepler held that these heavenly intervals determine what we experience as consonant and dissonant sounds. Remember that he was also an astrologer.

Without going into all the details of how the planets' motions combine to make musical statements, but to give a sense of this bizarre conception, here are the basic intervals produced by each planet and the Moon as they travel from aphelion to perihelion and back. The musical notation is that of the period and takes Saturn as the standard for notation of pitch:

Mercury begins at C and rises to E, a tenth higher. This would actually "sound" four octaves higher than written.

Venus varies so slightly that it never departs from the single note, E. Three octaves higher than written.

The Earth goes only a half-step from G to A-flat. Two octaves higher.

Mars, which has a more irregular path than the nearly circular orbits of Venus and the Earth, goes from F to C, a perfect fifth, and back. One octave higher than written.

Jupiter goes from G to B-flat, a minor third.

Saturn goes from G to B natural, a major third.

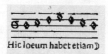 And the Moon goes from G to C, a perfect fourth.

Put these all together and you get the kind of "sounds" that Kepler believed was the basis of music that could touch the soul. Kepler was particularly enamored of the music of Orlando di Lasso and believed that di Lasso's music exhibited all these celestial harmonies, which made them particularly effective.

◆

Kepler's three laws constituted a significant breakthrough in planetary astronomy, leading as they did to a new, comprehensive formulation by Newton a generation later. But Kepler was a very strange man with some very off-beat ideas that he held just as strongly as he did the ones for which we remember him now. Kepler was one of the few believers in the Copernican system in the seventy years or so after the death of Copernicus. It should be easy enough to see that Kepler was not the sort of person any more than Copernicus was himself to convince the general public that their common sense had let them down, and Mother Earth was actually spinning around every day and whizzing through space at an incredible rate. It was going to take a different sort of personality for that.

Chapter Ten
Kepler's Celestial Harmony

For More Information

Alioto, Anthony M. *A History of Western Science,* 2nd ed. Englewood Cliffs, NJ: Prentice-Hall, 1987. Chapters 13.For more information

Caspar, Max. *Kepler.* Trans. and ed. by C. Doris Hellman. London & New York: Abelard-Schuman, 1959.

Cohen, I. Bernard. *The Birth of a New Physics,* Rev. ed. New York: Norton, 1985. Chapter 8.

_____. *Revolution in Science.* Cambridge: Harvard University Press, 1985. Chapter 7.

Hall, A. Rupert. *The Revolution in Science: 1500-1750.* London: Longman, 1983. Chapter 5.

Kearney, Hugh. *Science and Change: 1500-1700.* New York: McGraw-Hill, 1971. Chapter 4.

Kepler, Johannes. *Astronomia Nova*, trans. W.H. Donahue Cambridge: Cambridge University Press, 1992.

_____. *The Harmonies of the World.* Book 5. Translated by Charles Glenn Wallis. Chicago: Encyclopaedia Brittanica, 1952.

Koestler, Arthur. *The Sleepwalkers*, London: Arkana Books, 1989.

Kuhn, Thomas S. *The Copernican Revolution: Planetary Astronomy in the Development of Western Thought.* Cambridge: Harvard University Press, 1957.

MacLachlan, James. *Children of Prometheus: A History of Science and Technology,* 2nd ed. Toronto: Wall & Emerson, Inc., 2002. Chapter 9.

Ronan, Colin A. *Science: Its History and Development Among the World's Cultures.* New York: Facts on File, 1985. Chapter 8.

Spielberg, Nathan, and Bryon D. Anderson. *Seven Ideas that Shook the Universe,* 2nd ed. New York: Wiley, 1995. Chapter 2.

Wall, Byron E. "What the Copernican Revolution is All About," *The Nature of Science: Classical and Contemporary Readings.* Toronto: Wall & Emerson, Inc., 1990. Pages 45-54.

Westfall, Richard S. *The Construction of Modern Science: Mechanisms and Mechanics.* Cambridge: Cambridge University Press, 1977.

Galileo offers his telescope to the Muses in this illustration from the frontispiece of a collection of Galileo's works.

Chapter Eleven

Galileo, the Anti-Anti-Copernican

A very different sort of person from Johannes Kepler was Galileo Galilei, born in 1564, the same year as William Shakespeare and almost eight years before Kepler. Galileo was born in Pisa, son of a prominent musician, Vincenzio Galilei, author of a book on music theory which Kepler read while he was journeying to defend his mother from witchcraft charges and to which he then referred several times in *The Harmony of the World*.

Despite being a successful and respected musician, Vincenzio Galilei had very little money and wanted his son to follow a more financially reliable career. So young Galileo was sent to the University of Pisa to study medicine. When he got to the university he became particularly interested in mathematics and began to study it privately. Galileo left the University of Pisa before completing a degree, possibly because of financial hardship. But his studies of mathematics were enough that he was able to get a position at the Academy of Florence. In 1588, Galileo gave a lecture on the geography of Dante's *Inferno*, which discussed its mathematical aspects. This brought him to the attention of an influential Florentine, Guidobaldo del Monte, who helped Galileo attain the position of Professor of Mathematics at the University of Pisa the following year. Galileo was then twenty-five years old.

Mathematics was not high in the university pecking order in Galileo's time. Nothing makes this clearer than the salary levels. Galileo as Professor of Mathematics earned a salary that was one-tenth of what the Professor of Philosophy earned. How times have changed.

Galileo found this galling, especially as he needed money for his family. Thus began a life-long dislike that Galileo had for philosophers and philosophy. In Galileo's view, philosophers did not think, they just mouthed dogma. The dogma at that time was Scholastic Aristotelian philosophy. It's ironic that Aristotelian philosophy should have become divorced from experience in the Renaissance, since Aristotle himself was the great advocate of basing all knowledge on empirical observations. But in the nearly two thousand years since Aristotle lived, the search for truth had become a process of finding out what the great master had said and then using logic to build upon that instead of going back to basic experience.

Galileo Galilei.

Galileo's sharp mind could outwit almost anybody. He began to develop a habit of engaging philosophers on some point of doctrine and showing what fools they were. This may have been personally satisfying to Galileo, but it earned him a growing number of enemies who would have been happy to see Galileo get his comeuppance. It was during this period that Galileo made his famous demonstration from the Leaning Tower of Pisa. He took two metal balls of different weights and dropped them at the same time from the top of the tower. According to Aristotle, the heavier ball should fall much faster than the lighter ball because it will be more urgently seeking its natural place. But as Galileo showed, the balls hit the ground at virtually the same time. This was clearly not an experiment to see what the outcome would be. Galileo had determined that beforehand. This was a public demonstration to show that the philosophers were wrong. The colorful reports that he dropped the balls before the entire assembled university are considered false now, but the lesson is nevertheless clear. Conclusions about how the world works have to be checked out.

Galileo had always leaned toward the practical side of mathematics rather than the abstract subject in and for itself alone. In the late sixteenth and early seventeenth centuries, when Galileo lived, the practical uses of mathematics tended toward the military. A practical mathematician was a person who was good with instruments and calculations, such as were needed to operate artillery, for example. The term "engineer" was coming into use and had nearly the same meaning as practical mathematician. Galileo was a very competent engineer. He made instruments of precision and sold them to supplement his income. He also invented several instruments, such as one that aided artillery calculations. (He called it a "Geometrical and Military Compass.")

Chapter Eleven
Galileo, the Anti-Anti-Copernican

In 1591, his father died and Galileo assumed responsibility for the support of his entire family. He appealed once again to Guidobaldo del Monte, who used his influence to help Galileo obtain the position of Professor of Mathematics at the University of Padua, near Venice. The position at Padua paid three times the salary of Pisa and carried only light teaching duties, leaving Galileo considerable time for research. Padua, being part of the Venetian Republic, had a freer intellectual climate than Pisa and many other parts of Italy. Galileo's life was much improved, though his income was still not really adequate to meet all his financial obligations.

In addition to his instrument making, through which Galileo earned extra income, he was also able to give more time to some basic research into issues of fundamental concern to the physics of the earth, an earlier example of which was the experiment/demonstration of dropping balls off the Leaning Tower of Pisa. That work is a topic in itself and will form the subject matter of the next chapter. This chapter follows a different sequence of events.

Galileo and the Copernican System

Galileo was familiar with the ideas of Copernicus and appreciated the mathematical arguments, but Galileo was too much of an empiricist to accept the heliocentric (sun-centered) viewpoint without confirmation. Unlike Kepler, the mystical symmetry of the Copernican viewpoint was not enough for Galileo to decide it must be the truth. After all, the idea that the Earth is flying through space at a terrific rate is hard to swallow without some direct evidence. Then, according to one story, in 1595 or so, as he was visiting Venice and seeing water sloshing back and forth in the canals as vessels passed through them, he got an idea.

The Tides

Consider any container of water, being moved along at a steady pace. The surface of the water will be more or less level. However, if you start moving the container ahead suddenly at a faster pace, the water will rush toward the back of the container and the level of water will be higher there and lower in the front. Conversely, if the container slows down its forward motion, the water will rush forward and bunch up in the front, leaving the rear with a lower level.

Now consider the Copernican explanation of the motions of the Earth. Ignoring the effect of Copernicus' "third" motion, the main two motions attributed to the Earth are (1) the daily rotation, and (2) the annual revolution about the Sun. Consider, for purposes of illustration, a place on the equator of the Earth. At midnight at that place, the Sun is directly beneath one's feet. From the Copernican point of view, the Earth is turning in an easterly direction in its daily rotation, and from the vantage point of the chosen place, the Earth itself is traveling through space also in an easterly direction. The two motions are in the same direction. However, let 12 hours pass. It will now be noon, local time, and while the daily rotation continues in an easterly direction, the surface of the Earth has swung around and points in the opposite direction so that the annual motion around the sun is now toward the west. The two motions are in opposite directions.

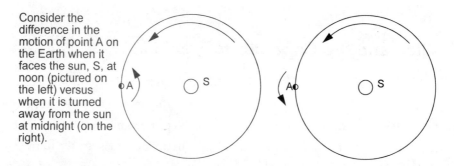

Consider the difference in the motion of point A on the Earth when it faces the sun, S, at noon (pictured on the left) versus when it is turned away from the sun at midnight (on the right).

Looked at from the vantage point of the Sun (which, in the Copernican system is the center of the universe and therefore a pretty special place), the spot on the Earth is speeding up and slowing down. Now suppose the spot we are talking about is on the shore of some large body of water. It would be just like being at one end of the container mentioned before. When the Earth "speeds up" (i.e., at night), the water rushes to the west and when it slows down (in the day), the water rushes to the east. The result is that the water level goes up and down a considerable amount at that place on shore. Does this phenomenon occur? Of course it does, these are the daily tides, a well-known feature of life in any seaport.

Chapter Eleven
Galileo, the Anti-Anti-Copernican

The Copernican system, thought Galileo, explains the tides. Here was a visible confirmation of the heliocentric viewpoint. It was not conclusive, but it made Galileo more open to Copernican ideas.

The Telescope

Several years passed during which Galileo busied himself on other matters, and then an event occurred which changed Galileo's life forever, making him a confirmed and very vocal supporter of the Copernican system. It also changed astronomy forever. This concerns the telescope.

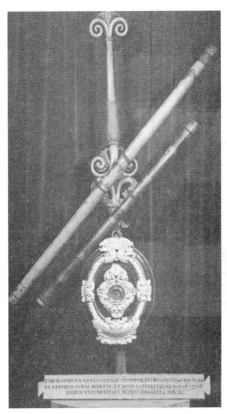
A Galilean telescope.

The origin of the original "spy glass" is a bit fuzzy, but it is known that in October, 1608, a patent application was submitted by a spectacles maker in the Netherlands for an instrument that could see at a distance. Two weeks later, another patent claim was made by another lens maker, and there were other applications in the Netherlands soon after, as well as reports of similar devices in other countries. None of this mattered until a Flemish merchant traveled to Italy in the spring of 1609 to try to sell the instrument to the Doge of Venice. Venice was a commercial center at the time, so it made much sense for anyone with a new product to try to promote it there.

Galileo heard of the instrument and had it described to him. He went back to his workshop and within a day or so he had made one of his own. Galileo's first "spy glass" had a magnification of three, but soon he improved the design and got the magnification up to ten, and later on up to thirty. Galileo had intended his new instrument, which was much better than the Dutch versions, for commercial applications. In a place like Venice where shipping is the lifeblood of the city, it would be extremely advantageous to know what ships were out at sea nearing port with their expected cargos. Galileo demonstrated how one could climb to a high spot, for example up in the Campanile in Piazza San Marco, and with his instrument make out the identifying flags of ships entering the Lagoon of Venice from the Adriatic. This so impressed the Venetians that his professorship was confirmed for life, his salary increased, and he was also able to add considerably to his income by manufacturing the device in his workshop and selling it to merchants.

The telescope (the name was coined later) was not intended as an instrument of research in the skies. After all, if the stars were immutable and just went round and round (even in the Copernican system) and the planetary orbits had been worked out, what else was there to know? It did not take the curious Galileo very long before he tried pointing one of his instruments up at the sky at night to see if it made a difference. What Galileo saw and recorded over the next few months was sufficient in his mind to confirm that Copernicus must have been right and Ptolemy was wrong. It also gave Galileo more ways to undermine the philosophers who were, of course, committed to the Ptolemaic viewpoint.

The Starry Messenger

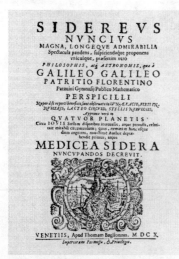

The title page of *The Starry Messenger*.

Galileo realized at once that he was seeing things that no one had ever seen before. Everything he saw fit and supported the Copernican system. To begin with, the telescope gave him a much more detailed view of the Moon. In the Aristotelian/Ptolemaic model, heavenly bodies were completely different from things on Earth. As Aristotle had it, everything in the heavens was formed of the fifth element, the quintessence, and was eternal and perfect. The perfect shape was the sphere, so therefore heavenly bodies were all perfectly smooth spheres. The planets could reflect light because they had a smooth surface, just as a mirror does. An irregular

Chapter Eleven
Galileo, the Anti-Anti-Copernican

surface such as the face of the Earth, would not be able to reflect light, or so the reasoning went. In the ancient view, all heavenly bodies went around the Earth. In the Copernican, everything went round the Sun—except the Moon. Why would one celestial body persist in going around the Earth while everything else went around the Sun? Was this not an indication that the Copernican system was jury-rigged?

When Galileo looked at the Moon, he could see that the slight discolorations on the face of the Moon (the "Man in the Moon") were actually shadows on the Moon caused by its very irregular surface. By careful measurement of the shadows over a period of time and a judicious use of the Pythagorean Theorem, Galileo was able to calculate that there were mountains on the Moon as large as four miles, a size that exceeds all but the highest mountains on Earth. (Mount Everest is 5 ½ miles high; the Alps, which Galileo would have been familiar with, run to nearly 3 miles high, though Galileo mistakenly thought they did not exceed one mile in height.) Other shadows led Galileo to believe that the Moon had oceans like the Earth. (These were the craters. Galileo had detected the indentations.) So, the Moon was a heavenly body, but it was rough and irregular just like the Earth. Therefore it was not so outrageous to say that the Earth was a planet.

On the question of reflecting light, Galileo saw that while we normally see only the part of the Moon that is in direct sunlight, at the time when the Moon is but a crescent in the sky, some of the dark part of it is also dimly visible through the telescope. This was possible, Galileo reasoned, because light was being reflected off the Earth onto the Moon. He called this "Earthshine" to make the parallel with Moonshine, which illuminates the Earth by reflected light. If the Earth, clearly not smooth, could reflect light onto the Moon, then there was no reason to conclude that heavenly bodies had to be smooth and perfect.

On the subject of the anomaly of the Moon circling the Earth while every other planet circled the Sun in the Copernican system, Galileo made the surprising discovery that there were four previously unnoticed "stars" near Jupiter. Becoming more curious, he watched these night after night and tracked their position. What he found was that they appeared to be in orbit around Jupiter. So the Earth was not alone in having an orbiting satellite.

Part II
Science Emerges

Orion's Belt and Sword as seen by Galileo with many more stars than had been previously charted.

And the telescope showed new things about the stars themselves. To begin with, there were many more of them than had been seen and recorded before. Aristotle had estimated the number of stars to be about seven thousand. Galileo showed there were many more, and moreover, he could see that the mysterious Milky Way and observed nebulae were just clusters of stars that the naked eye could not individually distinguish.

After several months of nocturnal observations with his new invention, Galileo was ready to tell the world what he had seen and what it meant. In 1610, he published a fifty-page pamphlet called *The Starry Messenger* (*Siderius Nuncius*) and dedicated it to the Duke of Tuscany, whose favor he was trying to court. The following excerpt from that booklet includes his observations of the Moon and his argument that it is, well, rather like the Earth.

> Great indeed are the things which in this brief treatise I propose for observation and consideration by all students of nature. I say great, because of the excellence of the subject itself, the entirely unexpected and novel character of these things, and finally because of the instrument by means of which they have been revealed to our senses.
>
> Surely it is a great thing to increase the numerous host of fixed stars previously visible to the unaided vision, adding countless more which have never before been seen, exposing these plainly to the eye in numbers ten times exceeding the old and familiar stars.
>
> It is a very beautiful thing, and most gratifying to the sight, to behold the body of the moon, distant from us almost sixty earthly radii, as if it were no farther away than two such measures—so that its diameter appears almost thirty times larger, its surface nearly nine hundred times, and its volume twenty-seven thousand times as large as when viewed with the naked eye. In this way one may learn with all the certainty of sense evidence that the moon is not robed in a smooth and polished surface but is in fact rough and uneven, covered everywhere, just like the earth's surface, with huge prominences, deep valleys, and chasms.

Chapter Eleven | 193
Galileo, the Anti-Anti-Copernican

Again, it seems to me a matter of no small importance to have ended the dispute about the Milky Way by making its nature manifest to the very senses as well as to the intellect. Similarly it will be a pleasant and elegant thing to demonstrate that the nature of those stars which astronomers have previously called "nebulous" is far different from what has been believed hitherto. But what surpasses all wonders by far, and what particularly moves us to seek the attention of all astronomers and philosophers, is the discovery of four wandering stars not known or observed by any man before us. Like Venus and Mercury, which have their own periods about the sun, these have theirs about a certain star that is conspicuous among those already known, which they sometimes precede and sometimes follow, without ever departing from it beyond certain limits. All these facts were discovered and observed by me not many days ago with the aid of a spyglass which I devised, after first being illuminated by divine grace. Perhaps other things, still more remarkable, will in time be discovered by me or by other observers with the aid of such an instrument, the form and construction of which I shall first briefly explain, as well as the occasion of its having been devised. Afterwards I shall relate the story of the observations I have made.

About ten months ago a report reached my ears that a certain Fleming had constructed a spyglass by means of which visible objects, though very distant from the eye of the observer, were distinctly seen as if nearby. Of this truly remarkable effect several experiences were related, to which some persons gave credence while others denied them. A few days later the report was confirmed to me in a letter from a noble Frenchman at Paris, Jacques Badovere, which caused me to apply myself wholeheartedly to inquire into the means by which I might arrive at the invention of a similar instrument. This I did shortly afterwards, my basis being the theory of refraction. First I prepared a tube of lead, at the ends of which I fitted two glass lenses, both plane on one side while on the other side one was spherically convex and the other concave. Then placing my eye near the concave lens I perceived objects satisfactorily large and near, for they appeared three times closer and nine times larger than when seen with the naked eye alone. Next I constructed another one, more accurate, which represented objects as enlarged more than sixty times. Finally, sparing neither labor nor expense, I succeeded in constructing for myself so excellent an instrument that objects seen by means of it appeared nearly one thousand times larger and over thirty times closer than when regarded with our natural vision.

It would be superfluous to enumerate the number and importance of the advantages of such an instrument at sea as well as on land. But forsaking terrestrial observations, I turned to celestial ones, and first I saw the moon from as near at hand as if it were scarcely two terrestrial radii away. After that I observed often with wondering delight both the planets and the fixed stars, and since I saw these latter to be very crowded, I began to seek (and eventually found) a method by which I might measure their distances apart...

Now let us review the observations made during the past two months, once more inviting the attention of all who are eager for true philosophy to the first steps of such important contemplations. Let us speak first of that surface of the moon which faces us. For greater clarity I distinguish two parts of this surface, a lighter and a darker; the lighter part seems to surround and to pervade the whole hemisphere, while the darker part discolors the moon's surface like a kind of cloud, and makes it appear covered with spots. Now those spots which are fairly dark and rather large are plain to everyone and have been seen throughout the ages; these I shall call the "large" or "ancient" spots, distinguishing them from others that are smaller in size but so numerous as to occur all over the lunar surface, and especially the lighter part. The latter spots had never been seen by anyone before me. From observations of these spots repeated many times I have been led to the opinion and conviction that the surface of the moon is not smooth, uniform, and precisely spherical as a great number of philosophers believe it (and the other heavenly bodies) to be, but is uneven, rough, and full of cavities and prominences, being not unlike the face of the earth, relieved by chains of mountains and deep valleys. The things I have seen by which I was enabled to draw this conclusion are as follows.

On the fourth or fifth day after new moon, when the moon is seen with brilliant horns, the boundary which divides the dark part from the light does not extend uniformly in an oval line as would happen on a perfectly spherical solid, but traces out an uneven, rough, and very wavy line as shown in the figure below. Indeed, many luminous excrescences extend beyond the boundary into the darker portion, while on the other hand some dark patches invade the illuminated part. Moreover a great quantity of small blackish spots, entirely separated from the dark region, are scattered almost all over the area illuminated by the sun with the exception only of that part which is occupied by the large and ancient spots. Let us note, however, that the said small spots always agree in having their blackened parts directed toward the sun, while on the side opposite the sun they are crowned with bright contours, like shining summits. there is a similar sight on earth about sunrise, when we behold the valleys not yet flooded with light though the mountains surrounding them are already ablaze with glowing splendor on the side opposite the sun. And just as the shadows in the hollows on earth diminish in size as the sun rises higher, so these spots on the moon lose their blackness as the illuminated region grows larger and larger.

Again, not only are the boundaries of shadow and light in the moon seen to be uneven and wavy, but still more astonishingly many bright points appear within the darkened portion of the moon, completely divided and separated from the illuminated part and at a considerable distance from it. After a time these gradually increase in size and brightness, and an

Chapter Eleven
Galileo, the Anti-Anti-Copernican

Galileo's Sketches of the Moon

Galileo made detailed observations of the new vistas that his telescope opened. Here, he drew sketches of the Moon at different times to show the irregularity of the shadow on its face, evidence which led him to believe that the Moon was far more like the Earth than had been previously imagined. Note the other dark patches, clearly visible on the left-hand drawings, that Galileo concluded were Lunar seas.

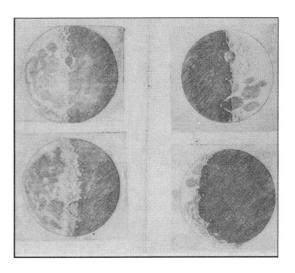

hour or two later they become joined with the rest of the lighted part which has now increased in size. Meanwhile more and more peaks shoot up as if sprouting now here, now there, lighting up within the shadowed portion; these become larger, and finally they too are united with that same luminous surface which extends ever further. An illustration of this is to be seen in the figure above. And on the earth, before the rising of the sun, are not the highest peaks of the mountains illuminated by the sun's rays while the plains remain in shadow? Does not the light go on spreading while the larger central parts of those mountains are becoming illuminated? And when the sun has finally risen, does not the illumination of plains and hills finally become one? But on the moon the variety of elevations and depressions appears to surpass in every way the roughness of the terrestrial surface, as we shall demonstrate further on.

At present I cannot pass over in silence something worthy of consideration which I observed when the moon was approaching first quarter, as shown in the previous figure. Into the luminous part there extended a great dark gulf in the neighborhood of the lower cusp. When I had observed it for a long time and had seen it completely dark, a bright peak began to emerge, a little below its center, after about two hours. Gradually growing, this presented itself in a triangular shape, remaining completely detached and separated from the lighted surface. Around it three other small points soon began to shine, and finally, when the moon was about to set, this triangular shape (which had meanwhile become more widely extended) joined with the rest of the illuminated region and suddenly burst into the gulf of shadow like a vast promontory of light, surrounded still by the three bright peaks already mentioned. Beyond the ends of the cusps, both above and below, certain bright points emerged which were quite detached from the remaining lighted part, as may be seen depicted in the same figure. There were also a great number of dark spots in both the horns, especially in the lower one; those nearest the boundary of light and shadow appeared larger and darker, while those more distant from the boundary were not so dark and distinct. But in all cases, as we have mentioned earlier, the blackish portion of each spot is turned toward the source of the sun's radiance, while a bright rim surrounds the spot on the side away from the sun in the direction of the shadowy region of the moon. This part of the moon's surface, where it is spotted as the tail of a peacock is sprinkled with azure eyes, resembles those glass vases which have been plunged while still hot into cold water and have thus acquired a crackled and wavy surface from which they receive their common name of "ice-cups."

As to the large lunar spots, these are not seen to be broken in the above manner and full of cavities and prominences; rather, they are even and uniform, and brighter patches crop up only here and there. Hence if anyone wished to revive the old Pythagorean opinion that the moon is like another earth, its brighter part might very fitly represent the surface of the land and its darker region that of the water. I have never doubted that if our globe were seen from afar when flooded with sunlight, the land regions would appear brighter and the watery regions darker. The large spots in the moon are also seen to be less elevated than the brighter tracts, for whether the moon is waxing or waning there are always seen, here and there along its boundary of light and shadow, certain ridges of brighter hue around the large spots (and we have attended to this in preparing the diagrams); the edges of these spots are not only lower, but also more uniform, being uninterrupted by peaks or ruggedness.

Near the large spots the brighter part stands out particularly in such a way that before first quarter and toward last quarter, in the vicinity of a certain spot in the upper (or northern) region of the moon, some vast prominences arise both above and below as shown in the figures reproduced below. Before last quarter this same spot is seen to be walled about with certain blacker contours which, like the loftiest mountaintops, appear darker on the side away from the sun and brighter on that which faces the sun. (This is the opposite of what happens in the cavities, for there the part away from the sun appears brilliant, while that which is turned toward the sun is dark and in shadow.) After a time, when the lighted portion of the moon's surface has diminished in size and when all (or nearly all) the said spot is covered with shadow, the brighter ridges of the mountains gradually emerge form the shade. This double aspect of the spot is illustrated in the ensuing figures.

There is another thing which I must not omit, for I beheld it not without a certain wonder; this is that almost in the center of the moon there is a cavity larger than all the rest, and perfectly round in shape. I have observed it near both first and last quarters, and have tried to represent it as correctly as possible in the second of the above figures. As to light and shade, it offers the same appearance as would a region like Bohemia if that were enclosed on all sides by very lofty mountains

Chapter Eleven
Galileo, the Anti-Anti-Copernican

arranged exactly in a circle.[1] Indeed, this area on the moon is surrounded by such enormous peaks that the bounding edge adjacent to the dark portion of the moon is seen to be bathed in sunlight before the boundary of light and shadow reaches halfway across the same space. As in other spots, its shaded portion faces the sun while its lighted part is toward the dark side of the moon; and for a third time I draw attention to this as a very cogent proof of the ruggedness and unevenness that pervades all the bright region of the moon. Of these spots, moreover, those are always darkest which touch the boundary line between light and shadow, while those farther off appear both smaller and less dark, so that when the moon ultimately becomes full (at opposition[2] to the sun), the shade of the cavities is distinguished from the light of the places in relief by a subdued and very tenuous separation.[3]

The things we have reviewed are to be seen in the brighter region of the moon. In the large spots, no such contrast of depressions and prominences is perceived as that which we are compelled to recognize in the brighter parts by the changes of aspect that occur under varying illumination by the sun's rays throughout the multiplicity of positions from which the latter reach the moon. In the large spots there exist some holes rather darker than the rest, as we have shown in the illustrations. Yet these present always the same appearance, and their darkness is neither intensified nor diminished, although with some minute difference they appear sometimes a little more shaded and sometimes a little lighter according as the rays of the sun fall on them more or less obliquely. Moreover, they join with the neighboring regions of the spots in a gentle linkage, the boundaries mixing and mingling. It is quite different with the spots which occupy the brighter surface of the moon; these, like precipitous crags having rough and jagged peaks, stand out starkly in sharp contrasts of light and shade. And inside the large spots there are observed certain other zones that are brighter, some of them very bright

---

1. This casual comparison between a part of the moon and a specific region on earth was later the basis of much trouble for Galileo.... Even in antiquity the idea that the moon (or any other heavenly body) was of the same nature as the earth had been dangerous to hold. The Athenians banished the philosopher Anaxagoras for teaching such notions, and charged Socrates with blasphemy for repeating them. [Translator's note.]
2. Opposition of the sun and moon occurs when they are in line with the earth between them (full moon, or lunar eclipse); conjunction, when they are in line on the same side of the earth (new moon, or eclipse of the sun). [Translator's note.]
3. Galileo was on to more than he realized. The "cavity" he described was a giant crater formed by the impact of a huge meteorite. His comparison of its appearance to Bohemia (which he had never seen from the ground, let alone from the air) was to establish that the moon was rocky and uneven, like parts of the earth. Bohemia was ringed by mountains and would, Galileo surmised, look much the same through a telescope from a comparable distance as the "cavity" on the moon did. Indeed it does, as we now know. But more than that, some scientists are now of the opinion that the "Bohemian Plateau" in western Czechoslovakia is itself a crater formed by the impact of a huge meteorite. See Robert Kunzig, "Back to Bohemia," *Discover* 10 (June 1989): 22—23.

> indeed. Still, both these and the darker parts present always the same appearance; there is no change either of shape or of light and shadow; hence one may affirm beyond any doubt that they owe their appearance to some real dissimilarity of parts. They cannot be attributed merely to irregularity of shapes, wherein shadows move in consequence of varied illuminations from the sun, as indeed is the case with the other, smaller, spots which occupy the brighter part of the moon and which change, grow, shrink, or disappear from one day to the next, as owing their origin only to shadows of prominences.[4]

After publishing *The Starry Messenger*, Galileo continued his telescopic studies and discovered more interesting things: The Sun, he found, had spots on it, and those spots moved in a regular way, suggesting that the Sun rotated too. Saturn, he discovered, had "ears"—these were the rings of Saturn, but Galileo's telescope could not make that out. He thought instead that Saturn had two stationary satellites around it. But for demonstrating that Copernicus was right, the most important discovery was that Venus did indeed show phases like the Moon, just as they should have in the Copernican arrangement. That made Copernicus' Ad Hoc assertion that Venus had its own light unnecessary.

The Starry Messenger itself had an enormous impact. The first printing was sold out immediately and word of the book quickly spread across Europe. Johannes Kepler sent congratulations to Galileo. A second edition was published in Frankfurt, and telescopes, particularly those made by Galileo, were in great demand. In his edition of *The Starry Messenger* and other related works, Stillman Drake reports the following story:

> An amusing illustration of his sudden fame is an event that occurred at Florence only two weeks after his book was published. The courier from Venice brought a package to one of his friends there, and neighbors at once surrounded him, demanding that it be opened at once. They were sure that a telescope must be inside. When instead the contents turned out to be a copy of the already famous book, they insisted that its new owner read aloud to them that very evening Galileo's account of his discovery of the Medicean stars. [That is, the satellites of Jupiter.][5]

Though Galileo now had a secure position at the University of Padua and a reasonable income that he could supplement by selling instruments from his workshop and tak-

---

4. From *Discoveries and Opinions of Galileo*, trans. and ed. by Stillman Drake, 27–38 (New York: Doubleday Anchor, 1957).
5. *Discoveries and Opinions of Galileo*, p. 59.

Chapter Eleven
Galileo, the Anti-Anti-Copernican

ing private students, he wanted more. He wanted to be completely free of all teaching duties and be able to devote himself full time to research. He also wanted to return home to Tuscany. Dedicating *The Starry Messenger* to the Duke of Tuscany and naming the satellites of Jupiter after him was all part of this plan. As soon as *The Starry Messenger* proved to be as great a success as Galileo had hoped, he pressed his advantage vigorously with the Duke and in the end obtained what he wanted, an appointment to the court at Tuscany as the Imperial Mathematician—a parallel position to Kepler's in Prague.

He was simultaneously appointed Professor of mathematics once again at the University of Pisa, but this time without any associated duties.

Unfortunately, by moving out of the Republic of Venice, Galileo also gave up the academic freedom and protection from outside meddling that independent Venice had afforded him. Galileo was vulnerable to mischief against him from two directions: the philosophers, whom he regularly treated with contempt and sarcasm for their muddleheadedness and obstruction of scientific investigation, and from religious leaders, whom he saw as making a terrible error by letting the Bible and other religious documents be interpreted by the dogmatic views about nature of a heathen philosopher who lived 2000 years ago. The former he thought had no business sticking their heads in science for which they had no training nor aptitude; the latter he thought were leading the church down a dangerous path by mixing religion and science. Both of these groups had seized on the views of Copernicus as either nonsense or heresy. They were the anti-Copernicans. Galileo set his sights on them more than on the astronomical issues themselves. He was of course pro-Copernican, but Copernicus' views were just the most conspicuous scientific controversy of the time.

Galileo Reinterprets the Bible

Martin Luther's Protestant Reformation shook the foundations of the Catholic Church in the early sixteenth century and undermined its authority. Many defections to the Lutheran cause decimated the ranks of Catholics in northern Europe, and in southern Europe unquestioned adherence to church doctrine was failing. In the middle of the sixteenth century, the Catholic Church fought back with the Council of Trent and the Counter-Reformation. It was a time when it was necessary to take sides either for or against church authority. This was the world in which Galileo lived. He was himself a devout Catholic with close relatives in high positions in the church; both of his daughters were nuns; and he had many friends who were bishops and cardinals.

Galileo was alarmed by the role that Aristotelian philosophy had come to play in church dogma. Where religious documents, such as the Bible itself, were vague or silent on questions about the natural world, church theologians had filled the gap by inserting an Aristotelian interpretation of events. By Galileo's time it was difficult to separate Christianity from Aristotelianism, and this troubled Galileo greatly since he believed that Aristotle was basically wrong about the natural world. Galileo wanted to urge good Catholics to separate their religion from matters of science. But the Counter-Reformation was not a good time to be arguing that any church dogma was wrong. It seemed nearly inevitable that someone as outspoken and feisty as Galileo would run afoul of the church's desperate attempts to hold on to its authority.

When Galileo moved to his new position as Imperial Mathematician to the Grand Duke of Tuscany, he was able to give his attention to his basic scientific research, but from time to time he lashed out against what he viewed as incompetent and unwarranted scientific pronouncements by armchair philosophers. Such outbursts merely added to the number of people in positions of influence who had a grudge against him.

If the philosophers could not best Galileo head on in an argument, they could put him in an untenable position by drawing him into theological issues. One of the most dangerous for Galileo was the interpretation of Scripture. Nevertheless, he felt it his duty as a good Catholic to express himself on the use of Biblical quotations as evidence in deciding matters of science. He wrote several times to friends and colleagues on the subject and when he learned that his letters had been recopied and circulated about by some of those who bore him ill will, Galileo decided that he had better think matters through completely and put them in one coherent statement. Because one of the people who had been drawn into discussions on these topics was the Grand Duchess of Tuscany, the mother of the Grand Duke, Galileo wrote his thoughts on the subject in the form of a letter to her. The *Letter to the Grand Duchess Christina* of 1615 was the result.

In brief, the thrust of Galileo's argument in the *Letter* was that the Bible was a book to be read and understood by ordinary people, not by scientists, and therefore it spoke in ordinary common-sense language that could not be taken as literal and technical scientific truth. For example, the Old Testament story about Joshua commanding the Sun to stand still (to make the day longer) did not imply that the Sun is what moved and the Earth was motionless, but that since this is the way it is ordinarily perceived, this is the way it would be expressed in the Bible.

> Since it is very obvious that it was necessary to attribute motion to the sun and rest to the earth, in order not to confound the shallow understanding of the common people and make them obstinate and perverse about believing in the principal articles of the faith, it is no wonder that this was very wisely done in the divine Scriptures.

Besides, he argued, if the Bible was being literally true and if the Aristotelian view was correct, it is not so much the Sun that moves as it is the celestial sphere that is constantly turning, dragging the Sun along. Therefore, the Bible should have said that Joshua commanded the celestial sphere to stop turning. And more generally, he argued that it is wrong to take a quotation out of the Bible, out of context, and think that it is a literal statement about the natural world.

The Dialogue

With this Galileo had taken a stand on the separation of science and religion. It was only a matter of time before he would be forced to back off or face the wrath of the church establishment. In a pre-emptive move, Galileo traveled to Rome to meet with Vatican officials to urge them not to make the mistake of tying church doctrine to Aristotle and to show them that the Copernican view was correct. His efforts were in vain, and the church, to the contrary, decided to ban Copernicus' work altogether. Moreover, Galileo was specifically instructed that he must not hold or defend the view that the Earth moves and is not in the center of the world. To Galileo, that meant that he was not to state that the Copernican system was correct.

Galileo believed that once all the facts were known and both the Copernican and Ptolemaic systems were explained and compared, that people would conclude on their own that the Copernican view had to be correct. But how to make the facts known when he was forbidden to hold or defend the heliocentric system. Then he had an idea. He would adopt the analytical style of Plato. He would *discuss* each viewpoint hypothetically in the form of a dialogue where each viewpoint was presented fairly. And to give it some structure, he would present it as a conversation among three men discussing the causes of the tides. Remember that Galileo had had the idea that the Copernican system explained the tides as being caused by the seas sloshing back and forth from the changing motion of the Earth, considering the daily and annual motions around the Sun.

In 1623, the sitting pope died and a friend of Galileo's, Cardinal Maffeo Barberini, who had encouraged Galileo in his attacks on philosophers, was elected the new pope, taking the name Urban VIII. The next year, Galileo called on the new pope in Rome and proposed the dialogue. The pope seemed supportive. Galileo returned to Florence and began writing in earnest. The manuscript was finished in 1629, and after obtaining the requisite imprimaturs of the censors, it finally appeared in 1632.

The dialogue would have three characters. One would present the Copernican viewpoint; one would present the Aristotelian/Ptolemaic view; and one would be a reasonable but neutral party who would put questions to the other two and help work through the implications of their answers.

Following the practice of Plato, Galileo chose real people for his interlocutors. The Copernican view would be defended by Salviati of Florence, a recently deceased friend whom Galileo described as a "sublime intellect." The role of the neutral interlocutor would be filled by Sagredo of Venice, also a recently deceased friend whom Galileo called "a man of noble extraction and trenchant wit." For the not-so-noble defender of Aristotle, Galileo chose the name of a famous ancient Aristotelian philosopher Simplicio, long since dead. The dialogue was set in the palace of Sagredo and took place over four days.

Since the purported purpose of their discussions was to consider explanations for the tides, Galileo had named the work, "Dialogue on the Tides," but his publisher objected, probably thinking that the title would mislead the public. As a compromise, the book originally came out with the simple title *Dialogue*. Later editions added a few explanatory words. The book is usually referred to now as *Dialogue on the Two Chief World Systems—Ptolemaic and Copernican*.

In order to reach the widest audience possible, the *Dialogue* was written in colloquial Italian rather than in Latin. Galileo wished to put his case to any literate reasonable

Chapter Eleven
Galileo, the Anti-Anti-Copernican

The frontispiece from the 1642 edition of Galileo's *Dialogue on the Two Chief World Systems*. On the left are the representatives of the old system, Aristotle and Ptolemy, and on the right is Copernicus, the herald of the new.

person, not just the scholars who had probably made up their minds anyway. It was an immediate success. It was bought, read, and discussed. We will never know whether, if left alone, it would have done the job for which it was intended, namely convinced the public of the correctness of the Copernican view.

The Inquisition

Within five months of publication, Galileo's enemies had caught up with him. They made representations to the Vatican that Galileo had deliberately violated the injunction not to hold or defend the heliocentric system by writing the *Dialogue*; that despite the supposed neutrality of the presentation, the work was indeed an argument for the truth of the Copernican view; and that, by doing so, he was meddling in "high matters," i.e., theological doctrine. All copies of the book were to be confiscated and destroyed. Galileo was summoned to appear before the Inquisition on a charge of heresy.

What happened to the support that Galileo expected from the new pope? Galileo had received encouragement from Pope Urban that he should write the work, and moreover he had obtained the approval of four censors whose names appeared in the book. How was it that despite these precautions, his enemies could prevail upon the Vatican to try him for heresy?

Historians of science have pored over this question again and again, finding much evidence of intrigue and backstabbing. However, a fairly compelling case can be made that the deciding factor was that Galileo lost the pope's support by what he wrote on the very last page of the *Dialogue*. At the end of the book when the summing-up is being done by each speaker, it is obvious that Salviati, presenting the Copernican view, has won every argument, convincing Sagredo by the force of his reasoning, and that Simplicio, the Aristotelian, has had every one of his arguments torn to shreds. Having reached this point, however, Simplicio has one last ace to play to win the day for the status quo: he intones that no matter what has been said, God is all powerful and could have done it some other way. This limp argument Simplicio attributes to "a most eminent and learned person, and before which one must fall silent." In fact that very argument had been made to Galileo by the pope himself. Now it was coming from the mouth of the fool Simplicio who just can't give up. The pope was furious and withdrew his protection of Galileo.

Despite being nearly seventy years old and in demonstrably poor health, he was forced to make the journey to Rome and stand trial. The Inquisition treated Galileo with respect and dignity, but it would not yield in its insistence that he must recant his views and admit his error. In the end, Galileo was convicted of "vehement suspicion of heresy." It's an odd charge; not enough to burn him at the stake as had been done to Bruno in

Chapter Eleven
Galileo, the Anti-Anti-Copernican

1600 for espousing Copernican views, but enough to warrant life imprisonment. That was Galileo's sentence, but in deference to his age and position, it was changed to house arrest. Galileo was permitted to return to his villa near Florence for the remainder of his life, but his movements were restricted, as were his visitors.

The Dialogue was banned in Italy and wherever else the arm of the Inquisition could reach, but it was readily available in Protestant countries. Being banned only made it in greater demand and eagerly read when it could be obtained.

Unlike Copernicus and unlike Kepler, Galileo's writing about astronomy was not highly technical and intended only for the mathematically astute. Neither, for that matter, *was* it mathematically astute. Galileo simplified Copernicus down to his essentials so that ordinary reasonable people could read and understand what the new astronomy was all about. He omitted many of the details that Copernicus included to refine his orbits, and he never even considered Kepler's elliptical orbits, though of course he knew of Kepler's work.

The result is that this argumentative, quick-witted man with a talent for writing and a feel for the dramatic is the person who got the views of Copernicus across to the general public. This was a great service to science. But actually his main role in the development of science concerns the work which has not been mentioned here except in passing. It is the subject of the next chapter.

For More Information

Alioto, Anthony M. *A History of Western Science,* 2nd ed. Englewood Cliffs, NJ: Prentice-Hall, 1987. Chapters 15.
Cohen, I. Bernard. *The Birth of a New Physics,* Rev. ed. New York: Norton, 1985. Chapter 4.
_____. *Revolution in Science.* Cambridge: Harvard University Press, 1985. Chapter 8.
Drake, Stillman. *Galileo at Work: His scientific biography.* New York: Dover, 1995.
Galileo. *Dialogue on the Two Chief World Systems.* Trans. Stillman Drake. Berkeley: University of California Press, 1988.
_____. *Discoveries and Opinions of Galileo.* Ed. & trans. by Stillman Drake. Garden City, NY: Doubleday Anchor, 1957.

Part II
Science Emerges

Hall, A. Rupert. *The Revolution in Science: 1500-1750.* London: Longman, 1983. Chapter 5.

Kearney, Hugh. *Science and Change: 1500-1700.* New York: McGraw-Hill, 1971. Chapter 5.

Kuhn, Thomas S. *The Copernican Revolution: Planetary Astronomy in the Development of Western Thought.* Cambridge: Harvard University Press, 1957.

MacLachlan, James. *Children of Prometheus: A History of Science and Technology,* 2nd ed. Toronto: Wall & Emerson, Inc., 2002. Chapter 9.

Ronan, Colin A. *Science: Its History and Development Among the World's Cultures.* New York: Facts on File, 1985. Chapter 8.

Spielberg, Nathan, and Bryon D. Anderson. *Seven Ideas that Shook the Universe,* 2nd ed. New York: Wiley, 1995. Chapter 3.

Wall, Byron E., ed. *The Nature of Science: Classical and Contemporary Readings.* Toronto: Wall & Emerson, Inc., 1990. Pages 45-54, 61-68.

Westfall, Richard S. *The Construction of Modern Science: Mechanisms and Mechanics.* Cambridge: Cambridge University Press, 1977.

Chapter Eleven
Galileo, the Anti-Anti-Copernican

Mural from a house in Rome showing telescopes in use in Venice.

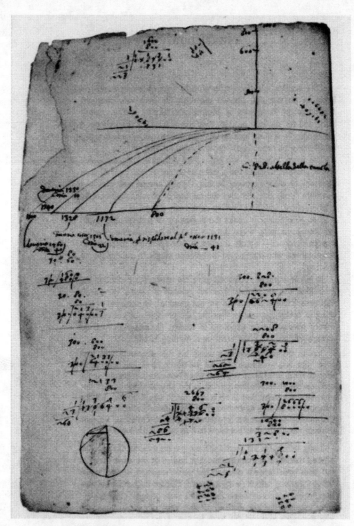

A page from Galileo's notebook showing his calculations of projectile motion.

Chapter Twelve

Galileo, Founder of Physics

Aristotle explained that motions on Earth were either natural or forced. A forced motion was caused by one thing pushing or pulling on another, i.e., direct contact. A natural motion was caused by an object striving to return to its natural place. This was just the sort of reasoning that Galileo was dead set against. Why does a heavy object fall and a light object (e.g., fire) rise? Because they seek their natural places. What are the natural places of heavy and light objects? On the ground for heavy objects, because that is their nature, and up into the sky for light objects, because that's their nature. What do we learn from this other than how the words heavy and light are used?

If Aristotle had left the matter there, it would only be a convoluted way of defining heavy and light. But Aristotle had in mind that this would be the foundation of an analysis of motion in the sublunar world. Going on from this, Aristotle also expounded on the rates at which various objects fell. Heavier objects surely fell more quickly because there was a greater extent to which they were out of their natural place. Where Aristotle could examine things with his own eyes, then reason from what he saw, he did so, but when that was impossible, he would use logic to deduce answers. Objects fall too fast to tell what is really happening, so he reasoned what *must* be happening.

The worst example of this is Aristotle's analysis of projectile motion being caused by antiperistasis—a heavy object thrown into the air continues to move after it has left contact with what threw it because it separates the air in front of it, which then goes around to the back of the object, closing the void that would have been produced otherwise and giving the object a bit of a slap forward. It then continues to move in this fashion until the object's natural motion takes over and brings it to the ground. This was so unsatisfactory that Aristotelian scholars for centuries after tried to find a better explanation. Here again, part of the problem is that objects flying through space move too quickly to be measured.

With the invention of the cannon in the late Middle Ages, it became a matter of considerable practical interest to know exactly how a cannon ball flies once it leaves the barrel of the cannon. Regardless of the theoretical formulation of natural versus forced motions, it was still important to know how to aim the cannon, that is, what should the angle of elevation of the barrel be to hit the target. These were among the problems of greatest interest to Renaissance "engineers."

Niccolo Tartaglia

An illustration from Tartaglia's *The New Science* showing how to aim a cannon.

Niccolo Tartaglia was an early 16th century Italian mathematician with an interest in practical problems. He was especially interested in understanding the mathematics of surveying, gunnery, and military engineering in general. In 1543, the year of the publication of Copernicus' *On the Revolutions*, Tartaglia published the first Latin edition of Archimedes' work and also an Italian edition of Euclid's *Elements*. Tartaglia's importance is that he made a systematic effort to understand the trajectory of projectiles, such as cannon balls, by firing cannon again and again and measuring where they hit—rather than try to solve the problem by reasoning from first principles. Thus, he started afresh from the phenomena rather than from preconceptions about motion. He published the results of his work in a short book entitled *The New Science*. The book did not succeed in discovering the relevant factors in cannon trajectories, but it did correctly report that maximum distance is reached if the cannon is aimed upwards at a 45° angle. One of Tartaglia's students was later one of Galileo's mathematics teachers.

Tartaglia's work heralded a shift that was on its way in scientific thinking. He wanted to know *how* things happened. In ancient times, the chief question about nature was *why*. Aristotle's four causes to explain anything emphasized the fourth cause, the "final" cause, or purpose. For Aristotle, that was the ultimate question: Why was some-

Chapter Twelve
Galileo, Founder of Physics

thing the way that it was? Tartaglia's military clients had little use for an explanation of why a cannon worked, but they certainly wanted to know in some detail how it worked. The more that questions about nature had practical implications, the more that people wished to know in detail how something occurred and cared little for why it happened. However, a careful recording and analysis of observations of natural phenomena may not be good enough to reveal the answers, as Tartaglia's efforts showed.

The Law of Free Fall

When, two generations later, Galileo dropped metal balls of different weight from the Leaning Tower of Pisa, it was to demonstrate the inadequacy of the Aristotelian explanation. It was not going to establish how objects actually did fall. That was going to require careful exact measurements of distance traveled and time elapsed, looking for a pattern. But a falling object has completed its fall in seconds. How can this be studied?

In working out the answer to this, Galileo started science down a path to knowledge that it has remained on ever since. Galileo began with the working assumption that nature operated in regular patterns that had an underlying simplicity. Half of the job of the scientist is to clear away the complications so that that simple structure can be detected.

The Cathedral and famous Leaning Tower of Pisa where Galileo performed the experiment described in Chapter 11 to test the theory regarding free-falling bodies.

One of the main problems to which Galileo applied himself for a major part of his life, when he was not distracted by holding and defending Copernicanism, was trying to understand falling objects. As Aristotle noted two thousand years before Galileo, heavy objects fall. Galileo did not find the rest of Aristotle's analysis of the matter convincing. In fact, at several times in his life, starting with the Leaning Tower demonstration, Galileo demonstrated that the Aristotelian account had to be wrong. But where to start to understand the phenomenon?

Galileo reasoned that whatever it was that made a heavy object fall, it was the same thing that made a ball roll downhill. But rolling downhill occurs at a considerably slower pace than falling through air. If you took a large round object, a ball or a wheeled cart, say, and set it loose at the top of a hill, you could follow it downward and study its descent and perhaps learn something. On the other hand, a hill introduces other complications. The ground may not be level. Even if it is, the surface may retard the rolling. Nature does not present itself in simple uncomplicated pieces, but all at once.

The Inclined Plane

Galileo's solution to these problems was to make an *artificial* environment. If it's falling that you want to study, try to eliminate everything else so you can just focus on that. Excellent craftsman that he was, Galileo made a device in his workshop to study falling in as ideal a setting as he could manage. He created the first laboratory equipment designed to study a particular problem under controlled conditions. It was the prototype scientific experiment.

He manufactured a number of small metal balls, making them as perfectly spherical and as smooth as he could. He then made a long shaft out of a very straight (i.e., plane) piece of hard wood. On that shaft he carved out a groove running the length of the shaft—as straight and even as he could. And he smoothed and polished the entire apparatus so as to minimize any friction. This shaft bore markings of distance along its length. It was attached to an apparatus that could raise or lower one end to any desired height so as to vary the steepness of the slope of the shaft. He then rolled the balls down the inclined plane to see what would happen.

Measuring Time

Well, of course, the balls rolled down more slowly when the shaft was inclined slightly than they did when it was inclined at a steeper angle. And he could also see that the rolling started off slow and picked up speed. This much Aristotle could have seen just as well. But what was the relation of angle to speed and was there a fixed rule by which

Chapter Twelve
Galileo, Founder of Physics

the balls increased their speed? These were new questions and to answer them something new had to be introduced into scientific analyses. That was time. Not long stretches of time, but small increments, on the scale of seconds or even less. It was an innovation in scientific thinking to study how things changed over tiny increments of time. Other people may have thought of doing this, but Galileo developed a way of making such measurements.

He actually used a number of devices. The most famous of these was the pendulum. Related to it is one of those legends that get attached to famous people to show that they were headed for great things at a tender age. Galileo was a choirboy in his local church in Pisa when he was a child. In the church, lamps hung down on long chains from the ceiling that were lit each night. The lamps hung at a height that was inconvenient for lighting, but they could be pulled to the side with a crook where they could be reached from an alcove. As the story goes, young Galileo noticed that when the lit lamp was released, it swung back and forth in a pendulum motion for a time before it came to rest, and as he watched it, he noticed that it took the same time each time to complete a cycle, back and forth, even though the angle that it swung out was getting smaller each time. This is the principle called *isochrony of the pendulum*, which is generally attributed to Galileo because he wrote about it in later life. It is also the principle on which pendulum clocks (e.g., "grandfather clocks") are constructed and how they keep regular time even though the weights that drive the pendulum are constantly falling. Whether Galileo noticed this as a child or later, it's fairly certain that he did note this feature. In fact, he made a thorough study of pendulums and could have used a swinging pendulum to measure short intervals of time.

Another device that Galileo used is a water clock. These were known in ancient times, so this is not a Galilean invention, but he may have been original in using them for small intervals of time. A large vessel, filled with water, that has a very small opening at the bottom, can be allowed to flow water at a regular rate. Weighing the water accumulated over a short period of time on a sensitive balance would be a way to get a fairly precise figure for time elapsed. Galileo wrote a detailed description of this method.

Still another method which has a certain appeal for its simplicity is just to count beats. Galileo's father was a famous musician. Galileo himself received a thorough musical training and could play several instruments competently. All musicians develop a fairly good sense of time in order to play at a constant rate. Galileo may have been able to measure short intervals fairly accurately simply by counting them in his mind. It would be much easier to do this while attending to a delicate experiment of short duration than to have to start and stop a water clock or keep one eye on a swinging pendulum.

The Odd-Number Rule

With a means of measuring time in small increments, a way of slowing down the fall of an object enough that its progress can be measured all along the way, and an environment that had all extraneous factors removed, Galileo could look for regularities in the way objects fall. He made a tremendous number of measurements, varying the parameters bit by bit—different balls, different steepness, different total distance, etc.—and then found a pattern that surprised him in its simplicity.

Time versus Distance

1st	1
2nd	3
3rd	5
4th	7
5th	9
6th	11
7th	13
8th	15

What Galileo found was that the balls rolling down the inclined plane always behaved in the following way: Start the ball rolling from a stop and let it roll for some arbitrary time, call that t. Mark the distance covered and call that d. Now start again and let it roll for twice the amount of time, that is, for $2t$. The distance it goes past the first marker will turn out to be $3d$. If you let it roll from stop for time $3t$, the distance past the second marker will be $5d$. In time $4t$ it covers $7d$ past the third marker. You will end up with having a series of marks on the inclined plane that represent the distance covered by the ball in each interval of time t. There is a pattern here. It is easier to see in a chart.

The left column, representing intervals of time, is just the ordinal numbers, while the right column, representing distance covered in that particular unit of time, is the list of odd numbers. This is called the *Odd-Number Rule*. Or, as the chart shows, as the ball rolls, it picks up speed in a constant fashion—it is constantly accelerating by two units of distance for each unit of time.

Chapter Twelve
Galileo, Founder of Physics

Here we have a general principle that emerged not from reasoning about the nature of matter and of motion, but as an induction from the observed facts. Aristotle surely would have approved of the method even if he reached different conclusions himself. What Aristotle himself believed was the rule governing falling bodies is not entirely clear. His comments on this are spread around in scattered statements in his book *On the Heavens* (*De Caelo*). What is clearer is what his followers took him to be saying, namely, that the speed of a falling body is proportional to its weight. That still does not explain the cause of acceleration unless a body increases in weight as it falls, which is one way to read Aristotle's meaning.[1]

The Times-Squared Law

The relationship shown by the Odd-Number Rule can be expressed more compactly in a surprisingly simple formula if instead of considering the distance traveled in a particular unit of time, the total distance traveled over a quantity of time is shown. Make another chart, similar to the previous one, but this time make the left column the total amount of time expressed in whatever unit is taken for the first increment, and the right column is the total distance traveled by the end of that time. To distinguish the increments of distance from the total distance, call the total distance *s*.

Time versus Total Distance

t	s
1	1
2	4
3	9
4	16
5	25
6	36
7	49
8	64

The pattern now is that the right column is the square of the left column. This can be expressed as a formula: the distance traversed, *s*, is equal to the distance traveled in the first unit of time multiplied by the square of the number of units of time. If *d* is the distance traveled in the first unit of time then the formula is

$$s = dt^2$$

---

1. See Stillman Drake, *A History of Free Fall: Aristotle to Galileo*. Toronto: Wall & Emerson, Inc., 1989, p. 7. Also reprinted in Galileo, *Two New Sciences* (Toronto: Wall & Emerson, Inc., 2000).

This formula applies to any object that is freely going down, rolling down a straight slope, or falling through empty space. The familiar version of this law for an object in free fall near the surface of the earth is $s = 4.9t^2$, where d and s are measured in meters and t in time, or $s = 16t^2$, where distance is measured in feet.

How, not Why

It is interesting to sit back here and look at the different kind of results we have in Galileo from, say, his Aristotelian contemporaries. Instead of reasoning from a general systematic philosophy to explain the causes of an event, he seeks to isolate what he wants to study from the rest of nature, in order to get as close as he can to the phenomena to be studied. Then without any assumptions (or with as few as possible) about the nature of what he is examining, he begins collecting measurements, looking for a pattern. When he finds a pattern, he tests further to make sure that it holds more generally, and then he expresses that pattern in a mathematical formulation. And then he is finished. For Aristotle, his result would at best be considered a statement of the efficient cause, but it would fall short of an explanation if it did not address the final cause, the purpose. Why does anything happen in this world? Galileo does not even attempt to answer that question, but he does try to find out how it happens in a very precise way.

This is a new and different goal for science. It has more limited objects than the open-ended quests of philosophy, but it gives far greater attention to getting the details right. It marks a major separation between science and philosophy.

Galileo's work exemplifies what we saw as the route to sure knowledge. He was explicit about this in some of his writings,[2] and it is implicit in the others. Science is figuring out the patterns and regularities of nature. It leaves metaphysics for others to deal with. Moreover, note the role of mathematics for Galileo. Mathematics is the compact language in which one expresses those patterns and relationships that are to be found in nature. It is *the* language of nature, but nature is something else. As a working hypothesis, nature is assumed to have rational and simple structures, but whatever is found out is what is.

---

2. Especially in the 1623 work, *Il Saggiatore*, which contains a detailed statement of the principles of experimental science.

Chapter Twelve
Galileo, Founder of Physics

Compare this view to his colleague Kepler, for example. Kepler's goal was to discover the hidden mathematical secrets as though nature itself was just a way of cloaking those secrets. The mathematics was all. For Kepler, to find some elegant mathematical relation was the ultimate purpose. For Galileo, finding out how nature worked was the purpose, and that could best be expressed in mathematical formulae.

Parabolic Projectiles

Once again, consider the troubling question of projectiles. Aristotle's analysis involved a combination of natural and forced motions, with the forced motions being both the original push that starts an object flying and the deduced continuation of that caused by the air whirling around the back of the object and continuing to push it forward, *antiperistasis*. The Aristotelian approach was never going to get anywhere with this because in its effort to make sense, i.e., fit in with the rest of the Aristotelian worldview, it continued to look in the wrong places. Projectiles eventually fall and hit the ground. Galileo had sorted out the falling part with his law of free fall. He knew how it happened in a fairly precise mathematical way. He could use that law as a starting place in his search to understand projectile motion.

To study projectile motion, Galileo once again set up the inclined plane and ran balls down it. But this time, the purpose of the plane was just to get balls rolling at a predetermined speed when they came to the bottom of the inclined plane. He could then run the ball off a table or other elevated precipice and let it fall to the ground, measuring where it hit. Judging from the working papers of Galileo that remain for scholars to pore over, he did such experiments again and again until he had found the pattern.

The pattern that Galileo found was that objects flying through the air fall at exactly the same rate whether they are moving horizontally or not. Take a ball at the end of Galileo's table and let it just slip over the edge. It will fall to the ground in exactly the same amount of time as another ball shot horizontally off the table with a considerable push. The vertical is not affected by the horizontal. The horizontal meanwhile has another surprising feature. It just does not change, except for the minor slowing caused by air resistance.

The classic example of this, frequently quoted in physics textbooks, is to consider a rifle shot perfectly horizontally off into the distance on a very flat plain. At the exact moment that the bullet leaves the barrel of the rifle, another bullet is dropped from a height equal to that of the rifle. Both bullets will hit the ground at the same time. Another feature of this law that Galileo discovered is that it accounts for Tartaglia's rule that a cannon shot at a 45° angle travels farther than the same cannon shot at any other angle.

This horizontal motion is what we would call the principle of inertia, the tendency of a body to remain at rest or in motion in a straight line until some force makes it move otherwise. This notion is never clearly expressed in Galileo. We attribute it to Descartes (see the next chapter) or to Newton. But whether made explicit or not in Galileo, it represents a major departure from Aristotle and shows what one of the problems was that Aristotle had in interpreting projectile motion. For Aristotle, motion is something that had to be caused. Natural motion (up and down) was caused by an object being out of its natural place and striving to return to it. Other motions were forced. When the force stopped, the motion stopped.

Galileo, through his experiments, saw otherwise. Motion continues, and does so at the same pace, until interrupted. His separation of a projectile's motion into the vertical and the horizontal made this clear. Galileo does not seem to have been terribly bothered by the extraordinary difference between his concept of a motion being natural and Aristotle's concept, though he does have his characters discuss it a bit in his dialogues.

Two New Sciences

Accounts of Galileo's place in the history of science often treat him as almost two different people: one, the supporter of Copernicus, who made the convincing arguments for the heliocentric system and suffered the wrath of the Catholic Church because of it; and two, the founder of physics or of experimental science, the person who first laid out explicit rules for how one sets up a science experiment and what sort of conclusions one can draw from it. That division has been followed here as well. But by proceeding in this fashion, it is sometimes difficult to keep straight what was happening when in Galileo's life.

Chapter Twelve
Galileo, Founder of Physics

From the time that Galileo started working in mathematics at Pisa when a young man, he studied physical phenomena and sought to describe them precisely and mathematically wherever possible. This work extended throughout his life and resulted in several minor publications and reams of working papers which are still available. His defense of Copernicus was a much more sporadic affair. The invention of the telescope set him off on reporting new observations, which supported the Copernican view. Once he became thoroughly convinced of the correctness of the sun-centered system, he sought to write and speak about it, both for the sake of leading the world (and in particular, his Church) to the truth, as well as for the sake of putting Aristotelian philosophers in their place. The culmination of those efforts was the *Dialogue on the Two Chief World Systems*, which explained the Copernican system so well to the general public and which got him in such trouble. He was seventy years old when he was silenced on the Copernican issue.

But that only gave him the time and the motivation to write up the results of the rest of his life work in another great work in the history of science. This he did while under house arrest and during a time in which he finally went completely blind. The work that he wrote, dictating it to his son in the end, was *Two New Sciences*, or, more fully, *Discourses and Mathematical Demonstrations Concerning Two New Sciences Pertaining to Mechanics and Local Motions*. He was not likely to get permission to publish this in Italy, given his conviction for suspected heresy, so the book was sent to a Dutch publisher, where it appeared first in 1638.

Two New Sciences is, as the title suggests, about two separate subjects. One of them is what this chapter has been reviewing, that is, Galileo's work on motion. The other topic is what would now be called strength of materials in engineering terms. Galileo's work on strength of materials was considered significant. It laid down some of the groundwork for engineering studies that would be important for structural designs of all sorts. Among the topics was the relation of thickness of beams to resistance to fracture when under stress.

For pure science, it is the topic of motion that is more important. Galileo's account of his work with falling bodies and projectiles lays out how science should proceed and makes the case that the objects of science are and should be different from philosophy.

DISCORSI
E
DIMOSTRAZIONI
MATEMATICHE,
intorno à due nuoue scienze

Attenenti alla
MECANICA & i MOVIMENTI LOCALI,

del Signor
GALILEO GALILEI LINCEO,
Filosofo e Matematico primario del Serenissimo
Grand Duca di Toscana.

Con vna Appendice del centro di grauità d'alcuni Solidi.

IN LEIDA,
Appresso gli Elsevirii. M. D. C. XXXVIII.

The title page of the original 1638 edition of Galileo's *Two New Sciences*.

The work is in the same format as his *Dialogue on the Two Chief World Systems*, and stars the same three interlocutors, Salviati, Sagredo, and Simplicio. This time it is not so much a debate between two antithetical worldviews as a question of approach and methodology. Simplicio still represents the Aristotelian position, but in this case that is more a matter of minimizing the importance of mathematics than it is clinging to any particular interpretation. The work is a mixture of Italian and Latin. The conversation is

Chapter Twelve | 221
Galileo, Founder of Physics

in Italian, as it was in the *Dialogue*. But a significant part of the book is Salviati reading aloud mathematical propositions in Latin that are identified as being the work of Galileo himself. As in the *Dialogue*, dogmatic Aristotelian positions are shown to be untenable as Salviati leads Simplicio into a blind alley. But altogether the work is much more technical and difficult and is clearly intended for working scientists more than the general educated public.

In the following passage, Galileo goes after the same issue that was the point of the Leaning Tower demonstration, Aristotle's conviction that heavier bodies fall faster than lighter bodies. Here, instead of a physical demonstration, he has his interlocutors Salviati and Sagredo engage Simplicio, the Aristotelian, on a discussion of the subject, resulting in Simplicio getting into a logical muddle. The issue is whether a lighter stone tied to a heavier stone falls faster or slower than the heavier stone itself.

> SALVIATI: ... I seriously doubt that Aristotle ever tested whether it is true that two stones, one ten times as heavy as the other, both released at the same instant to fall from a height, say, of one hundred braccia,[3] differed so much in their speeds that upon the arrival of the larger stone upon the ground, the other would be found to have descended no more than ten braccia.
>
> SIMPLICIO: But it is seen from his words that he appears to have tested this, for he says "We see the heavier..." Now this "We see" suggests that he had made the experiment.
>
> SAGREDO: But I, Simplicio, who have made the test, assure you that a cannonball that weighs one hundred pounds (or two hundred, or even more) does not anticipate by even one span the arrival on the ground of a robinet ball weighing only half [as much], both coming from a height of two hundred braccia.
>
> SALVIATI: But without other experiences, by a short and conclusive demonstration, we can prove clearly that it is not true that a heavier moveable is moved more swiftly that another, less heavy, these being of the same material, and in a word, those of which Aristotle speaks. Tell me, Simplicio, whether you assume that for every heavy falling body there is a speed determined by nature such that this cannot be increased or diminished except by using force or opposing some impediment to it.

---

3. A braccio, meaning an "arm," was a measure of length of 58.4 cm, about an inch less than two feet. So, a hundred braccia was about 58 meters, or 64 yards. This is very close to the height of the Leaning Tower of Pisa.

SIMPLICIO: There can be no doubt that a given moveable in a given medium has an established speed determined by nature, which cannot be increased except by conferring on it some new impetus, nor diminished save by some impediment that retards it.

SALVIATI: Then if we had two moveables whose natural speeds were unequal, it is evident that were we to connect the slower to the faster, the latter would be partly retarded by the slower, and this would be partly speeded up by the faster. Do you not agree with me in this opinion?

SIMPLICIO: It seems to me that this would undoubtedly follow.

SALVIATI: But if this is so, and if it is also true that a large stone is moved with eight degrees of speed, for example, and a smaller one with four [degrees], then joining both together, their composite will be moved with a speed less than eight degrees. But the two stones joined together make a larger stone than that first one which was moved with eight degrees of speed; therefore this greater stone is moved less swiftly than the lesser one. But this is contrary to your assumption. So you see how, from the supposition that the heavier body is moved more swiftly than the less heavy, I conclude that the heavier moves less swiftly.

SIMPLICIO: I find myself in a tangle, because it still appears to me that the smaller stone added to the larger adds weight to it: and by adding weight, I don't see why it should not add speed to it, or at least not diminish this [speed] in it.

SALVIATI: Here you commit another error, Simplicio, because it is not true that the smaller stone adds weight to the larger.

SIMPLICIO: Well, that indeed is beyond my comprehension.

SALVIATI: It will not be beyond it a bit, when I have made you see the equivocation in which you are floundering. Note that one must distinguish heavy bodies put in motion from the same bodies in a state of rest. A large stone placed in a balance acquires weight with the placement on it of another stone, and not only that, but even the addition of a coil of hemp will make it weigh more by the six or seven ounces that the hemp weighs. But if you let the stone fall freely from a height with the hemp tied to it, do you believe that in this motion the hemp would weigh on the stone, and thus necessarily speed up its motion? Or do you believe it would retard this by partly sustaining the stone?

We feel weight on our shoulders when we try to oppose the motion that the burdening weight would make; but if we descended with the same speed with which such a heavy body would naturally fall, how would you have it press and weigh on us? Do you not see that this would be like trying to lance someone who was running ahead with as much speed as that of his pursuer, or more? Infer, then, that in free and natural fall the smaller stone does not weigh upon the larger, and hence does not increase the weight as it does at rest.

SIMPLICIO: But what if the larger [stone] were placed on the smaller?

SALVIATI: It would increase the weight if its motion were faster. But it was already concluded that if the smaller were slower, it would partly retard the speed of the larger so that their composite, though larger than before, would be moved less swiftly, which is against your assumption. From this we conclude that both great and small bodies, of the same specific gravity, are moved with like speeds.

SIMPLICIO: Truly, your reasoning goes along very smoothly; yet I find it hard to believe that a birdshot must move as swiftly as a cannonball.

SALVIATI: You should say "a grain of sand as [fast as] a millstone." But I don't want you, Simplicio, to do what many others do, and divert the argument from its principal purpose, attacking something I said that departs by a hair from the truth, and then trying to hide under this hair another's fault that is as big as a ship's hawser. Aristotle says, "A hundred-pound iron ball falling from the height of a hundred braccia hits the ground before one of just one pound has descended a single braccio." I say that they arrive at the same time. You find, on making the experiment, that the larger anticipates the smaller by two inches, that is, when the larger one strikes the ground, the other is two inches behind it. And now you want to hide, behind those two inches, the ninety-nine braccia of Aristotle, and speaking only of my tiny error, remain silent about his enormous one.

> Aristotle declares that moveables of different weight are moved (to the extent this depends on heaviness) through the same medium with speeds proportional to their weights. He gives as an example moveables in which the pure and absolute effect of weight can be discerned, leaving aside those other considerations of shapes and of certain very tiny forces, which introduce great changes from the medium, and which alter the simple effect of heaviness alone. Thus one sees gold, which is most heavy, more so than any other material, reduced to a very thin leaf that goes floating through the air, as do rocks crushed into fine dust. If you wish to maintain your general proposition, you must show that the ratio of speeds is observed in all heavy bodies, and that a rock of twenty pounds is moved ten times as fast as a two-pound rock. I say this is false, and that in falling from a height of fifty or a hundred braccia, they will strike the ground at the same moment.

SIMPLICIO: Perhaps from very great heights, of thousands of braccia, that would follow which is not seen at these lesser heights.

SALVIATI: If that is what Aristotle meant, you saddle him with a further error that would be a lie. For no such vertical heights are found on earth, so it is clear that Aristotle could not have made that trial; yet you want to persuade us that he did so because he says that the effect "is seen."

After *Two New Sciences* was published, Galileo continued to work at his villa, under house arrest as always. Much of Galileo's work in physics would be incorporated into the new comprehensive mathematical physics of Isaac Newton. Galileo lived another four years, dying in 1642, the year that Newton was born.

For More Information

Alioto, Anthony M. *A History of Western Science,* 2nd ed. Englewood Cliffs, NJ: Prentice-Hall, 1987. Chapters 15.

Cohen, I. Bernard. *The Birth of a New Physics,* Rev. ed. New York: Norton, 1985. Chapter 5.

_____. *Revolution in Science.* Cambridge: Harvard University Press, 1985. Chapter 8.

Drake, Stillman. *Galileo at Work: His scientific biography.* New York: Dover, 1995..

Galileo. *Discoveries and Opinions of Galileo.* Ed. & trans. by Stillman Drake. Garden City, NY: Doubleday Anchor, 1957.

_____. Two New Sciences. Ed. & trans. by Stillman Drake, with Drake's *A History of Free Fall.* Toronto: Wall & Emerson, 2000.

Hall, A. Rupert. *The Revolution in Science: 1500-1750.* London: Longman, 1983. Chapter 7.

Kuhn, Thomas S. *The Copernican Revolution: Planetary Astronomy in the Development of Western Thought.* Cambridge: Harvard University Press, 1957. Kearney, Hugh. *Science and Change: 1500-1700.* New York: McGraw-Hill, 1971. Chapter 5.

MacLachlan, James. *Children of Prometheus: A History of Science and Technology,* 2nd ed. Toronto: Wall & Emerson, Inc., 2002. Chapter 10.

Ronan, Colin A. *Science: Its History and Development Among the World's Cultures.* New York: Facts on File, 1985. Chapter 8.

Spielberg, Nathan, and Bryon D. Anderson. *Seven Ideas that Shook the Universe,* 2nd ed. New York: Wiley, 1995. Chapter 3.

Wall, Byron E., ed. *The Nature of Science: Classical and Contemporary Readings.* Toronto: Wall & Emerson, Inc., 1990. Pages 45-54, 61-68.

Westfall, Richard S. *The Construction of Modern Science: Mechanisms and Mechanics.* Cambridge: Cambridge University Press, 1977.

Chapter Twelve
Galileo, Founder of Physics

A portrait of Galileo by Francesco Villamena.

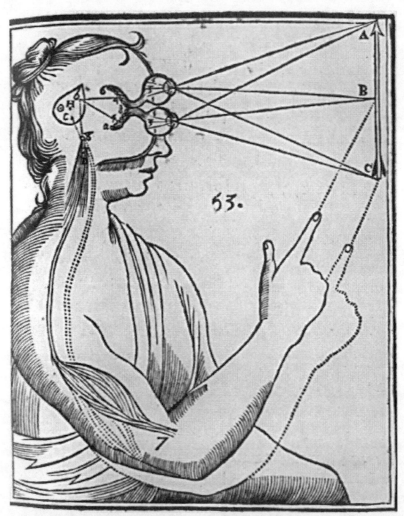

Descartes' idea of the interaction of *res cogitans* and *res extensa* through the pineal gland.

Chapter Thirteen

The Res Extensa of Descartes

A younger contemporary of both Galileo and Kepler was René Descartes, who was also one of those few convinced of the basic correctness of the Copernican system. Though Galileo and Descartes also had many other convictions about the world in common, their general philosophical outlook was about as far apart as possible. Galileo took every opportunity to malign the efforts of philosophers and show how great systems, such as Aristotle's, led to dogmatic generalizations, while close attention to physical phenomena might reveal some of the workings of nature with certainty. Descartes, on the other hand, while seeing major faults with Aristotle, sought to reform philosophy with a major overhaul going back to first principles and deducing what must follow from that.

Galileo's emphasis on discovering how physical processes behave by making experiments to measure them exactly and then expressing the process mathematically dictated a role for science that was quite separate from philosophy. Metaphysical questions would be banished from science and left to philosophers who could argue themselves in circles about them for all that Galileo cared. Descartes thought that he could go back to those metaphysical questions and straighten them out, then build a whole new system of philosophy from them, which would encompass all the issues of science.

Descartes' major work belongs in philosophy rather than in science, where most of his ideas are still examined and debated. However, Descartes played a crucial role in science as well, for two reasons. First, his conception of the physical world provided a very useful model that guided scientific thinking along very productive lines. Second, Descartes developed very useful analytical tools for the physical world in the form of an entire new branch of mathematics.

A Jesuit-trained Soldier with a Law Degree

René Descartes was born in 1596 into an aristocratic family—Kepler was then publishing his *Cosmographical Mystery* and Galileo was a professor at the University of Padua. His mother died soon after his birth and Descartes was raised by an indulgent father. The boy was in very poor health throughout his childhood and early adolescence. At the age of eight he was sent to a Jesuit college for training in the classics, and because of his frail health was permitted to stay in bed until eleven o'clock each morning, a habit he kept all his life whenever circumstances allowed it. At the Jesuit college he received a thorough training in Aristotelian philosophy and a grounding in the classics and in science. After a period of desultory living in Paris, his Jesuitical training took hold again, and he began to think more seriously about his life and about how one acquires knowledge. He went to the University at Poitiers and took a degree in law. Then, he decided that real knowledge comes not from book learning, but from the experiences of real people in situations that tested them. So, he opted for a career as a soldier.

First he joined the Dutch army; then later he switched sides and joined the Bavarian army. What they were fighting for did not seem to matter to him. During this time an interest which he had earlier in mathematics was reawakened and he began to study it in earnest. He soon decided that soldiers did not have any special route to knowledge and became bored by the military life. He retired from the army and began to work more seriously on the philosophical problems that had been in his mind since his days as a scholar in the Jesuit college: how do we know anything? Where do we begin and how do we build on that? The next thirty years or so of his life were given over to thinking about that and related problems, traveling around Europe, and writing various works. He settled finally in Holland and did most of his serious work there.

Chapter Thirteen
The Res Extensa of Descartes

The Principles of Philosophy

Descartes believed that Aristotle's systematic philosophy had somewhere gotten off on the wrong foot. It had taken the wrong things for granted, and by minimizing mathematics had lost its ability to be precise. He believed the Aristotelian system was wrong, but unlike Galileo, he did not believe that systems of knowledge were in principle wrong. Descartes set out to reform philosophy. In effect, he intended to replace Aristotle with himself.

Where does one begin? Descartes believed that you had to start with the only things that you knew for certain, and go from there. He put this out in 1644 in a work entitled *The Principles of Philosophy*. The book is laid out almost like a mathematics text with each assertion numbered, as Euclid did his propositions, followed by the justification for the statement. He begins by asserting that in order to find truth, we must begin by doubting everything until we find something that cannot be doubted. That one thing that cannot be doubted is one's own existence, since how else could one be thinking?

> I. THAT in order to seek truth, it is necessary once in the course of our life, to doubt, as far as possible, of all things.
>
>> As we were at one time children, and as we formed various judgments regarding the objects presented to our senses, when as yet we had not the entire use of our reason, numerous prejudices stand in the way of our arriving at the knowledge of truth; and of these it seems impossible for us to rid ourselves, unless we undertake, once in our lifetime, to doubt of all those things in which we may discover even the smallest suspicion of uncertainty.
>
> II. That we ought also to consider as false all that is doubtful.
>
>> Moreover, it will be useful likewise to esteem as false the things of which we shall be able to doubt, that we may with greater clearness discover what possesses most certainty and is the easiest to know.
>
> III. That we ought not meanwhile to make use of doubt in the conduct of life.
>
>> In the meantime, it is to be observed that we are to avail ourselves of this general doubt only while engaged in the contemplation of truth. For, as far as concerns the conduct of life, we are very frequently obliged to follow opinions merely probable, or even sometimes, though of two courses of action we may not

perceive more probability in the one than in the other, to choose one or other, seeing the opportunity of acting would not unfrequently pass away before we could free ourselves from our doubts.

IV. Why we may doubt of sensible things.

Accordingly, since we now only design to apply ourselves to the investigation of truth, we will doubt, first, whether of all the things that have ever fallen under our senses, or which we have ever imagined, any one really exist; in the first place, because we know by experience that the senses sometimes err, and it would be imprudent to trust too much to what has even once deceived us; secondly, because in dreams we perpetually seem to perceive or imagine innumerable objects which have no existence. And to one who has thus resolved upon a general doubt, there appear no marks by which he can with certainty distinguish sleep from the waking state.

V. Why we may also doubt of mathematical demonstrations.

We will also doubt of the other things we have before held as most certain, even of the demonstrations of mathematics, and of their principles which we have hitherto deemed self-evident; in the first place, because we have sometimes seen men fall into error in such matters, and admit as absolutely certain and self evident what to us appeared false, but chiefly because we have learnt that God who created us is all-powerful; for we do not yet know whether perhaps it was his will to create us so that we are always deceived, even in the things we think we know best: since this does not appear more impossible than our being occasionally deceived, which, however, as observation teaches us, is the case. And if we suppose that an all-powerful God is not the author of our being, and that we exist of ourselves or by some other means, still, the less powerful we suppose our author to be, the greater reason will we have for believing that we are not so perfect as that we may not be continually deceived.

VI. That we possess a free-will, by which we can withhold our assent from what is doubtful, and thus avoid error.

But meanwhile, whoever in the end may be the author of our being, and however powerful and deceitful he may be, we are nevertheless conscious of a freedom, by which we can refrain from admitting to a place in our belief aught that is not manifestly certain and undoubted, and thus guard against ever being deceived.

VII. That we cannot doubt of our existence while we doubt, and that this is the first knowledge we acquire when we philosophize in order.

Chapter Thirteen
The Res Extensa of Descartes

> While we thus reject all of which we can entertain the smallest doubt, and even imagine that it is false, we easily indeed suppose that there is neither God, nor sky, nor bodies, and that we ourselves even have neither hands nor feet, nor, finally, a body; but we cannot in the same way suppose that we are not while we doubt of the truth of these things; for there is a repugnance in conceiving that what thinks does not exist at the very time when it thinks. Accordingly, the knowledge, I think, therefore I am (in the original Latin, cogito ergo sum), the first and most certain that occurs to one who philosophizes orderly.

So, Descartes concluded that anyone who thinks must conclude that they exist, but only because they are thinking. There is therefore a world of thought that exists. Having gotten that far, Descartes asserts that the mind—the world of thought—is different from the body. He then goes on to explore at length how the world of the mind operates. That world he calls, in Latin, *res cogitans*, the world of thought. That world is completely different from the world that we perceive with our senses, the world of the body, which has as its defining characteristic that it takes up space; it is therefore the world of "extension," *res extensa*.

It's a bit reminiscent of Plato's Divided Line. For Plato, the world of thought, the upper part of the Divided Line, was where the forms were and was really the only world worth bothering about. What existed below the major division of the line was the world that could be perceived by the senses, and it was of little interest to the philosopher. Descartes does not seem to despise the world of extension, but his separation indicates that he views them as entirely different from each other.

And this has consequences for science. Because if *res cogitans* is in a different realm from the physical world, it does not impinge on it. Therefore, *res extensa* has its own structure, its own rules, and knowledge of it comes by other means. This sharp division called "dualism," that is characteristic of Descartes' thought, means that his work in philosophy and his work in science can virtually be separated from each other. Many of those issues of philosophy that troubled Galileo don't arise with Descartes, because they are in a different category altogether.

The Principles of Material Objects

After the care with which Descartes establishes the reasons that the world of thought must exist, his reasons for asserting that there indeed is a corporeal world—*res extensa*—and what its characteristics are may seem Ad Hoc and peremptory. For example, he argues the reason that there must really be a world of extension is that we perceive it and therefore there must be something other than the mind. Furthermore, it must be really out there because otherwise God would be deceiving us, and that would be repugnant to God's nature.

Nevertheless, whatever his reasons, Descartes lays out what to him are the necessary attributes of the natural world. Unlike Aristotle's world in which everything was in some sense alive and had a purpose, Descartes' *res extensa* is totally lifeless and mechanical. Everything operated blindly and without meaning. Only when *res cogitans* was able to influence *res extensa* did anything happen that was not already predetermined.

Take motion, for example. Descartes says that all motion is caused by God and there is a fixed amount of it in the world. Then, what he calls his first law of nature is that any body remains in the same state, at rest or in motion, and if in motion it will continue in motion until brought to a stop. Notice how this disposes of the Aristotelian *antiperistasis* problem. Aristotle had to provide a reason that a projectile will continue to move when it is no longer in contact with whatever moved it. Descartes says that the object just naturally stays in motion.

The second law of nature is that all movement is, of itself, along straight lines. For Aristotle, natural motion in the sublunar world was also in straight lines, though that motion was all vertical, either up or down to the natural place. Descartes names no natural places, nor is there any natural end to the motion, as there is in Aristotle. But more importantly, Descartes did not have a separate set of rules for the heavens as Aristotle did. For Aristotle, the motions of the planets were naturally in circles, since circles are never ending and the heavenly motions are eternal. For Descartes, the heavenly bodies, being part of *res extensa*, would naturally move in straight lines. Therefore, if they actually moved in circles, there must be something pushing them back to the circular position.

Chapter Thirteen
The Res Extensa of Descartes

And here we see in Descartes the same tendency that appears in Aristotle to use logical deduction to fill the gap where direct observation fails to provide an answer. Descartes says it is a law of nature that the planets move in straight lines. Therefore, if they don't, something is acting on them. But the world of extension is a purely mechanical world so there can be no such thing as forces that cause a body to do something. (Descartes' expression is that forces are occult—belonging to the spiritual rather than the corporeal world.) Aristotle's natural motion to a natural place cannot be the explanation because that implies a purposefulness that is not part of *res extensa*. How does Descartes solve this problem? By his infamous theory of the *vortices*.

A vortex is a swirling whirlpool. It can be set up by independent bodies rushing past each other in opposing directions and colliding, as for example, water running down a drain or air turning into a tornado. Descartes did not believe that empty space could exist, therefore even out among the planets there were particles jam-packed against each other. The universe he believed to be spherical. As a planet, for example, sped along in a straight line, it would be headed toward the edge of the world, but it would continually be running into these invisible particles pushing back at it, with the result that the planet stayed in orbit round the sun.

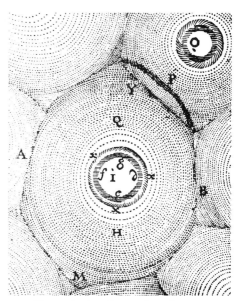

Descartes' vortex system, from *Principles of Philosophy* 1644.

The World as Machine

Aristotle's characterization of the world was as an organism. Everything in it was alive, at least in principle. Descartes' characterization was that the world was a machine. Nothing was alive. Life, in the sense of a spirit or force, belonged in *res cogitans*. Even plants and animals to Descartes were just complicated mechanisms, without souls, without feelings.

But, back to *cogito ergo sum*, Descartes did attribute souls to humans, though they were in that other world. Since humans had souls, could think and feel, and take voluntary actions, how could their bodies, which were in the world of extension, be connected to their souls? If humans could will their actions, how did they do it?

Once again, Descartes does not fail to leave us without an answer. It seems that anatomists in his time had found a tiny gland in the human brain that they could not understand. This was the *pineal gland,* so named because it is shaped like a pine cone. Moreover, this gland was not found in any other species. (It has since been found in several other animals.) Therefore, it must be the interaction between the mind and the body. Sensations travel through the nerve channels in the body into the brain where they somehow connect with the pineal gland. In the pineal gland a transfer is made between the body and the spiritual mind. The mind can then direct the body to do something in response. A favorite example is eye-hand coordination, with the visual sensation being directed to the pineal gland and after processing in the non-corporeal mind, the pineal gland would send signals to the motor nerves controlling the hand. This, like the vortex system, stems from Descartes' unwillingness to leave unanswerable questions unanswered. It had an ancient feel to it.

Nevertheless, the framework of a totally mechanical physical world was a very suggestive and useful model. It particularly showed its usefulness when combined with the mathematical tools that Descartes developed to deal with it.

Analytic Geometry

The second way that Descartes played a crucial role in the development of science was in the mathematics he developed that helped him study that mechanical world of extension that he had isolated. Descartes contributed a lot to mathematics, but the most important was creating the subject that we now call analytic geometry. Analytic geometry is really a combination of algebra and geometry. The ancient Greeks excelled at the study of figures in two- and three-dimensional space. What we now call algebra developed slowly over a long period of time going back to ancient Egypt. The name "algebra" tells us that it came into European thought in the Renaissance from the Arab countries along with many other scientific subjects. Nevertheless, algebra was not in general use in math-

Chapter Thirteen | 235
The Res Extensa of Descartes

ematics in Europe until Descartes brought the subject matter together and gave it coherence. Galileo, for example, primarily used the mathematics of Euclid in places where algebra might have simplified some problems.

Consider now that projectile motion problem that Galileo worked on. When Galileo finally figured out how the projectile moved, if he had known Descartes' system (he didn't), he could have written down the exact path of the projectile as it ran off the table and fell as a series of pairs of numbers that we would now call Cartesian coordinates. The reason that this would have been valuable is that once you can write a path down as a series of positions in x and y, you can then look for a mathematical formula—an algebraic formula—that describes that path more compactly. Descartes' greatest contribution was to show how a marriage of geometry and algebra was of tremendous use in studying any object in motion in *res extensa*. The key point that he added to the subject was the coordinate system, which we still name in his honor, Cartesian coordinates. Descartes showed that any place whatsoever in the physical world can be uniquely identified by three numbers that specify its position from a given starting point. For example, sitting at a rectangular desk, consider the starting point to be one of the corners of the desk, say, the near left-hand corner. Call the line formed by the front edge of the desk the x-axis. Call the line formed by the left edge of the desk the y-axis, and call a line that one could make starting at the corner and going vertically upward the z-axis. Think of each line as marked off in centimeters. A series of numbers (such as 13, 9, 26) would then correspond to a point that is 13 cm to the right of the starting point, along the front edge of the desk, then 9 cm directly across the desk in a line parallel to the left edge, then 26 cm straight up from there. With those three axes being defined and the way of assigning distances along each of them also defined, every point in space in the entire world of *res extensa* can be identified with a unique series of three numbers, which are called the coordinates of the point.

In the case of Galileo's projectile, he ran balls off the end of a table and found that they fell some distance away onto the ground. The motion had two directions, down and horizontally across. If the horizontal direction is called the x direction and the down direction is called the y direction, then at every point in the flight of the ball, it is at some x position and some y position and that can be expressed as two numbers. Now, it turns out that Galileo found out that the path of his projectiles was the shape of a parabola, one of the conic sections written about by the ancient mathematician Apollonius.

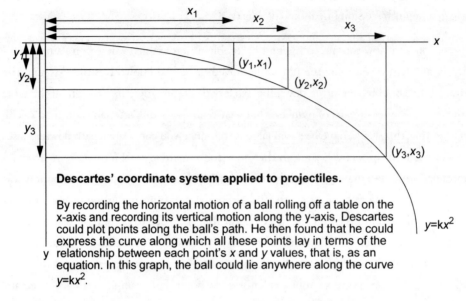

Descartes' coordinate system applied to projectiles.

By recording the horizontal motion of a ball rolling off a table on the x-axis and recording its vertical motion along the y-axis, Descartes could plot points along the ball's path. He then found that he could express the curve along which all these points lay in terms of the relationship between each point's x and y values, that is, as an equation. In this graph, the ball could lie anywhere along the curve $y=kx^2$.

Descartes' analytic geometry showed that a parabola is represented by the algebraic equation $y = kx^2$, where k can be any specified number. Every pair of numbers that can be put into this equation and make the left side equal the right side make up the points along a parabola. Galileo's law of projectile motion could be expressed quite compactly as a simple algebraic equation. Given an equation in two variables, x and y, or in three variables with an extra z, a graphical model of the equation can be drawn on paper, if it is 2-dimensional, or constructed if it is in three dimensions. Scientific studies of motion, for example, can be plotted on graphs and then a simplifying equation found that describes that graph. Or, starting with the equation, you can picture the result.

Here's another example using two dimensions. Take a piece of paper with a horizontal and vertical axis drawn. We now have "graph paper" to make this easy, thanks to Descartes. Mark the intersection of the axes as O, for "origin." Mark off units of measure on the axes in all directions. With a compass draw a circle of radius 10, for example, with O as the center. You can measure the radius on one of the axes to get it right. Now take any point whatsoever on the circle and mark it. Call its coordinates (x, y). Draw a vertical line from the point chosen to the x-axis. The length of that line will be y (or $-y$ if the point chosen was below the *x*-axis). The length of the line from O to the line just drawn where it intersects the *x*-axis is simply x (or $-x$, if the point was to the left of the *y*-axis). Connect the line from O to the point (x, y). You have drawn a right triangle where the radius, 10, is the hypotenuse, and the sides of the triangle are x and y. From the Pythagorean theorem, you know that $x^2 + y^2 = 10^2$. Every point on that circle has an x and a y that fit the above equation. And there are no x and y that do fit the equation that are not on that circle. Therefore we can say that $x^2 + y^2 = 100$ is the equation of a circle of radius 10 centered at the origin, O. And other geometric figures studied by Euclid have simple algebraic formulae as well.

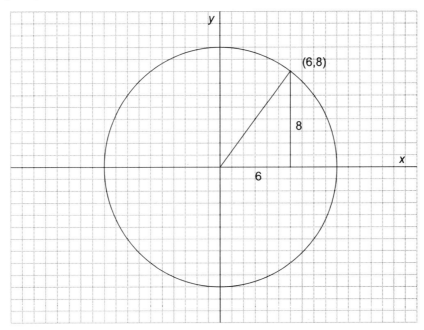

The graph of the circle $x^2 + y^2 = 100$.

Both Descartes and Galileo took as goals to study the phenomena of nature and then describe them in mathematical formulae. To be able to represent every point in space as Cartesian coordinates made the step of describing phenomena mathematically much simpler.

Descartes published much of his mathematical work in a book called simply *Geometry (La Géométrie)*. This and two other short works on optics and meteorology were to be bound together and preceded by a work called *Discourse on Method*, which would explain Descartes' method of doing science. The *Discourse* has become separated from the other works and is now the single most widely read of Descartes' works.

◆

Throughout most of Descartes' productive years he maintained contact with the rest of the scientific world through extensive correspondence with the French priest Marin Mersenne, who kept in touch with almost everyone doing scientific work in 17th century Europe and who had been his friend from childhood.

Like Kepler and Galileo, Descartes was a believer in the Copernican system. He even had written a book in which he explained his conception of the world, which involved the Copernican system, (*Le Monde.*) However, when he heard of Galileo's troubles with the Inquisition, he stopped writing it, not wishing to become another martyr to Copernicanism. The incomplete manuscript was finally published after his death.

Descartes was comfortably settled in Holland and doing some of his best work when he was enticed by the young Queen Christina of Sweden. The Queen brought him to Sweden ostensibly as a scholar in residence who would be given free reign to do his work and bring honor to her court. One of his few duties was to be the personal mathematics tutor of the queen, and here the clash of styles was disastrous. Descartes maintained that his lifelong habit of staying in bed until 11 a.m. enabled him to do some of his best thinking while lying there awake. He even claimed that it was the only way he could do good work in mathematics and preserve his health. The Queen, on the other hand, liked to get up very early in the morning and was oblivious to the harsh Swedish climate. She wished to have her daily lessons at 5 a.m. Descartes had to get up in the morning and walk to the palace to tutor the queen. He lasted at this only a few months before he caught pneumonia and died.

For More Information

Alioto, Anthony M. *A History of Western Science,* 2nd ed. Englewood Cliffs, NJ: Prentice-Hall, 1987. Chapters 14.

Cohen, I. Bernard. *Revolution in Science.* Cambridge: Harvard University Press, 1985. Chapter 9.

Descartes, René. *The Principles of Philosophy.* Translated by John Veitch.

Hall, A. Rupert. *The Revolution in Science: 1500-1750.* London: Longman, 1983. Chapter 7.

Kearney, Hugh. *Science and Change: 1500-1700.* New York: McGraw-Hill, 1971. Chapter 5.

Kuhn, Thomas S. *The Copernican Revolution: Planetary Astronomy in the Development of Western Thought.* Cambridge: Harvard University Press, 1957.

MacLachlan, James. *Children of Prometheus: A History of Science and Technology,* 2nd ed. Toronto: Wall & Emerson, Inc., 2002. Chapter 10.

Ronan, Colin A. *Science: Its History and Development Among the World's Cultures.* New York: Facts on File, 1985. Chapter 8.

Westfall, Richard S. *The Construction of Modern Science: Mechanisms and Mechanics.* Cambridge: Cambridge University Press, 1977.

Isaac Newton investigating the phenomena of light.

Chapter Fourteen

Let Newton Be!

Nature, and Nature's Laws lay hid in Night.
God said, Let Newton be! and All was Light.

Most summaries of Isaac Newton's life and work cannot resist quoting the above epitaph by Alexander Pope. It characterizes the effect of Newton on the development of science and sums up the esteem in which he was held in Britain. Take almost any book that surveys the whole history of science and open it right in the middle. Chances are, the pages that lie there are about Newton's work or something just before or after him. Newton took all that came before him (in the physical sciences anyway) and made it into a grand and comprehensive system with tremendous power. Science after Newton was, at first, exploring the as yet undiscovered implications of the Newtonian synthesis, solving new problems with methods either taken directly from Newton or developed from Newton's thought. Then, later on, it was finding out the limitations of the Newtonian system and discovering a deeper structure to the world with Newton's views as a valuable first approximation. It would be difficult to overemphasize the pivotal role of Newton's work.

Newton himself, struck by the enormity and complexity of nature, tended to minimize his own work, saying that it was possible only because he stood "on the shoulders of giants" that came before him. And late in life he compared his scientific accomplishments to that of a boy playing on the seashore who had found some pretty pebbles while "the great ocean of truth" lay undiscovered before him. This is not to say that Newton was modest about his abilities compared to his contemporaries. He disparaged almost anyone else who was trying to understand the same subjects as he, jealously guarded his findings, and accused others of stealing his ideas and claiming them for their own. His vitriolic accusations were ruinous to more than one promising career. He was a complicated person.

As is befitting someone who was made into a public icon, his early life had a few unusual characteristics that have been blown into harbingers of his greatness to come. For starters, he was born on Christmas Day in the year 1642, which almost made him a semi-god right there. Of course, it was only Christmas Day on the Julian calendar, which England had been slow to adopt. On the European continent, it was January 4, 1643, which does not sound nearly as auspicious. Next, he was a "posthumous child." The meaning of this curious term is that his father had died before he was born. He was therefore already "in heaven" as it were, looking after the affairs of his son. His home was in Lincolnshire County, where his father had been a well-off landowning farmer. Newton's mother remarried when he was three and left the care of her son to his grandmother. There are stories of dubious validity about young Isaac being a dreamy child who liked to construct mechanical toys, such as windmills.

What is known well enough is that when he reached an appropriate age, he was sent to Cambridge University to study law. At Cambridge, he discovered the glorious world of mathematics, which pushed all other subjects aside. In 1665, Newton graduated with an ordinary degree, and planned to stay on at Cambridge for further study. But in 1666 the bubonic plague that had devastated Europe in the mid 14th century returned, and, as a precaution against the epidemic spreading, the university was closed for a period of about 18 months. Newton returned home to the family estate in Lincolnshire to wait it out.

1666, the Annus Mirabilis

Many great scientists did their best work in a few spurts of feverish activity, often at a fairly young age. Newton may be the most extraordinary example of this. In 1666, Newton was just the right age, had just the right amount of learning under his belt, and had put his mind to some difficult problems that he was going to need solitude to think through. The plague was the best thing that happened to science in the 17th century.

The period of time that Newton spent back at home waiting out the plague was the most productive in his life. Almost every one of the major scientific accomplishments of his career was at least set in motion during that period. Historians have called the time Newton's "miracle year"—his *annus mirabilis*.

In that time, Newton investigated the nature of light, founded a completely new branch of mathematics—the Calculus, and, if we can believe his own account from years

Chapter Fourteen | 243
Let Newton Be!

later, had the critical insight that linked the falling of objects on Earth with the orbits of the planets. It took Newton the next twenty years to work out all the details and ramifications of what he thought of in that plague year.

Light

There is hardly a more basic and essential phenomenon of nature to understand than light itself. Light is the chief means through which we perceive the world. Light comes chiefly from the Sun, which brings us warmth and makes life possible. Questions to be addressed include what light is, how it travels—if it does, what form it takes, how we see, what colors are, why fire, for example, makes light. Light has been as subject of great interest to people trying to understand nature throughout recorded history. It was a natural topic to interest Newton.

He began with the attempt to understand color. His approach is interesting for its ingenuity in isolating the phenomenon to study and then once he had a hypothesis, manipulating his experiment in order to rule out other interpretations. To perform the experiment that he wished, he needed glass prisms and a room that he could darken except for a small opening through which he could direct sunlight, which he would take apart and study with the help of the prisms. When he was ready to do the experiment, he traveled back to Cambridge in order to use his rooms at Trinity College.

Newton set up his equipment, let just a crack of light in through the windows, placing the prism in the path of the light. Light entered the prism and was refracted (i.e., bent) as it went through the prism and produced the familiar rainbow spectrum on the opposite wall. He identified the colors as Violet, Indigo, Blue, Green, Yellow, Orange, and Red.

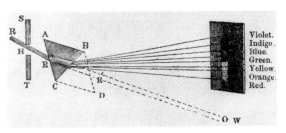

Newton diagram of the refraction of light into seven colors.

The number of colors is curious. Newton saw seven colors. Anyone who has seen the visible spectrum, whether in a rainbow or produced by a prism would see several colors gradually moving from a deep blue or purple to red. But the colors shade into each other. How many distinct colors one sees is a question of individual perception. In Newton's published papers and his later book on optics, this discovery of seven colors is presented as an objective experimental result. In a draft of a lecture published only after Newton's death, he admitted that both orange and indigo were added to the main perceived colors "in order to divide the image into parts more elegantly proportioned to one another."[1]

What are these more elegant proportions that Newton was trying to perceive? They were what corresponded to the mathematical proportions of the just intonation scale—one of the two main tuning methods for the seven note musical scale. Newton was trying to find Pythagorean musical ratios in ordinary light. This is highly reminiscent of Kepler's efforts to find celestial harmonies and is a clue to an aspect of Newton's work.

White Light is a Composite of Colors

At the time of Newton's work, the chief issue was the relationship between white and colored light. Since the spectrum could be produced by bending white light, were colors brought into existence by disturbing white light somehow? For example, if light was some sort of disturbance of an invisible medium, does the process of refraction alter that disturbance and produce colors? Or, if light consisted of streams of particles, did bending light rays set the particles spinning at different rates, each spin rate corresponding to a certain color? These were possible interpretations and were being discussed around the time of Newton's experiment.

Newton had a different explanation: he considered the different colors of light to be the simple forms and white light to be a composite produced when different colors were mixed together.

> Whiteness is the usual colour of Light: for Light is a confused aggregate of Rays imbued with all sorts of Colours.[2]

---

1. Quoted in Penelope Gouk, "The Harmonic Roots of Newtonian Science." In John Fauvel et al., eds. *Let Newton Be!*, p. 118.
2. From H. W. Turnbull, ed., *Correspondence of Isaac Newton* (Cambridge, 1959), Vol I, p. 98.

Chapter Fourteen
Let Newton Be!

Newton's characterization of white light as a "confused aggregate" suggests that it is an accidental result of a blind mechanism at work in the world. This would be in keeping with Descartes' prescription for how *res extensa* must be analyzed.

Colors are perceived. For Descartes, perception is a cognitive activity and therefore is part of *res cogitans*. Is a color a thought or a thing? For many philosophers, color was clearly a thought because it was a quality of light. Newton's view changed this around. Colored light was fundamental; white light was the perception. Therefore, in the Cartesian *res extensa*, white light did not exist—only mixtures of colored light. Newton took colors into the mechanical world. They still had to be perceived—Newton said that red light, for example, isn't actually red, but is perceived as red. Nevertheless, red light and blue light are different things.

The Crucial Experiment

One way that Newton established that colored light existed on its own, rather than being a transformation of white light, was to do an experiment that would come out one way if he was right and another way if he was wrong. This is an example of a *crucial experiment*, that is, it decides which alternative at a crossroads is the correct one and eliminates the other. This is a key tool in modern science. It is interesting to see how Newton reasoned it through and then set up the experiment.

Newton's crucial experiment was to show that once white light had been separated into its constituents by refraction, a color, red, for example, would remain red no matter how much it was refracted again. He used one prism to produce a spectrum and then let one color only of the spectrum pass through a small hole in a barrier. On the other side of the barrier, he placed another prism that then refracted the light of the color that was allowed to pass through the hole. When it could not be transformed into anything other than the light that it was, he concluded that the colors were primary. Among Newton's papers is a drawing in his own hand showing the set-up of this experiment.

When the plague had passed and normal life resumed, Newton returned to Cambridge and continued his studies. He did some further experiments on light to confirm what he had determined back in 1666 and extended his studies into more detailed analysis of refraction and the differences among colors. Newton's work in optics was quite ex-

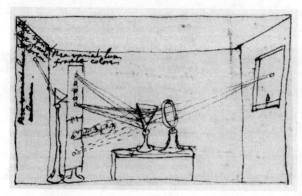

Newton's diagram of the crucial experiment.

tensive, but he kept most of it to himself, not publishing his major study *The Opticks* until 1704, when his chief critic, Robert Hooke, had died.

Newton's Letter to the Royal Society of London

Before closing himself off from his critics, Newton did make public his original work on light and the way he did so was a landmark for the history of science. He sent a letter in 1669 to the editor of *the Philosophical Transactions of the Royal Society of London*. When that letter was finally published in 1672, it represented in effect the first published paper representing the results of experimental work in science. The following is an excerpt from that letter:

Newton's Letter to the Editor of the Philosophical Transactions of the Royal Society of London

(3075) Numb.80
PHILOSOPHICAL TRANSACTIONS
February 19. 1671/72.
The CONTENTS.

A Letter of Mr. Isaac Newton,

Mathematick Professor in the University of Cambridge; containing his New Theory about Light and Colors: Where Light is declared to be not Similar or Homgeneal, but consisting of difform rays, some of which are more refrangible that others: And Colors are affirmed to be not Qualifications of Light, derived from Refractions of natural Bodies, (as tis generally believed;) but Original and Connate properties, which in divers rays are divers: Where several Observations and Experiments are alledged to prove the said Theory...

A Letter of Mr. Isaac Newton, Professor of the Mathematicks in the University of Cambridge; containing his New Theory about Light and Colors: sent by the Author to the Publisher from Cambridge, Febr. 6. 1671/72; in order to be communicated to the R. Society.

Sir,

Chapter Fourteen
Let Newton Be!

To perform my late promise to you I shall without further ceremony acquaint you that in the beginning of the Year 1666 (at which time I applyed my self to the grinding of Optick glasses of other figures than Spherical) I procured me a Triangular glass-Prisme, to try therewith the celebrated Phænomena of Colours. And in order thereto having darkened my chamber, and made a small hole in my window-shuts, to let in a convenient quantity of the Sun's light, I placed my Prisme at his entrance, that it might be thereby refracted to the opposite wall. It was at first a pleasing divertisement, to view the vivid and intense colours produced thereby; but after a while applying my self to consider them more circumspectly, I became surprised to see them in an oblong form; which, according to the received laws of Refraction, I expected should have been circular.

And I saw... that the light, tending to (one) end of the Image, did suffer a Refraction considerably greater than the light tending to the other. And so the true cause of the length of that Image was detected to be no other, than that Light consists of Rays differently refrangible which, without any respect to a difference in their incidence, were, accordingly to their degrees of refrangibility transmitted towards divers parts of the wall...

Then I placed another Prisme...so that the light...might pass through that also, and be again refracted before it arrived at the wall. This done, I took the first Prisme in my hand and turned it to and fro slowly about its Axis, so much as to make the several parts of the Image...successively pass through...that I might observe to what places on the wall the second Prisme would refract them.

When any one sort of Rays hath been well parted from those of other kinds, it hath afterwards obstinately retained its colour, notwithstanding my utmost endeavors to change it ...

I have refracted it with Prismes, and reflected with it Bodies which in Daylight were of other colours; I have intercepted it with the coloured film of Air interceding two compressed plates of glass; transmitted it through coloured Mediums, and through Mediums irradiated with other sorts of Rays, and diversely terminated it; and yet could never produce any new colour out of it.

But the most surprising, and wonderful composition was that of Whiteness. There is no one sort of Rays which alone can exhibit this. Tis ever compounded, and to its composition are requisite all the aforesaid primary Colours, mixed in a due proportion. I have often with Admiration beheld, that all the Colours of the Prisme being made to converge, and thereby to be again mixed, reproduced light, intirely and perfectly white.

Hence therefore it comes to pass, that Whiteness is the usual colour of Light; for Light is a confused aggregate of Rays indued with all sorts of

Colours, as they are promiscuously darted from the various parts of luminous bodies.

Calculus

The branch of mathematics that we call the Calculus developed, as the name suggests, as a collection of rules for calculating, that is, as a set of shortcuts that would achieve the same result as a much longer and more tedious process of breaking a problem into many parts and calculating the results of them one by one. Though the subject was expanded greatly in the 18th and 19th centuries and given a formal and rigorous structure, its origins in the 17th century were far less concerned with mathematical proof and much more concerned with problem solving. As mentioned in Chapter 10, Kepler had worked out some of the principles of the Calculus to aid him in calculating the area of irregular figures, but he was still left with a very slow and cumbersome process. Other people, most notably Descartes, had found other shortcuts and useful restatements of problems that made calculations easier.

Newton was familiar with this work and extended it much further into a system that he called *fluxions*. It is primarily a set of techniques for handling the algebra of infinite series. Compared to the methods of Calculus used today, it also was difficult to use and to understand, but it did make an enormous advance over the methods of calculation that were available at the time. Newton worked out the basics of his version of the Calculus in his *annus mirabilis* of 1666 and then worked on it further in the next couple of years. In 1669, he had completed a monograph to be entitled *De analysi* and, in 1671, another longer work to be entitled *De methodis serierum et fluxionum*, but both the Royal Society and Cambridge University Press rejected the works! In 1684, Newton's arch rival, the German philosopher Gottfried Leibniz, who had independently discovered many of the same principles, published his work *Nova methodis pro maximis et minimis*, which then became the first publication on the Calculus. This led to a lifelong feud between the two over who invented the Calculus, with Newton accusing Leibniz of hearing about and then stealing his idea.

Newton's method of fluxions had greater depth and insight than the ideas expressed by Leibniz, and Newton's techniques could solve more difficult problems. Nevertheless, Leibniz's work was easier to grasp and work with. Leibniz introduced the notation that has

Chapter Fourteen | 249
Let Newton Be!

remained with us, *dy/dx* for differentiation, the integral sign $\int$, and he introduced the concept of a function—all basic terminology in Calculus today.

The reason that the Calculus was such a valuable advance in mathematics was that it could handle many problems that defeated ancient mathematics, for example, the mathematics summed up in Euclid's *Elements*. Ancient Greek mathematics was very effective at analyzing the properties of geometric shapes that were combinations of straight lines and arcs of circles, and also of motions that did not vary from a set path in a straight line or a circle and did not change in speed. But people such as Kepler and Galileo had found that other geometric shapes were better at describing nature, and motions did indeed vary in speed constantly. For example, Kepler determined that the planets moved around the Sun in elliptical orbits and sped up when closer to the Sun. Galileo found that falling bodies increase the velocity of fall at a constant rate over time and projectiles travel in the path of an ellipse. These shapes and paths could be put into algebraic terms using Descartes' analytic geometry, but it was still not easy to calculate such quantities as the area of a figure bounded on one side by an ellipse or a parabola, nor the speed of an accelerating object at any particular moment in its motion. Calculus provided the tools to solve these problems and then to use those solutions to find the answers to problems that could not even be easily expressed in ancient mathematics.

Instantaneous Velocity

To get a better idea of the power of the Calculus, just consider just one of these questions, that of the instantaneous speed of a falling object. To begin with, what is speed?[3] We are so familiar with the notion that it may be difficult for us to realize that it's a concept that presented problems for the ancient mind. A highway speed of 100 km/h means that if you continue driving at precisely those speeds for one hour, then at the end of the hour you will have traveled 100 kilometers. But we also recognize the concept as meaning speed at a given moment, a moment, for example, when the speedometer in the car registers 100. We always express instantaneous speed in terms of average speed over a specified

---

3. I will use the terms speed and velocity virtually interchangeably. The difference is that velocity is speed in a particular direction. Ten meters per second is a speed; ten meters per second straight down is a velocity.

distance. For highway speeds, kilometers per hour, for falling objects, meters per second. (Or miles per hour and feet per second in Imperial units.)

In antiquity, even the average speed formulation would have been problematic. Our expression divides units of one kind by units of a different kind. For the ancients, this made no sense, since the only ratios that they allowed were between magnitudes of the same kind. A speed was viewed as an intensity. A faster speed was a greater intensity than a slower speed. It made for convoluted analysis, but then they rarely thought of changing speeds, so it was not as great a problem as it became in the Renaissance.

But what about instantaneous speed? How fast does something move in an instant? An instant is a frozen moment of time. It has no "width," that is, it does not consume any amount of time, and during that instant, nothing is moving. Remember Zeno's paradoxes. The arrow can never reach its target because it is always still at any instant of time and therefore does not move. The formula, distance divided by time, in an instant becomes 0/0, which has no defined meaning in mathematics.

We know that arrows do reach targets. We know there is a problem with this analysis, and we know that the concept of speed is a good and useful one. We can measure average speed over any finite distance, but we are stymied in this logical morass if we try to apply the same notion to instantaneous speed.

Calculus solves this problem and gives a way to calculate instantaneous speed. It overcomes the logical difficulty and the prohibition in mathematics of dividing by zero by the simple expedient of turning a blind eye. It is part of Newton's genius that he was able to discover mathematical principles and natural laws by holding conflicting ideas in his mind at the same time until he found some new synthesis.

We can illustrate this best by taking one of the principles of nature recently discovered in Newton's era that involved changing speeds and show how the Calculus provides an answer to the question of instantaneous speed. This will be much easier to follow if we use the formulation of the Calculus of Leibniz rather than of Newton.

Chapter Fourteen | 251
Let Newton Be!

Galileo's law of free fall determined that an object increases its velocity at a constant rate over time. Using the convenient formulation of Descartes' analytic geometry, we can express this relationship as

$$s = 4.9t^2$$

where s is the total distance fallen over time t, and s is measured in meters and t is measured in seconds, and all of this takes place near the surface of the Earth with any air resistance ignored.

So at any moment in the fall of an object, we could stop it, measure the time elapsed, t, and the distance fallen, s, and we would find that the numbers fit the equation. From that you can take the s and divide it by the t (since it does not bother us to divide one kind of unit by another) and get a figure for the average velocity over the whole journey. Since this is an equation that applies to any moment in the fall, it also applies at a point that is just a tiny bit later than the time that we have called t. Call that tiny bit of extra time Δt ("delta" t). During that extra bit of time, the object will have fallen an extra distance, which we will call Δs. The equation expressing the total distance must also be true at this point, which we can represent as the time $t + \Delta t$ and the distance $s + \Delta s$. So the equation can also be written as

$$s + \Delta s = 4.9(t + \Delta t)^2$$

The right-hand side can be multiplied out and rewritten (as Newton had shown) as $4.9t^2 + 9.8t(\Delta t) + 4.9(\Delta t)^2$, making the whole equation

$$s + \Delta s = 4.9t^2 + 9.8t(\Delta t) + 4.9(\Delta t)^2$$

Equations have the wonderful property that since the left side of any equation equals the right side, equations can be added and subtracted from other equations using the common notions of Euclid. So we can now subtract that first equation, $s = 4.9t^2$, from the equation above as follows:

$$s + \Delta s = 4.9t^2 + 9.8t(\Delta t) + 4.9(\Delta t)^2$$
$$\text{minus } s = 4.9t^2$$
$$\text{yields } \Delta s = \mathbf{9.8t(\Delta t) + 4.9(\Delta t)^2}$$

This tells us that the tiny increment of distance, Δs, is equal to an amount that can be calculated if you know the amount of time that had elapsed at time t plus the size of the small increment Δt. These are still just distances and times. The only speed we know is the average speed up to time t, which is s/t.

Now, the reasoning in Calculus (we could say Newton's reasoning, though here we are looking at how Leibniz might have expressed it) is that the speed that something is traveling at any instant is very close to the average speed that it is traveling over a very small interval of time beginning at the desired instant. In other words $\Delta s/\Delta t$ is very close to the instantaneous speed at time t. In fact it *would be* the instantaneous speed if the increment of time Δt were equal to zero. But then, as we know, the increment of distance Δs would also be zero and we would be back at trying to evaluate 0/0. But there is another way. Go back to that last equation, $\Delta s = 9.8t(\Delta t) + 4.9(\Delta t)^2$. So long as the increment of time Δt is not zero, we can divide both sides of the equation by it. If we do we get

$$\Delta s/\Delta t = 9.8t + 4.9\Delta t$$

Now, forget about the 0/0 problem and think that if Δt were allowed to go to zero, or at least get as close to it as we like, the expression $\Delta s/\Delta t$ would be as close to the instantaneous speed as we need. What would happen to the right side of the equation if Δt were zero? We would get

$$9.8t + 4.9(0) = 9.8t$$

Therefore, in the limit, as Δt approaches zero, $\Delta s/\Delta t$ approaches $9.8t$. The instantaneous speed can consequently be taken to be that limit $9.8t$. So long as we don't worry about the problem of division by zero, or what amounts to the same thing here, the contradiction of a body having a speed in a moment of time, we have an answer to the question. With that answer, Newton was able to solve many problems involving changing speeds. More generally, with the Calculus, a great array of physical relationships could be calculated that otherwise would have been either extremely difficult or impossible to solve.[4]

The Calculus is a very effective set of tools, but in its original formulation(s) it lacked the sort of rigor that had become the standard since Euclid. Its effectiveness led

---

4. The idea for using the example of Galileo's law of falling bodies to illustrate the derivation of instantaneous velocity is taken from Alioto, *A History of Western Science*.

mathematicians and physicists of the 17th century to forgive its lack of mathematical rigor. When mathematicians of the 18th and 19th centuries put their minds to finding a logically defensible foundation for Calculus, the subject became much more complicated and inelegant.

The Falling Apple

The other major insight that Newton had during the famous *annus mirabilis* of 1666 was that which led to his conception of universal gravitation. In Newton's own recollection, he says he was sitting in the garden of Woolsthorpe Manor, the family home to which he had retreated, when an apple falling from a nearby tree got his attention and started him thinking that whatever it was that made the apple fall was the same thing that kept the Moon in its orbit around the Earth. How far he got with this idea then is difficult to say. We only have Newton's account of it to go by. But this line of thinking was the driving force behind his great work, *The Principia,* which was published over twenty years later.

Newton's summation of what he accomplished in that miracle year, written in an autobiographical note fifty years later was as follows:

> In the beginning of the year 1665 I found the Method of approximating series & the Rule for reducing any dignity of any Binomial into such a series. The same year in…November [I] had the direct method of fluxions & the next year in January had the Theory of Colours & in May following I had entrance into ye inverse method of fluxions. And in the same year I began to think of gravity extending to ye orb of the Moon &…from Kepler's rule of the periodical times of the Planets…I deduced that the forces wch keep the Planets in their Orbs must [be] reciprocally as the squares of their distances from the centers about wch they revolve: & thereby compared the force requisite to keep the Moon in her Orb with the force of gravity at the surface of the earth, & found them answer pretty nearly. All this was in the two plague years of 1665-1666. For in those days I was in the prime of my age for invention & minded Mathematicks & Philosophy more then [sic] at any time since.[5]

That idea that Newton had, that gravity extended to the Moon, was his single most fruitful idea. It is the key concept that enabled him to bring together all the various

---

5. Quoted in D. T. Whiteside, "Newton's marvelous year: 1666 and all that." *Notes and Records, Royal Society of London* 21:32-41.

strands of thought that were going their separate ways. It took twenty more years to figure out how it all worked together. The result of his labors was the most important book in the history of science, *The Principia*.

For More Information

Alioto, Anthony M. *A History of Western Science,* 2nd ed. Englewood Cliffs, NJ: Prentice-Hall, 1987. Chapters 16.

Cohen, I. Bernard. *The Birth of a New Physics,* Rev. ed. New York: Norton, 1985. Chapter 7.

———. *Revolution in Science.* Cambridge: Harvard University Press, 1985. Chapter 10.

Fauvel, John et al. *Let Newton Be.!: A New Perspective on His Life and Works.* Oxford: Oxford University Press, 1988.

Hall, A. Rupert. *The Revolution in Science: 1500-1750.* London: Longman, 1983. Chapter 12.

Kearney, Hugh. *Science and Change: 1500-1700.* New York: McGraw-Hill, 1971. Chapter 6.

Kuhn, Thomas S. *The Copernican Revolution: Planetary Astronomy in the Development of Western Thought.* Cambridge: Harvard University Press, 1957.

MacLachlan, James. *Children of Prometheus: A History of Science and Technology,* 2nd ed. Toronto: Wall & Emerson, Inc., 2002. Chapters 10-11.

Manuel, Frank. *A Portrait of Isaac Newton.* New York: Da Capo Press, 1990.

Ronan, Colin A. *Science: Its History and Development Among the World's Cultures.* New York: Facts on File, 1985. Chapter 8.

Spielberg, Nathan, and Bryon D. Anderson. *Seven Ideas that Shook the Universe,* 2nd ed. New York: Wiley, 1995. Chapter 3.

Stillwell, John. *Mathematics and Its History.* New York: Springer-Verlag, 1989. Chapter 8.

Westfall, Richard S. *The Construction of Modern Science: Mechanisms and Mechanics.* Cambridge: Cambridge University Press, 1977.

———. *Never at Rest: A Biography of Isaac Newton.* New York: Cambridge University Press, 1980.

Chapter Fourteen
Let Newton Be!

Trinity College, Cambridge University, in the 17th century.

PHILOSOPHIÆ
NATURALIS
PRINCIPIA
MATHEMATICA.

Autore *JS. NEWTON,* *Trin. Coll. Cantab. Soc.* Matheseos
Professore *Lucasiano,* & Societatis Regalis Sodali.

IMPRIMATUR·
S. PEPYS, *Reg. Soc.* PRÆSES.
Julii 5. 1686.

LONDINI,

Jussu *Societatis Regiæ* ac Typis *Josephi Streater.* Prostat apud
plures Bibliopolas. *Anno* MDCLXXXVII.

The title page from the first edition of Newton's *Principia*.

Chapter Fifteen

The Mathematical Principles of Natural Philosophy

Isaac Newton's special place in the history of science exists because he took all the major loose ends of the study of the physical world, which themselves were going off in different directions and becoming less and less compatible, and brought them all back under one system that provided a secure framework for building the vast edifice that we now call science.

The Pieces that Newton Put Together

Newton's role in science is akin to the role played by Aristotle for philosophy in general. Like Aristotle's works, Newton's analysis of a scientific problem became the starting place for further investigation, either to continue Newton's work, adding more detail, or to find its shortcomings and go on to a new formulation. Aristotle's work included much about the physical world. What he said remained the starting place for studies of nature until something else could replace it, despite the shortcomings of his theories that became manifest during the Scientific Revolution.

The established and recognized way to build a secure framework of knowledge in mathematics was to be seen in Euclid's *Elements*. The Copernican system had usurped the Aristotelian cosmos, but left a great many questions unanswered. Johannes Kepler found amazing new regularities in the positions and motions of the planets. Galileo had discovered simple laws that described the motions of falling bodies and projectiles on Earth. Descartes had sought for a new systematic framework for all of knowledge and in doing so laid down some principles of motion that contradicted Aristotle. He also developed a lot of useful mathematics for studying the natural world.

Newton had studied and absorbed all of this. The awe that he inspired in his lifetime and in the generations that followed was engendered by what he did with it.

Aristotle's Approach

From the modern perspective, one of the most extraordinary doctrines of ancient philosophy was that the Earth and the heavens are entirely separate worlds with different defining principles. This view was codified by Aristotle and made part of the general understood and tacitly accepted framework on which every other concept about nature was built. In the world below the spherical shell that comprised the "orb" of the Moon, the sublunar world, natural motions were in straight lines, starting from where an object was (when released) and heading as directly as possible to its "natural place," while forced motions were caused by direct contact with a moving body. Everything was finite. Things came into existence, had a life, and then ended in death. That description fit not just living things, but was extended to everything in the sublunar world—large, apparently inanimate bodies like mountains, as well as motions, which also had beginnings, middles, and ends.

Beyond the Moon, in the superlunar world, everything was eternal. The stars (including the wandering ones) were forever and so were their motions, which were unending because they went in circles. Even the material of the superlunar world was different. Instead of combinations of earth, air, fire, and water as in the sublunar world, everything in the heavens was formed from the quintessence, the fifth element, which was invisible or at least transparent everywhere except where it was formed into stars.

Even after Galileo succeeded in turning public opinion toward the Copernican view and thereby put the Earth out among the stars, these distinctions still remained in people's minds. Galileo's own work can be divided into two compartments, that which concerned the heavens and that which concerned matters on Earth.

Euclid's Axiomatic System

The most successful and enduring contribution of ancient thought to science was Euclid's *Elements*. Not only was Euclid's work the standard textbook of mathematics for thousands of years, it also demonstrated the method for building a secure and powerful system that depended solely on a few specifiable assumptions. The axiomatic system of Euclid is built up from a set of definitions of how certain technical terms are to be used, a small number of postulates that are specific to the subject matter under study (e.g., geometry, number theory), and then some "common notions," that clarify some generally understood ideas (e.g., if equals be added to equals, the sums are equal). Those specified starting points and a handful of implicit principles of logic were all Euclid needed to build a tremendous edifice of mathematics, all of which is true and certain if his stated assumptions are true.

Kepler's Laws

Johannes Kepler could be viewed as an amazingly clever mental gymnast who spent his life finding curious mathematical coincidences in astronomy and astrology. We could dismiss him as some crazed mystic in the Pythagorean and Hermetic traditions who prattled about silly relationships among music, geometry, and the Copernican planetary system, and was of no consequence for science when put next to serious scientists like his contemporary Galileo.

We could do this, but we don't. Instead we hail him as the discoverer of his three laws of planetary motions. We do so primarily because Newton found much of interest in Kepler's works and did not despise him because of his numerological mysticism. Newton himself was not above a bit of "seeing" patterns in nature that corresponded magically to other realms. For example, "seeing" seven colors in the spectrum, which corresponded to the seven notes in the musical scale.

Kepler's celebrated "three laws" are the bits of Kepler's work that Newton incorporated into his own system.

Galileo's Laws and His Scientific Method

Galileo made a convincing argument for the general public in favor of the Copernican system, but he did so by simplifying the system rather than examining its minute details as Kepler did. Hence, for someone like Newton, Galileo's astronomical work was of limited value. But Galileo's work on terrestrial physics pointed Newton directly toward his eventual conception of universal gravitation. Galileo had also, through his experiments with inclined planes and projectiles, shown Newton the way to study a problem: simplify it to its essentials, make measurements under controlled conditions, and express his findings in mathematical terms.

The Lucasian Professor of Mathematicks

When the plague finally abated, Newton returned to Cambridge and that autumn was elected to a fellowship at Trinity College. When he was an undergraduate, his mathematics professor had been Isaac Barrow, who was the first holder of the Lucasian Professorship of Mathematicks at Cambridge. Barrow had doubtless taken note of the extraordinary abilities of young Newton and they became friends.

When Newton returned to Cambridge, his primary interest was alchemy to which he devoted himself for a couple of years until his attention was suddenly brought back to mathematics by the publication of a book[1] that briefly presented some of the techniques that Newton had developed in greater detail in his Calculus work. Newton, fearing being upstaged and losing credit for having developed these methods himself, promptly composed a treatise, *De analysi*, which explained his findings in detail. This was one of the works, mentioned in Chapter 14, which was rejected for publication, ultimately allowing Gottfried Leibniz to publish the first treatise on the Calculus. Newton relied on help from Isaac Barrow to help establish his priority in these mathematical studies.

Very soon after this, Barrow resigned his position as Lucasian Professor in order to devote himself to theology. Barrow recommended that Newton be chosen to replace him, which he was. Newton held the position for twenty-seven years. Newton was the second Lucasian Professor. Ever since he held the post, it has been occupied by a series of notable mathematical physicists. Stephen Hawking is the present Lucasian Professor at Cambridge.

Newton had only modest duties. Every year he was supposed to give some lectures on "some part of Geometry, Astronomy, Geography, Optics, Statics, or some other Mathematical discipline." Newton chose to lecture on whatever interested him at the time, beginning with optics. He performed his duties to the letter of his contract, but he was not known to have made any efforts to be understandable. His personal assistant at Trinity, a man named Humphrey Newton, not related to Isaac, commenting on the lectures said

> so few went to hear Him, & fewer y^t understood him, y^t oftimes he did in a manner, for want of Hearers, read to ye Walls.[2]

---

1. Nicholas Mercator's *Logarithmotechnia*, 1668.
2. Cited in Richard S. Westfall, *Never at Rest*, p. 209.

Chapter Fifteen
The Mathematical Principles of Natural Philosophy

The Reflecting Telescope

Newton's work on colors had brought home to him the problems of using glass lenses in telescopes. Since light of different colors refracts differently, which is why we get a spectrum, light that passes through a lens to magnify an image will also spread out into the colors of the spectrum, and the image produced will not be in focus. After studying the matter for some time, Newton concluded that ordinary telescopes with lenses could never be made much more powerful without completely losing the benefit of magnification because of blurring. Putting his mind to this, he found a solution. Instead of magnifying an image by refraction, he could magnify it by reflection. He invented a new kind of telescope that worked by aiming a parabolic mirror at the object to be seen. That mirror would then focus the light that hit it on a point some small distance from the mirror, where Newton had placed a flat mirror at a 45° angle, which would direct the image out to the side where it could be viewed through an eyehole. Since this design did not have the problem of refracting into colors, it could be made larger and more powerful.

Newton presented one of his reflecting telescopes—made entirely by himself—to the Royal Society of London, where it was greeted with enthusiasm, and shortly thereafter Newton was elected to membership in the Society.

Edmund Halley's Fateful Visit to Newton

When Newton ventured to publish the results of his optical researches in 1672 (see Chapter 14), a storm of protests resulted, some claiming that Newton's results were wrong, others claiming that Newton had stolen the ideas from them. One of the latter was Robert Hooke, Curator of the Royal Society, who remained one of Newton's lifelong enemies. After replying to his critics for a time, Newton grew impatient and decided that he was better off not making his work public and then getting embroiled in distracting controversy. So from about the middle of the 1670s, Newton withdrew into his own studies, keeping most of his results to himself. Newton worked with great intensity, as he always did, but his interest turned to alchemy, theology, and religious history, spending little time on mathematics and physics.

Meanwhile ideas similar to those which struck Newton during the plague years were being talked about in academic circles. In London, a discussion took place in 1684 between the astronomer Edmund Halley (after whom the comet is named) and the architect Sir Christopher Wren about the force that held the planets in their orbits around the Sun. They thought there must be an inverse square relationship between the force and the distance of a planet from the Sun which would produce the elliptical paths described by Kepler. But they could find no way to prove it. They discussed this with Robert Hooke at the Royal Society, who said he knew the proof but wasn't going to tell them. Halley wrote,

> Mr Hook said that he had it, but that he would conceale it for some time so that others, triing and failing might know how to value it, when he should make it publick.[3]

Halley decided to ask Newton and traveled to Cambridge to do so. When Halley put the question to Newton, he turned it around. He asked Newton what curve would be produced by an object, such as a planet, subject to a force of attraction to the Sun that was as the reciprocal of the square of the distance from it.

> Sir Isaac replied immediately that it would be an Ellipsis, the Doctor struck with joy & amazement asked him how he knew it, why, said he I have calculated it, whereupon Dr Halley asked him for his calculation without any further delay. Sir Isaac looked among his papers but could not find it, but he promised him to renew it, & then send it him.[4]

Three months later, Newton sent Halley a nine-page proof of the solution to the problem that Halley had put to him. Halley urged Newton to have it published immediately, but Newton refused, having now seen the scope of where his analysis led. So, for the next 18 months he worked feverishly on developing the ideas further, and in April of 1686, Newton sent the first of three parts of his manuscript to the Royal Society for publication. The Royal Society had, however, squandered all its available funds for publications on an elaborate edition of *The History of Fishes* and had no money left to pay for the printing of Newton's work. Halley came to the rescue and undertook to have the work printed at his own expense.

---

3. In *The Correspondence of Isaac Newton* (Cambridge: Cambridge University Press), 1959-1977, Vol II, p. 442.
4. Cited in I. B. Cohen, *Introduction to Newton's "Principia,"* (Cambridge: Cambridge University Press, 1971), p. 297.

Chapter Fifteen
The Mathematical Principles of Natural Philosophy

Even with that, Halley's role in getting Newton's masterwork into print was not at an end. Newton was correct when he said to Halley in 1684 that he had previously calculated that the resulting orbit was an ellipse. He had done so back in 1680 after an exchange of letters with Robert Hooke. This proved to be trouble later. It was Hooke with whom Newton had tangled over his theory of colors, and it was Hooke whose death Newton waited for before he would publish is *Opticks*. Hooke now was objecting that he had been the originator of the inverse-square law of attraction, not Newton. This sent Newton into a rage and nearly resulted in his withdrawal of Book III from publication. Finally, it all came together after much coaxing from Halley, and in July of 1687, the work finally appeared.

Philosophiæ Naturalis Principia Mathematica

The work was written in Latin and bore the title *Philosophiæ Naturalis Principia Mathematica*, or in English, *The Mathematical Principles of Natural Philosophy*. It is generally referred to as *The Principia* for short.

What is immediately obvious upon first looking at *The Principia* is the extent to which it is modeled upon the style and structure of previous works and is thereby a rejoinder to them. Start with the title. Descartes' work that was intended to make a fresh start to philosophy and replace Aristotle was entitled *The Principles of Philosophy*. Newton's title is the same, but with the addition of two modifying and limiting words. Newton's stated subject matter is "natural" philosophy, not all of philosophy, which, in Cartesian terms would mean that it concerns itself only with *res extensa*, the physical world. It is as though Newton either was rebuking Descartes for having bitten off more than he could chew, or, more modestly, Newton was saying that he had something to say only about the natural world. The other word added is "mathematical." Descartes had proposed to lay out all the principles of knowledge, but Newton confines himself to the mathematical only. Was Newton merely describing his own approach or was he tacitly asserting that the mathematical was the only route to knowledge of the external world?

Names aside, the format of the book is clearly copied from Euclid's *Elements*. Newton begins with Definitions, goes on to Axioms, and then clarifies his procedure by laying out Rules of Reasoning. Then, just like Euclid, he proceeds to propositions that build one upon another. Like Euclid (that is, like the Latin translations of Euclid), his proofs end with the initials Q.E.D. for *quod erat demonstrandum*, [that] which was to be proved. It is in every respect intended to have the same logical rigor and mathematical certainty as Euclid's *Elements*, but to be describing the physical world, not just the abstract world of mathematics. It could be argued (and has been) that Newton's usage of the Euclidean structure is a bit of a cheat, because it suggests that there is more logical necessity and certainty in his results than are warranted by his actual demonstrations. But on the other hand, it asserts that rigor of this sort is the ideal to which scientific treatises about the natural world should aspire.

The axiomatic structure of *The Principia* means that Newton simply begins with what he wants his readers to grant him, and he will show them what is implied by it. There may be some reasons given as to why his axioms are true, but the argument is more a justification for choosing the particular starting place than attempt to prove the assertions. Compare this with the beginning of Descartes' *Principles* where Descartes endeavors to prove every step as logically necessary, starting only with the (undoubtable) assertion that there exists someone to be doing the thinking.

In Descartes' *Principles* after a long argument in which Descartes "deduces" what the characteristics of the external world are, he states—as though it is now proven—the first and second laws of nature, which define natural movement in the world of extension. (See Chapter 13.) Newton, on the other hand, states those same two "laws of nature" not as proven, but as assumptions on which he will build his system. Newton, therefore, identifies his argument as belonging to the world of mathematics where logical validity is preeminent and can be verified, not the world of philosophy where arguing from first principles can go astray and can be subject to different interpretations.

The Axioms, or Laws of Motion

Descartes' first and second laws of nature are embodied in Newton's first Axiom of Motion. There are three stated Axioms of Motion, and these are often called Newton's Laws of Motion (as Newton himself called them). But it is important to remember that these are the starting points of *The Principia*, not the end result. These are the assumptions on which Newton builds the edifice of his system. They are not proven. Their confirmation lay in the vast amount of observable results that can be accounted for ("saved"

Chapter Fifteen
The Mathematical Principles of Natural Philosophy

in the Platonic sense) by assuming them true. This is a very great shift in scientific thinking.

Here are the axioms as given in the beginning of *The Principia*. The translation is from 1729, by Andrew Motte.

> LAW I. Every body perseveres in its state of rest, or of uniform motion in a right line, unless it is compelled to change that state by forces impressed thereon.
>
>> PROJECTILES persevere in their motions, so far as they are not retarded by the resistance of the air, or impelled downwards by the force of gravity. A top, whose parts by their cohesion are perpetually drawn aside from rectilinear motions, does not cease its rotation, otherwise than as it is retarded by the air. The greater bodies of the planets and comets, meeting with less resistance in more free spaces, preserve their motions both progressive and circular for a much longer time.
>
> LAW II. The alteration of motion is ever proportional to the motive force impressed; and is made in the direction of the right line in which that force is impressed.
>
>> If any force generates a motion, a double force will generate double the motion, a triple force triple the motion, whether that force be impressed altogether and at once, or gradually and successively. And this motion (being always directed the same way with the generating force), if the body moved before, is added to or subtracted from the former motion, according as they directly conspire with or are directly contrary to each other; or obliquely joined, when they are oblique, so as to produce a new motion compounded from the determination of both.
>
> LAW III. To every action there is always opposed an equal reaction; or the mutual actions of two bodies upon each other are always equal, and directed to contrary parts.
>
>> Whatever draws or presses another is as much drawn or pressed by that other. If you press a stone with your finger, the finger is also pressed by the stone. If a horse draws a stone tied to a rope, the horse (if I may so say) will be equally drawn back towards the stone: for the distended rope, by the same endeavour to relax or unbend itself, will draw the horse as much towards the stone as it does the stone towards the horse, and will obstruct the progress of the one as much as it advances that of the other.
>>
>> If a body impinges upon another, and by its force change the motion of the other, that body also (became of the quality of, the

mutual pressure) will undergo an equal change, in its own motion, towards the contrary part. The changes made by these actions are equal, not in the velocities but in the motions of bodies; that is to say, if the bodies are not hindered by any other impediments. For, because the motions are equally changed, the changes of the velocities made towards contrary parts are reciprocally proportional to the bodies. This law takes place also in attractions...

Galileo's Laws and Kepler's Laws Subsumed

The power of *The Principia* is that it accounts for widely different known phenomena in the same system, that is, all following logically from the same set of assumptions. Newton shows this right at the start by demonstrating that with arguments deriving from the Axioms of Motion, he is able to account for Galileo's law of free fall and of projectile motion, *and* Kepler's laws of planetary motion. Consider the importance of this. Both Galileo's and Kepler's laws are empirical, that is, they are mathematical formulae that describe actual observed events in nature. Therefore, Newton demonstrates right at the outset that his book is not some abstract piece of mathematical virtuosity, but a description of the physical world. Second, and even more important, these laws of Galileo's and Kepler's refer to events in the different and separate Aristotelian realms of sublunar and superlunar. For Aristotle and his followers, there were entirely different rules for the heavens and for the Earth. Newton shows that his system will encompass both in a single set of laws.

In a "scholium," which means a more general discussion of the significance and implications of the preceding sections, Newton asserts that Galileo's laws follow from the first two axioms and the first two (of six) corollaries that Newton showed were implied by the axioms.

COROLLARY I.

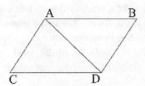

A body by two forces conjoined will describe the diagonal of a parallelogram, in the same time that it would describe the sides, by those forces apart.

COROLLARY II.

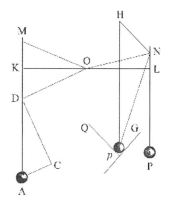

And hence is explained the composition of any one direct force AD, out of any two oblique forces AC and CD; and, on the contrary, the resolution of any one direct force AD into two oblique forces AC and CD: which composition and resolution are abundantly confirmed from mechanics.

SCHOLIUM.

Hitherto I have laid down such principles as have been received by mathematicians, and are confirmed by abundance of experiment. By the first two Laws and the first two Corollaries, Galileo discovered that the descent of bodies observed the duplicate ratio of the time, and that the motion of projectiles was in the curve of a parabola; experience agreeing with both, unless so far as these motions are a little retarded by the resistance of the air. When a body is falling, the uniform force of its gravity acting equally, impresses, in equal particles of time, equal forces upon that body, and therefore generates equal velocities; and in the whole time impresses a whole force, and generates a whole velocity proportional to the time. And the spaces described in proportional times are as the velocities and the times conjunctly; that is, in a duplicate ratio of the times. And when a body is thrown upwards, its uniform gravity impresses forces and takes off velocities proportional to the times; and the times of ascending to the greatest heights are as the velocities to be taken off, and those heights are as the velocities and the times conjunctly, or in the duplicate ratio of the velocities. And if a body be projected in any direction, the motion arising from its projection as compounded with the motion arising from its gravity.

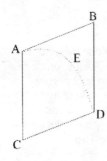

As if the body A by its motion of projection alone could describe in a given time the right line AB, and with its motion of falling alone could describe in the same time the altitude AC; complete the parallelogram ABDC, and the body by that compounded motion will at the end of the time be found in the place D; and the curve line AED, which that body describes, will be a parabola, to which the right line AB will be a tangent in A; and whose ordinate BD will be as the square of the line AB. ...

To express this more simply, Newton shows that a constant force in one direction produces a constant acceleration over time, which is what Galileo's law of free fall asserts for falling objects. That constant force is gravity. Note here that the term is used freely in its ordinary meaning of "heaviness," not in the technical sense of universal gravitation that Newton gave it later on. So long as gravity is a constant force, it will, due to the second axiom of motion, cause a constant acceleration on whatever it pulls. For a projectile, there are motions in two directions to consider: the motion in the direction that the object was thrust and the motion downward due to gravity. The corollaries assert that these act together to take the object to a place defined by the addition of the two motions in their respective directions. The scholium then shows how that combination of inertial motion in one direction and accelerated motion in another direction produces a parabola, as in Galileo's law of projectiles.

Kepler's laws take a bit more showing, but they follow readily enough from these assumptions, with a few other mathematical propositions demonstrated first. Newton's first proposition of Book I is a demonstration that Kepler's second law, that planets sweep out equal areas in equal times, is just an extension of Euclid's area law for triangles (in particular, that triangles with equal bases and heights have the same area, regardless of shape). The triangles in this case are produced by a body moving inertially past a point while a force is drawing the body toward the point. At first, the force pushing the body to the point operates only at finite, equally spaced instants of time. Then those instants become closer and closer together until in the limit they are constant. Before making the instants infinitesimally close to each other, Newton shows that a series of triangles are created as the object moves along, being subjected to these spurts of centripetal force (force directed to a single point). Each of these triangles, he shows, has equal area. Therefore, in the limiting situation of infinitesimally small spaces between the spurts of centripetal force, the object will sweep out equal areas in equal times.

PROPOSITION I. THEOREM I.

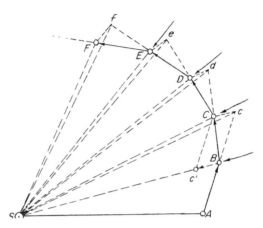

The areas which revolving bodies describe by radii drawn to an immovable center of force do lie in the same immovable planes, and are proportional to the times in which they are described.

> For suppose the time to be divided into equal parts, and in the first part of that time let the body by its innate force describe the right line AB. In the second part of that time, the same would (by Law I), if not hindered, proceed directly to c, along the line Bc equal to AB; so that by the radii AS, BS, cS, drawn to the center, the equal areas ASB, BSc, would be described. But when the body is arrived at B, suppose that a centripetal force acts at once with a great impulse, and, turning aside the body from the right line Bc, compels it afterwards to continue its motion along the right line BC. Draw cC parallel to BS, meeting BC in C; and at the end of the second part of the time, the body (by Cor. I of the Laws) will be found in C, in the same plane with the triangle ASB. Join SC, and, because SB and Cc are parallel, the triangle SBC will be equal to the triangle SBc, and therefore also to the triangle SAB. By the like argument, if the centripetal force acts successively in C, D, E, &c., and makes the body, in each single particle of time, to describe the right lines CD, DE, EF, &c., they will all lie in the same plane; and the triangle SCD will be equal to the triangle SBC, and SDE to SCD, and SEF to SDE. And therefore, in equal times, equal areas are described in one immovable plane: and, by composition, any sums SADs, SAFs, of those areas, are to each other as the times in which they are described. Now let the number of those triangles be augmented, and their breadth diminished in

infinitum; and (by Cor. IV, Lem. III) their ultimate perimeter ADF will be a curved line: and therefore the centripetal force, by which the body is continually drawn back from the tangent of this curve, will act continually; and any described areas SADs, SAFs, which are always proportional to the times of description, will, in this case also, be proportional to those times. Q.E.D.

The Clockwork Universe

Galileo's laws, Kepler's laws, and everything else that can be deduced from Newton's system describe events that can be seen in nature. But instead of starting with nature and generalizing from it, Newton starts with some abstract assumptions about motion and shows that these observable phenomena match up with what would be predicted by his system. In other words, *The Principia* is a *model* of the world, created on paper, just as one might create a model in the form of a machine with moving parts. A model is not the thing itself; it is a representation of the thing that more or less behaves like the thing modeled. In this case, what is modeled is the universe itself. The universe is much too difficult to comprehend directly, but a simplified model might be. The value of the model lies in how well it corresponds to reality.

Newton's model derives from deterministic mechanical laws. The world he describes is, in effect, a machine, a very sophisticated and complex machine. In Newton's age, the most complex machinery known were clocks, so it is natural that the analogy with a clock would come to mind, and Newton's model would be dubbed the Clockwork Universe.

The clock metaphor was a good one for Newton's system, especially if two aspects of clocks (especially clocks of that age) were kept in mind: (1) they did not keep perfect time, and (2) they required some kind of power to make them work. Newton's description of the world in *The Principia* did not predict all events perfectly, as he himself knew very well. Not even all aspects of the orbits of the planets were accounted for. And on Earth, the whole of the living world lay beyond the reach of his system. Newton was a deeply religious man. In fact, he probably spent more of his time studying theological and Biblical issues than he did scientific ones. If Newton's system was not able to account something, it was then evidence for Newton of the hand of God. This could refer to correcting planetary orbits or to selecting the configuration of limbs and organs in living beings. These

Chapter Fifteen
The Mathematical Principles of Natural Philosophy

were not failings in *The Principia;* these were places where the supernatural superseded the clockwork system.

In Book III, Newton added a "General Scholium", wherein he discusses many of the ramifications of his theory, including what it answers and what it does not. It is there, for example, where he says,

> This most beautiful system of the sun, planets, and comets could only proceed from the counsel and dominion of an intelligent and powerful Being.

Universal Gravitation

Newton's openness to influences from outside his "mechanical" system may have been what allowed him to see what Descartes could not. Descartes, convinced that everything in the physical world operated by blind push and pull mechanisms, could not fathom how any body could influence another over empty space. That would imply a "force" and forces, for Descartes, were *occult*, that is, they were mystical hocus pocus that did not belong in a scientific explanation. So Descartes rejected the idea of the Sun attracting the planets in the sort of magnetic pull that Kepler had speculated about, and as a result ended up with his labored vortex theory.

Newton, on the other hand, was willing to tolerate the idea of a force that he could not understand, and he then tried to analyze it mathematically. Newton saw the similarity between objects falling on Earth, and the Moon orbiting it, and sought a common pattern. He did not have to first answer the question of what could be causing the attraction.

The Principia in fact does not say what the cause of gravity is. What it does do is show that *if it were the case* that all bodies attracted each other by a force that is proportional to the product of their masses divided by the square of the distance between them, *then* objects on the Earth would fall either straight down or as projectiles as they have been seen to do, the Moon would remain in its orbit around the Earth, and the planets would go around the Sun, all as they have been observed and calculated to do. Gravitation, in *The Principia*, is a calculated effect. Its cause remains unknown.

Newton, in fact, wanted to be sure that his readers understood that he was not claiming that he had *explained* universal gravitation, only that he had discovered and calculated the effect, that is, discovered that heaviness on Earth and what bound the planets in their orbits had exactly the same characteristics and could both be calculated by the same simple formula. He also, by the way, showed that the gravitational force that he calculated between the Earth and the Moon accounted nicely for the tides. (Galileo had been barking up the wrong tree on that one.)

Newton Frames No Hypotheses

Newton endeavors to explain that he doesn't know what gravity is, only how it works. Elsewhere, in less formal writings, he speculated that gravity might be caused by a stream of "gravity particles" or by some disturbance of an otherwise undetected pervasive medium, but these musings he kept out of *The Principia*, in which he wanted to show how science should be done. In a much quoted passage in the General Scholium of Book III, Newton says that on the cause of gravity he "frames no hypotheses" (*Hypotheses non fingo*), meaning that he does not explain gravity by attributing to it some unknown cause, only noting that it exists and has been measured.

> Hitherto, we have explained the phenomena of the heavens and of our sea by the power of gravity, but have not yet assigned the cause of this power. This is certain, that it must proceed from a cause that penetrates to the very centres of the sun and planets, without suffering the least diminution of its force, that operates not according to the quantity of the surfaces of the particles upon which it acts (as mechanical causes used to do) but according to the quantity of the solid matter which they contain, and propagates its virtue on all sides to immense distances, decreasing always in the duplicate portion of the distances. . .
>
> Hitherto I have not been able to discover the cause of those properties of gravity from the phenomena, and I frame no hypotheses; for whatever is not deduced from the phenomena is to be called an hypothesis; and hypotheses, whether metaphysical or physical, whether of occult qualities or mechanical, have no place in experimental philosophy. In this philosophy particular propositions are inferred from the phenomena, and afterward rendered general by deduction. Thus it was that the impenetrability, the mobility, and the impulsive force of bodies, and the laws of motion and of gravitation, were discovered. And to us it is enough that gravity does really exist, and acts according to the laws which we have explained, and abundantly serves to account for all the motions of the celestial bodies, and of our sea.

Chapter Fifteen
The Mathematical Principles of Natural Philosophy

This is very much the same attitude as that expressed by Galileo, who sought to determine *how* nature works and to leave the question of *why* alone. Newton's endorsement of this attitude reinforced that this was the role of science.

The Scientific Revolution comes to an end with the work of Newton, because by the end of his career, science has found its identity. It is not the philosophy of knowledge about the natural world, though it is still called that. It is a discipline that studies nature very carefully and endeavors to describe it exactly, in the most precise language possible, which is often the language of mathematics. In that description, unifying and simplifying features are always sought. The simpler the explanation of phenomena, the better. For explanations to have any merit, they must account for more than the data that the explanation originally fit. Preferably, the explanation should be able to predict some otherwise unknown phenomenon that can be then verified. And, from the time of *The Principia* onwards, scientific explanations should be consistent with each other; in particular, they should fit nicely into the model of the world implied by the Newtonian synthesis.

◆

The Scientific Revolution began with Copernicus finding fatal flaws in the ancient cosmology. It continued through discoveries, theoretical formulations, and refutations that destroyed all confidence in the authority of Aristotle to explain the natural world. At the end of the Scientific Revolution, science had a new mandate, and fresh set of tools for the study of nature that was producing results at an impressive rate. It also had a new authority to contend with: the Newtonian system.

Part II
Science Emerges

For More Information

Alioto, Anthony M. *A History of Western Science,* 2nd ed. Englewood Cliffs, NJ: Prentice-Hall, 1987. Chapter 16.

Cohen, I. Bernard. *The Birth of a New Physics,* Rev. ed. New York: Norton, 1985. Chapter 7.

———. *Revolution in Science.* Cambridge: Harvard University Press, 1985. Chapter 10.

Dobbs, Betty Jo Teeter, and Margaret C. Jacob. *Newton and the Culture of Newtonianism.* Atlantic Highlands, NJ: Humanities Press, 1995.

Fauvel, John et al. *Let Newton Be.!: A New Perspective on His Life and Works.* Oxford: Oxford University Press, 1988.

Hall, A. Rupert. *The Revolution in Science: 1500-1750.* London: Longman, 1983. Chapter 12.

Kearney, Hugh. *Science and Change: 1500-1700.* New York: McGraw-Hill, 1971. Chapter 6.

Kuhn, Thomas S. *The Copernican Revolution: Planetary Astronomy in the Development of Western Thought.* Cambridge: Harvard University Press, 1957.

MacLachlan, James. *Children of Prometheus: A History of Science and Technology,* 2nd ed. Toronto: Wall & Emerson, Inc., 2002. Chapter 12.

Manuel, Frank. *A Portrait of Isaac Newton.* New York: Da Capo Press, 1990.

Ronan, Colin A. *Science: Its History and Development Among the World's Cultures.* New York: Facts on File, 1985. Chapter 8.

Spielberg, Nathan, and Bryon D. Anderson. *Seven Ideas that Shook the Universe,* 2nd ed. New York: Wiley, 1995. Chapter 3.

Westfall, Richard S. *The Construction of Modern Science: Mechanisms and Mechanics.* Cambridge: Cambridge University Press, 1977.

———. *Never at Rest: A Biography of Isaac Newton.* New York: Cambridge University Press, 1980.

Chapter Fifteen | 275
The Mathematical Principles of Natural Philosophy

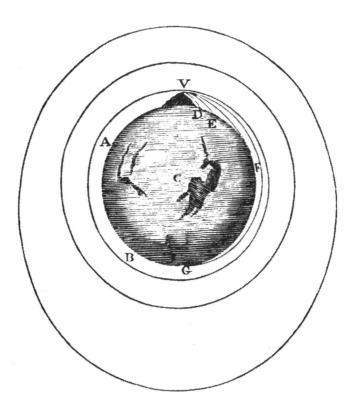

Newton's illustration of the relationship between a projectile's parabolic motion and the elliptical path of an object in orbit.

The first Solvay Conference in Brussels in 1911. Most of Europe's top physicists were there, including Albert Einstein, standing second from the right, and Marie Curie, seated.

Part Three

From Certainty to Uncertainty

The German philosopher Immanuel Kant called the 18th century the Enlightenment. It was a time of unbridled optimism and confidence that human understanding would come to comprehend all there was to know. Newton's amazing results had fostered the idea that a complete understanding of nature was within human capability, and not all that far off in the future. In Europe, the circumstances of everyday life were in a process of radical change with the advent of the Industrial Revolution. Machines were becoming more and more part of everyday life. For people of all levels of education and personal circumstances, the idea that the world was a vast machine that could be understood and made to obey was gaining strength.

Newton's physics had provided a guide in two ways. In physics itself, it was a systematic framework of physical laws. *The Principia* was so successful at accounting for the general shape and motions of the universe that every other explanation of the physical world had to fit into the Newtonian picture or get a very difficult reception.

Doing scientific work became synonymous with finding the "mechanism" that would account for phenomena. Curiously, even though the greatest triumph of Newtonianism was the discovery of the law of universal gravitation, it was rather forgotten that there was no "mechanism" here. Gravitational attraction was just as mysterious after Newton extended it into the heavens as it was when it just meant "heaviness" on Earth. Attraction at a distance had been forbidden by Descartes in his thoroughgoing mechanist model for *res extensa*. Newton had merely calculated how gravity worked and proclaimed it a universal force. This was not a mechanism in the Cartesian sense at all. But those were details that did not slow the rush to the mechanical viewpoint.

In France, where Descartes' physical views still had a following, mechanism was extended to every realm of the natural world. A physician named Julian Offroy de la Mettrie dispensed with Descartes' *res cogitans,* claiming that the soul was really just brain activity and that human beings were just as thoroughly mechanical as Descartes thought other living creatures were. His book, *Man a Machine (L'homme machine),* caused a sensation in France. He was pilloried as an atheist and had to flee from France. Nevertheless, he had just taken mechanism to its logical conclusion.

Another Frenchman, the mathematician and astronomer Pierre Simon de Laplace, developed new techniques in the Calculus and with them found that some of those anomalies that Newton thought could not be explained actually did fit in and were completely accounted for with Newtonian principles. In his work *Celestial Mechanics (Méchanique céleste),* Laplace showed that the irregularities in the speeds in orbit of Jupiter and Saturn that Newton could not explain were caused by the gravitational interaction between the two planets, which Newton had been unable to calculate.

Another phenomenon that Newton found no reason for was why the planets all circle the Sun in the same direction and in approximately the same plane. Here Laplace, relying on better observations of the heavens with much improved telescopes, had been able to make out that certain blotches in the sky were groups of stars revolving around each other. Laplace calculated that this would be the natural result if massive objects in space got close enough to each other for their mutual gravitational attraction to pull them nearer each other and ultimately into orbits around each other, picking up other objects that came by due to their increased gravitational pull. These *nebulae* that he saw were, he concluded, stars whose random motions, happening to take them near each other, resulted in a predictable and stable spiral motion once an orbital pattern had been initiated. This explanation would also account for the solar system, Laplace thought. Stray heavenly bodies, passing near enough around the Sun, would be caught by gravitational pull and begin to circle the sun. Once a pattern had been established, other planets passing into this system would also be caught by the mutual gravitational pull and begin to orbit. The combined effects of the gravitational attraction to the planets already caught in the web would cause these orbits to form into a spiral that would flatten out in time. Hence the solar system could have arisen from totally natural causes, all accounted for by Newtonian physics.

Since Newton had taken these unexplained phenomena to be evidence of the intervention of God, Laplace cheerfully announced that he had no longer any need for God.

The other way that Newtonianism guided intellectual activity in the Enlightenment was due to the form of the axiomatic system that he adopted so successfully in *The Principia.* Euclid had, of course, shown how effective it was to arrange logically dependent propositions in an order that showed exactly what had to be assumed in order to derive results. But Euclid's work was just for mathematics. Newton showed that the same arrangement was highly effective in forming an understanding of the world of nature itself. The axiomatic system became the model for all serious pursuits of knowledge. This ranged from related disciplines in the sciences, such as chemistry, to subjects of human interaction such as psychology, sociology, economics, even theology. The search was on for the "natural laws" of society. Both Adam Smith's *Wealth of Nations* and Karl Marx's *Das Capital,* though reaching opposite conclusions, set out to argue their case using an axiomatic presentation, where certain assumptions were made about human nature and conclusions then drawn from them.

As science progressed, great strides were made and many mysteries were clarified. But inevitably, the more scientists learned, the more they found they did not know. Even worse, more and more aspects of nature were uncovered that just did not seem to fit into the world as described by Newton.

By the end of the nineteenth century, there were so many cracks in the edifice that scientists were beginning to rethink the fundamental basis of the Newtonian system. The first half of the twentieth century saw another revolution in physics as Newton's authority was compromised by being subsumed into systems of greater generality.

On the other hand, Newton was not *replaced* by another authority (at least not yet) because as yet no completely satisfactory new theory has emerged that can account for every-

Part III
From Certainty to Uncertainty

thing. Instead we have different worldviews, depending on which way we look at things. If we take a very wide perspective, considering the vast distances and vast speeds in the universe as a whole, we get the formulations of general relativity and its related cosmological viewpoints. If, on the other hand, we try to examine nature at its smallest and most fundamental level, we get quantum mechanics and particle physics. But these different viewpoints are not entirely compatible with each other. Indeed, they are in many ways not compatible with themselves. We may see farther now than Newton ever did, but our view is much cloudier. Physical science has advanced out of confident certainty into doubting uncertainty.

In Part Three, we follow some of these paths that led from Newton to new understandings: Energy and Thermodynamics, Electricity and Magnetism, Relativity, Quantum Mechanics, and briefly, Cosmology. Each of these topics can be horrendously technical, but here we try to look at the major issues without getting bogged down in details. This treatment is therefore necessarily superficial, but I hope not without some value.

Part III | 281
From Certainty to Uncertainty

The role of science in the Enlightenment is optimistically portrayed in this engraving of the work of the new French Academy of Science.

An 18th century British coal mine with steam-driven locomotives hauling bins of coal in the yard.

Chapter Sixteen

Energy, A New Form of Being

The Industrial Revolution came roaring in to the Britain in the middle of the 18th century. Industrialization had been proceeding for some time before. Cottage industries were well established all around Britain, manufacturing textiles on foot-powered spinning wheels and looms. Along with other owner-operated businesses, such as blacksmiths, fullers, bakers, cheesewrights, green grocers, millers, and every other small enterprise necessary for town and urban life, these industries flourished across the British Isles and made a well-organized infrastructure for the growth of commerce. Some entrepreneurs wishing to capture more of the power of nature built larger operations located near fast running water and powered their businesses with waterwheels.

Growing populations and growing industry made greater demands for coal for heating, both domestically and in industrial furnaces and kilns, and metal ore to be smelted into useful metals. The mining industry had a ready market for whatever it could produce. Britain, sometimes described as a huge slab of coal, had the resources if only it could get them out of the ground.

Mining has an inherent natural obstacle that has been a hindrance to it from the time of the earliest civilizations that dug into the ground for minerals: Nearly everywhere that one might dig, there is groundwater not very far down. A mining shaft quickly fills up with water, making it difficult for miners to go down and extract ore. Inventive genius has been applied to the task of keeping the mines dry for as long as there have been mines.

A 16th century idea for a machine to pump out mines. From Agricola, *De re metalica*, 1555.

The simplest way to empty a mineshaft of water would be a suction pump operating from the entrance to the mine. This indeed can work for shallow mines, but there is a limitation here that makes this method useless in deeper mines. Suction can lift water only about 10 meters, after which no amount of pumping will get it any higher. This fact had been known for centuries and was a mystery to people. Galileo had thought there was some sort of internal tension at work in the water that could not hold a greater weight, rather like the tension limits on a piece of rope. The correct insight came after Newton's mechanist model gave people the idea that air had weight, and we lived, in effect at the bottom of a sea of air. Then it was understood that the reason for the 10-meter limit for sucking water up was that a 10-meter column of water exerted the same downward pressure as the pressure of the atmosphere. Creating a vacuum removed the weight of the air above the column of water and, as a result, the water rose up 10 meters, bringing it and the atmosphere into equilibrium.

Steam Engines

The discovery of atmospheric pressure also suggested other ways to use that pressure for human benefit. Aristotle had held that nature abhorred a vacuum, and from that argued that vacuums did not exist. The explanation of the water pump showed that what nature does is put a great deal of force, coming from the weight of the atmosphere, around any vacuum that might exist and closes it up very quickly. Atmospheric pressure could then be used to do work if the process of closing vacuums could be harnessed to human ends.

The key to such usage was water again, but this time in the form of steam. When water boils and turns to steam, it expands enormously and will occupy a space vastly larger than the same amount of water in a liquid state. If you could fill a closed chamber with steam and then cool that steam until it condensed back to water, you would have a chamber that for all practical purposes was a vacuum. The pressure of the atmosphere would push very hard against anything that could close up that space. This gave ideas to several inventive minds in the late 17th and 18th centuries who were looking for better ways to keep the mines dry.

The Savery Steam Pump

In 1698 a British engineer named Thomas Savery invented a machine that combined suction with condensing steam to make a pump for mines. The principle, in brief, was to boil water to make steam, let that steam into a large closed vessel that had an opening with a valve to a pipe running down into the mine. When the vessel was filled with steam, a valve shutting off the steam supply was closed and the steam was allowed to cool and then condense back to a small amount of water leaving a vacuum in most of the space. The valve leading to the mine opened causing the water from down in the mine to be drawn into the chamber until it filled. Then steam would be pushed into the chamber once again, driving the water into another pipe connected to the top of the vessel where it was pushed up and out of the mine.

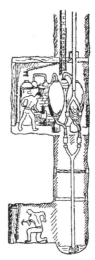

A Savery Steam Pump.

It was a very inefficient machine, and a dangerous one too, since it had to be located down in the mines where fires were always a hazard. And, the best it could do was suck water up 10 meters and then push it another 10. Mines were much deeper than that. Nevertheless, it proved the value of producing a vacuum by condensing steam.

The Newcomen Atmospheric Engine

A few years later, in 1712, a British ironmonger named Thomas Newcomen produced a machine with a different design that overcame some of the design flaws of the Savery pump. Newcomen retained the principle of condensing steam to produce a vacuum, but then used the atmospheric pressure that was exerted against the vacuum to propel a piston that would slide up and down in the steam chamber. The Newcomen pump overcame the 20-meter limit and did not have to be located down in the mine because the action of the piston pulled down a large pivoted beam, the other side of which raised a long arm which could be extended down into the mine as deep as required and used to lift containers of water (e.g., buckets) up.

This machine was also incredibly inefficient, but it worked, and so long as the cost of fuel to boil the water in the furnace was not an issue, this was a practical solution. This machine, however, was very large and cumbersome, and its jerky motion made it practical only for such tasks as pumping water. About 500 of these giant engines were built and used in Britain in the 18th century, all as pumps.

The Watt-Boulton Steam Engine

In 1764, a Scottish instrument maker named James Watt working at the University of Glasgow was asked to repair a working model of a Newcomen engine that was used for teaching purposes and was, well, *not* working. In fact, it was not working because the design of the engine was not one that could be successfully scaled down. Watt, however, examining the model decided that there were other design flaws that made the machine unnecessarily inefficient. The main flaw was that the chamber with the moving piston had to be filled with steam, then cooled enough to condense that steam, then reheated with the next batch of steam and so forth. Watt reasoned that much heat was unnecessarily lost in heating and cooling the walls of the chamber in each cycle. So, he devised a plan where the steam chamber with the moving piston would be kept hot at all times and a valve would be opened between that chamber and another *condensing* chamber which would be kept cold at all times with running water. As soon as the valve was opened, steam would rush into the condensing chamber where it would condense, form a vacuum, and draw the rest of the steam in behind it. In fact this worked much faster than the system which had the condensing take place in the same chamber.

Chapter Sixteen | 287
Energy, A New Form of Being

Watt's design was four times more efficient than Newcomen's and it worked faster too. In 1769, Watt obtained a patent for his new design and planned to market these engines to the mining industry, competing with Newcomen. This was an expensive proposition requiring a lot of capital and an entrepreneur to run it. Watt joined with an industrialist named Matthew Boulton, making a partnership, which was one of the great driving forces of the Industrial Revolution. Some historians, looking for a technological event to mark the beginning of the Industrial Revolution pick Watt's patent filing in 1769 as the launching date.

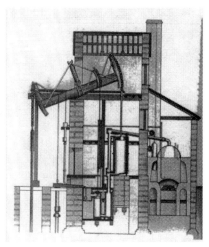

Watt's Steam Engine.

Watt and Boulton continued to improve their engine, adding a governor and a fly wheel and automatic valves to smooth out the action and turn it into an all-purpose source of continuous rotary motion that could be put to any industrial purpose. The Industrial Revolution gained momentum very quickly as factories could be located anywhere so long as there was room for a steam engine to provide a constant source of power. The textile industry, in particular, became mechanized, with large power looms and spinning machines replacing the cottage businesses.

Railroads and Steamships

Watt and Boulton held in effect all the patents on steam engines in Britain until their expiry in 1800. Though they improved their engines considerably and found more and more uses for them, they opposed the introduction of engines based on a different principle that were already being tried out on the Continent. These were high-pressure engines. They worked, not by filling a chamber with steam and then condensing it to make a vacuum, but by heating the steam to still higher temperatures, producing much larger pressures than atmospheric pressure, and then using that pressure to push pistons outward, instead of letting the pressure of the atmosphere collapse the piston inward. The great advantage of these high-pressure engines was that they could be made much smaller and still deliver a lot of power. The disadvantage was that the high pressures they produced taxed the strength of the walls of the boilers in which the steam was produced. Watt and Boulton objected on the grounds of safety, not without some justification. But metallurgy had come a long way in a few years, and boilers were made much stronger than they had been when Watt and Boulton first set up shop.

So, in 1800, with the expiry of the existing patents, high-pressure steam engines were introduced very quickly. Their smaller size made them portable, so steam-driven vehicles began to appear. Steam-driven road vehicles eventually were made practical, but long before then came the steam locomotives on the railroads and steamboats and steamships.

Heat Causes Motion, Does Motion Cause Heat?

Steam was everywhere by the early 19th century. Industry ran on it. But science had a problem with all this, namely, what was going on here? According to Newton, there were basically two entities in the universe: matter and motion. The amount of matter was taken to be fixed, though it may be transformed any number of ways by chemistry. And the amount of motion, or, more properly speaking, the amount of momentum (mass × velocity), was also fixed. So how is it that one takes a pile of fuel, coal, for instance, burns it to make heat, the heat causes water to change to steam, and the expansion of steam causes a piston to move, which is capable of doing work? Is this capable of scientific analysis?

You start with some lumps of coal, you produce motion which is captured to do useful work, at the end you have ashes and what resulted from the motions you produced. Is there the same amount of matter at the end? Is there the same amount of momentum? If so where is it all? Science was not doing very well at analyzing this.

Work in chemistry in the 18th century had supported the conviction that the amount of matter seemed to be unchanged when fuels were burned, provided that you took into account all the gases and residues. But the other variables were problematic. Chemists had begun to think of heat as an element. A chemical reaction would require so much of this substance, so much of that substance, and so much heat. Heat was added and subtracted just like any other element. So, a flammable object that was set on fire could be thought of as releasing heat that had been trapped in it. But this analogy broke down in many cases, and the idea that heat was some sort of stuff was no longer viable.

Friction provided a good counter example. A famous case in 1798 involved the American Benjamin Thompson, who, being a monarchist, fled the United States during the Revolution, moved to Germany and was so popular there that he was made Count Rumford by the Elector of Bavaria. Rumford had developed a technique for making superior cannon shafts by taking a solid metal cylinder and boring out the insides (rather than casting a hollow cylinder, for example). His technique involved using a blunt borer and immersing the cylinder in a barrel of water. What he discovered was that the boring generated enough heat to boil the water in the barrel. The amount of heat that could be produced was inexhaustible. Hence heat could not be some finite fluid trapped in the material. Since it was motion, in the form of friction, that produced the heat, perhaps heat was some form of motion.

The Mechanical Equivalent of Heat

Science in the post-Newtonian age demanded specific measurable quantities and mathematical formulae. If heat and motion were related, there must be some precise equivalence between the two. Also, there was the question whether the process was reversible. Clearly heat can be used to produce motion. That's how the steam engine worked. And given the effects of friction, it seemed that motion could produce heat.

The British physicist James Joule decided that what was needed was a precise way to measure the conversion of motion to *heat*. Joule sought an answer to a question that might be casually expressed as, "If you stir water, does it heat it up, and if so, by how much?" If you literally stir water, for example, stirring a hot beverage in a cup, far from heating up, it cools down. But that's because the stirring exposes more of the surface to the air and heat is carried away. To measure this phenomenon, a precisely controlled laboratory experiment would be required.

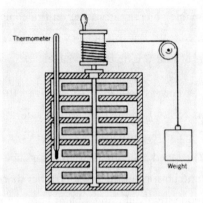

A cross-section of Joule's Churn.

Joule made an instrument, which we now call Joule's Churn, that would apply a precisely measured amount of effort to stirring a quantity of water and made very accurate measurements of the temperature of the water. The effort would be measured as *work*, a technical term introduced into physics that meant force applied over a distance. He made a cylinder into which he fitted a wheel with multiple paddles on a vertical axis and filled with water. When the wheel turned, the paddles would pass by very close fitting vanes that allowed the water to be swirled around but just barely. To the axis of the paddle wheel he attached a spool around which he wound a length of cord, which was draped over a pulley and attached to a heavy object. With the cord wound up, the object would be up near the pulley. When released, the object would descend a fixed distance to the table. The amount of work represented by this is measured by the weight of the object multiplied by the distance it descended.

Joule wanted to measure the amount by which heat was produced in the water, so he had inserted into one side of the churn a very sensitive thermometer that could register small changes of temperature. Repeating the experiment many times with different fluids and correcting for any possible systematic error, he determined that there was a fixed relationship between the amount of work expended and the amount of heat produced. That relationship is called the *Mechanical Equivalent of Heat*.

The Conservation of Energy

Heat can produce motion, and motion can produce heat. Moreover, there is something in flammable materials that allows them to produce heat. Take the steam engine for example. Start with coal, from that make fire, from that boil water to make steam, use the expansion of steam to drive a piston to produce work. There are a lot of things here producing other things. Is there something that remains the same?

Chapter Sixteen | 291
Energy, A New Form of Being

Science progresses by finding the things that are *invariant*, that is, do not change, and then explains everything else in terms of them. Newton pulled heaven and Earth together by showing that the same gravitational force worked in both realms. Newton had declared that the amount of matter in the world was fixed. The chemical experiments of Lavoisier in the 18th century confirmed that indeed matter did seem to remain the same through chemical processes. Newton also declared that momentum remained the same, but the work with heat engines seemed to be contradicting this. Was Newton right on one count and not on the other?

As often happens in the history of science, when an idea that served a purpose is found lacking, it is often replaced by another idea that includes the original one as a special case. The ancients thought all the planets moved in circles because circles have no end and repeat themselves. Kepler found that the correct path is an ellipse, which has the same characteristics. In fact all circles are ellipses—the special case where the two foci of the ellipse are at the same point.

Newton's idea of conservation of momentum was too restrictive. What is there that can replace it? In the mid-19th century several people reached much the same conclusion at about the same time. The two most notable of these are James Joule, whose work with his "churn" is described above, and Julius Mayer, a German physician. Joule was a physicist and paid attention mostly to physical interactions between heat and motion. Mayer, the physician, also thought about the relation between heat and bodily processes. They both concluded that all of these things: heat, momentum, forces, chemical potential, etc. were all manifestations of the same thing, which was to be called *energy*.

That is, to go back to Newton's categories, the two entities that exist in the universe would then be *matter* and *energy*. Moreover, just as Newton had declared that the amount of matter in the universe was fixed, Joule and Mayer claimed that the amount of energy was fixed. That is, the amount of energy is *invariant*. This is called the principle of the *Conservation of Energy*.

So what is this entity that exists in the universe in a fixed, invariant amount and that is the handmaiden of matter? Energy takes all of those forms that have been included under it, but energy itself is just an abstract concept. It is a quantity which can be stated precisely in mathematical terms for any isolated system and which obeys exact laws. But it remains an abstraction.

Unavailable Energy

Just because the total amount of energy is constant, that does not mean it is all useful. The history of the steam engine showed that well enough: coal is burned to boil water. But some of the heat produced goes to heating up the walls of the boiler. Some of it dissipates into the atmosphere. The steam produced goes into a steam chamber. Some of that is wasted heating up the sides of the steam chamber, more escapes through the spaces around the piston. Part of the thrust of the piston is wasted in overcoming the friction in the axles of the moving parts. And so on. The Newcomen Atmospheric Engine was calculated to have an efficiency of one percent. That is, ninety-nine percent of the energy that was released from the fuel was wasted. Even the much better Watt-Boulton engine had an efficiency of only four percent. Was this just bad design or was there something inevitable about this?

Available energy is the ability to do work. That ability arises from a degree of organization. In a machine that is highly efficient, energy is converted from one form to another smoothly with minimal loss. Actually, the steam engine is a particularly telling example of this loss of available energy because it is a heat engine. It produces heat early on in its process and then uses that heat to produce motion. Energy comes in many different forms and those forms can be converted into each other in very efficient ways with little loss of available energy, except when heat is involved. Heat, it was realized, is actually the lowest form of energy since heat is a state of maximum disorder.

It is because of this that perpetual motion machines are impossible. A perpetual motion machine is one that takes energy in one form and converts it to another form and then ultimately back again to the first form which then converts to the second and so on. But the principle of increasing unavailability means that when the first form of energy is reached again, there will be less energy available to continue the process, the entropy will rise, and eventually the machine will run down. Nineteenth century physicists determined that once energy is in the form of heat a certain portion of it is permanently unavailable for use, because it can never be recovered into another form. Moreover, every energy transformation inevitably produces at least some heat. The result of this is that ultimately every transformation of energy from one form to another leads to a bit more of the energy becoming unavailable for use. The amount of energy that is unavailable can be expressed in terms of the *entropy* of the system. Entropy is a technical term in use in chemistry and other disciplines. Here I am just using it as a measure of disorder. The more disorder, the more entropy.

Thermodynamics

In the 19th century, through the work of several scientists these principles were subjected to intense study and made into a separate branch of physics, devoted to the study of the forces that are involved with energy transfer in general, and heat in particular. The importance with which this was regarded is an indication of the importance of the steam engine for the economy and lifeblood of the industrialized nations. This new branch of physics was called *thermodynamics.*

The guiding insights in this field became its chief laws. First was the principle of the conservation of energy. *The first law of thermodynamics* is that in any closed system the total amount of energy is constant. Since the universe itself may be considered a closed system, this is equivalent to saying that there is a fixed amount of energy in the universe.

The increasing unavailability of energy is the second principle. *The second law of thermodynamics* is that in any processes involving energy exchange in a closed system, some of the energy becomes unavailable for later use. Another way of saying this is that in any closed system the entropy inevitably increases.

Temperature versus Heat

Heat is a form of energy. It can be measured; it can be transferred from one body to another. But it is not the same as temperature. It takes much more heat to raise the temperature of a standard volume of water one degree than to raise the temperature of the same volume of air one degree. If heat is thought of as some sort of molecular vibration, then the temperature of a body is a figure that represents the average level of that vibration.

Maxwell's Demon

There is something very different about the concepts and the laws that are stated for the science of thermodynamics; different, that is, from Newton's mechanics, which was the model for physical science. Newton's laws applied equally to each and every particle in the universe. They did not apply to just "most" particles, or just to the "typical" or "average" particle. Yet, thermodynamics speaks of temperature as an average energy level and states, in its second law, that the overall state of a system can only go in a given direction.

What does it mean to have laws that apply only to the overall state of a system and not to each and every bit of it? The second law says that some energy is made unavailable, but it has no way of determining which bits of energy will be the ones lost from use. This represents a change in thinking of what a scientific law is, and it was a change that did not sit well with all scientists.

In particular, it bothered James Clerk Maxwell, one of the most important physicists of the 19th century, the central player in the theory of electromagnetism (see Chapter 17). Maxwell, wishing to point out the limitations of these new kinds of laws, asked his colleagues to imagine the following theoretical situation:

Suppose, he said, that there is a closed box, divided into two chambers—a left chamber and a right one, with a door between them. The door is open. The air in both chambers is at equilibrium and since the door is open, both chambers register the same temperature. What that means is that the air molecules in each chamber have on average the same momentum as they careen about randomly, but only "on average." In both chambers there are some molecules moving quickly, the hot ones, and some moving slowly, the cold ones. Altogether both rooms are lukewarm. This would be a situation of maximum entropy since nothing can be done with this energy in this closed situation.

Chapter Sixteen
Energy, A New Form of Being

Now, imagine that standing at the door between these chambers, is a "demon"—some sort of tiny gremlin who is small enough to be able to see individual molecules of air. This demon takes it upon himself to watch the molecules of air that are nearing the door from either side, and when a fast moving molecule approaches the door from the left side, he holds the door open so it can pass to the right. But when a fast molecule approaches from the right, he slams the door so that it will bounce off it and stay in the right chamber. Conversely, he traps all the slow molecules in the left chamber. After a time, there will be a preponderance of fast molecules in the right chamber and slow ones in the left, and there will therefore be a temperature difference between them.

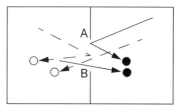

Maxwell's Demon creates a one-way transfer of heat by permitting fast moving molecules, represented here by black circles and solid lines, to move through door AB from the chamber on the left to the chamber on the right but not from that on the right to that on the left, while simultaneously allowing the slow-moving molecules, shown as hollow circles and dotted lines, to cross AB from the right to the left but not from the left to the right. Over time this will result in the chamber on the right having more fast moving molecules and thus being hotter than the left-hand chamber

Now, a temperature difference implies a lower entropy, meaning that there is useful energy where there was not before. To see this, imagine that the wall between the two chambers was a moveable piston that could slide back and forth. The right chamber, with the higher temperature, will be exerting a greater pressure on the dividing wall than the left chamber will, so if the wall were a moveable piston it would move toward the left chamber. That is measurable work that the chambers could not have done before.

Maxwell's point is that statistical laws have exceptions, in rare cases, while the universal laws of science that existed up until then did not. Something was happening to science and he was not sure he approved of it.

Absolute Zero

If heat is ultimately molecular motion and temperature is a measure of the average of that motion, then, if all molecular motion ceased, there must be some lowest possible temperature. Temperature scales always had arbitrary benchmarks. For example, in the Celsius scale, zero is the temperature at which water freezes and 100 is the temperature at which water boils, at standard atmospheric pressure, that is. But as we know, temperatures can go well below 0° C. Is there a point beyond which temperatures cannot go?

This thought occurred to William Thomson, a British physicist, who calculated the relationship between molecular motions and temperatures and then worked backwards to a theoretical utter absence of molecular motion. He found that this corresponded to what would be minus 273° Celsius. This temperature is then the absolute lowest possible, what we could call absolute zero. Though we continue to use the relative scales, Celsius and Fahrenheit, that have arbitrary zero points, it is useful for science to work with a scale where zero has an actual physical meaning of no motion.

The scale that science now uses beginning with absolute zero is named after Thomson. He was raised to a peerage by the queen for his contributions to science, and at that point he took the name Lord Kelvin. The absolute zero scale we use is called the Kelvin scale. The size of the degrees are identical to those in the Celsius scale, but the zero is at the theoretically lowest point while +273 degrees is the freezing point of water.

The Third Law of Thermodynamics

The Kelvin scale is more than a curiosity. It has physical significance. Since it represents total lack of molecular activity, it is also impossible to get to. Imagine trying to slow down molecular motion to nothing. Molecular motion can be reduced by compression, just as a refrigerator cools by compressing the coolant that circulates through it. The more that the molecular motion is retarded, the more energy it takes to retard it even more. Ultimately there is an unreachable limit where the amount of energy it takes to reduce the temperature further becomes infinite.

This is the *third law of thermodynamics*, that absolute zero cannot be reached, except through an infinite number of steps.

The Heat Death of the Universe

Since all of the laws of thermodynamics apply to any closed system, they apply to the universe as a whole. In the universe, there is, from the point of view of thermodynamics, quite a lot of order. That is, there are lots of very hot spots (stars), which contain vast amounts of energy, and there is a lot of very cold and empty space in between. This means that the entropy of the whole universe is low.

But the second law applies here, so that over time, the stars will all dissipate and burn out, having radiated their energy far and wide and heated up the space around them. As the stars burn out, they become cooler and space becomes warmer and the entropy of the system rises. Where does it all end?

This is the sort of question that scientists liked to put their minds to in the late 19th century and they thought they knew enough about the universe that they could make some informed estimates. What they came up with was the principle called the "Heat Death" of the universe. When all the energy of the stars has been thoroughly dissipated, they calculated, the universe will be all dark and the temperature will be uniform everywhere at something less than ten degrees on the Kelvin scale.

Not with a bang but a whimper. At least, that's the way it was seen then.

For More Information

Alioto, Anthony M. *A History of Western Science,* 2nd ed. Englewood Cliffs, NJ: Prentice-Hall, 1987. Chapter 22.
Asimov, Isaac. *The History of Physics.* New York: Walker & Co., 1984. Chapters 14-15.
MacLachlan, James. *Children of Prometheus: A History of Science and Technology,* 2nd ed. Toronto: Wall & Emerson, Inc., 2002. Chapter 17.
Motz, Lloyd, and Jefferson Hane Weaver. *The Story of Physics.* New York: Avon, 1989. Chapters 7 & 11.
Ronan, Colin A. *Science: Its History and Development Among the World's Cultures.* New York: Facts on File, 1985. Chapter 9.
Spielberg, Nathan, and Bryon D. Anderson. *Seven Ideas that Shook the Universe,* 2nd ed. New York: Wiley, 1995. Chapters 4-5.

Benjamin Franklin demonstrating that lightning is a form of electricity.

CHAPTER SEVENTEEN

Electromagnetism and the Æther

The basic Newtonian world consisted of matter moving in empty space. The universe was a huge empty box in which particles collided with each other and were knocked off in different directions, like a three-dimensional billiards table. It was not actually necessary that space be empty, but there was nothing in Newton's physics that referred to anything there. Newton himself had speculated about an all-pervasive medium, which would carry the gravitational attraction, but those were speculations, not part of his system.

Where others might have chosen a plenum, i.e., a filled space, with no voids anywhere, Newton accounted for phenomena with particles in motion. Consider the difference between Newton and Descartes on how the planets move. They both agree on the basic inertial motion of any object at a constant speed in a straight line. Since the planets do not move in such a path, Descartes declares that the heavens are full of invisible particles all pushing on each other within a spherical universe. The planets, while heading in straight lines, run into these particles that push them back toward the middle. Newton, reaching this point, calculates the gravitational attraction that would hold the planets in place and then declares that he "frames no hypotheses" to explain them.

Light

Newton's account of light represents a determined commitment to a particle explanation when others of about the same period, notably Christiaan Huygens, had chosen a different view. If the universe is a plenum, then light could be some sort of wave disturbance of the all-pervading medium, that produced the effect that we perceive as light. Then, the differences in these wave motions could also account for changes in color.

Newton did consider such explanations, and discussed what he took to be a crucial experiment to rule out a wave explanation. He reported on this in *The Principia* and argued that light would not behave the way it actually did if it was a wave motion. Newton's argument was that if light were a wave motion, it would spread out in all directions from its source, in much the same way that wave crests would form if a stone was dropped into a still pool of water. Similarly, continuing the water analogy, if light were let in through a small opening, it would begin to spread out again in all directions as soon as it was through the opening, just as waves are seen to do when they enter a protected area such as a harbor. Therefore light would be seen to diffuse and brighten up a wide area as it passed through even small openings. But instead, Newton says, if you take a point source of light, marked *A* in his diagram, and let it shine through a narrow slit in a barrier, marked *B* and *C*, then the light continues straight through and would strike a wall opposite in a band, marked by the endpoints *P* and *Q*, and not outside that band (i.e., that *A, B,* and *P* are in a straight line and so are *A, C,* and *Q*). That, says Newton, shows that light must travel in straight-line rays, as it would if it consisted of particles.

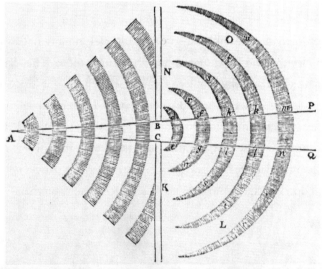

Newton's diagram of what would happen to light if it were a wave motion. From *The Principia,* Book II, Section VIII, Proposition XLI. Theorem XXXII.

The power of the Newtonian synthesis and the plausibility of this argument was enough to convince the majority of scientists, in Britain at least, that Newton was correct and that light must be a stream of particles—corpuscles, as Newton called them. But not everyone was convinced, especially those who had already thought through a wave theory of light, especially scientists on the European continent.

Chapter Seventeen
Electromagnetism and the Æther

Thomas Young's Two-Slit Experiment

Ironically, the death blow to Newton's corpuscular theory of light came not from the Continental scientists who had remained supporters of wave explanations, but from a British subject, a physician named Thomas Young, living a century after Newton (1773-1829). Young's interest in light was more concerned with issues of perception than of physics, nevertheless he reopened the case for waves by showing that Newton's crucial experiment was flawed. Newton had drawn conclusions about how light should behave if it were a wave, and then declared that it did not do so. The flaw in Newton's reasoning was

Thomas Young.

that it was not that light did not do as he said it must if it were a wave motion, but that he could not detect that it did. Remember the objections to the Copernican system: because of the Earth's motion around the Sun, there should have been an observable stellar parallax, but there wasn't. Copernicus had replied that the reason for that was not that it did not exist but that it was too small to notice. On this he was proven correct, but not until 300 years later when telescopes were made powerful enough.

It is the same with Newton's optics. If light was a wave motion, then when light passed through a narrow slit, it should spread out and diffuse past the slit, a process called *diffraction*. But Newton was not in a position to say by how much. In fact in the experiment that Newton described, light would spread out, but by a very small amount that Newton was unable to measure.[1]

What Thomas Young did was to conceive a different but related experiment that would show the diffraction of light unmistakably. Young set up a light source and a barrier, just as Newton had described, but instead of making one small slit in the barrier, Young made two slits spaced a bit apart.

---

1. Perhaps Newton could have been more careful. In 1665 (prior to Newton's optical work) the results of an experiment by an Italian physicist, Francesco Grimaldi, were published which showed that light passing through an aperture did spread out and showed color fringes at the edges. Newton was aware of these results and repeated them, getting and noting the color fringes, but viewed them as of minor significance and therefore ignored them.

Thomas Young's experiment showing interference patterns characteristic of wave motion.

The light from the source reached the two slits simultaneously and began to spread outward from each, producing crests and troughs of waves. Where the crests of waves from each slit came together, they reinforced each other producing brighter light. Where the crests of one intersected the troughs of the other, the waves cancelled each other out, leaving a dark spot. As these wave fronts traveled past the barrier and on to the wall beyond they were alternately reinforcing and canceling each other so that what shone on the wall beyond was a series of dark and light bands.

This was an effective refutation of Newton's crucial experiment and lent much support to a wave theory of light. This discovery earned Young a place in history of science books, but it did not do his career in Britain much good since he was daring to cast aspersions on the great man Newton. However it did bring the wave theory back to life.

Augustin Fresnal.

Soon after Young made his results known, a young French engineer named Augustin Fresnel worked out a mathematical theory that accounted for Young's results as well as some of Fresnel's own. Young and Fresnel began to work together devising new experiments and new explanations that made the corpuscular theory untenable. But the wave theory brought with it strange implications.

Transverse Waves

Young and Fresnel had assumed that light was a wave motion through an imponderable fluid that pervaded all space. Initially they thought that these would be *longitudinal*

Chapter Seventeen | 303
Electromagnetism and the Æther

waves, meaning that the wave undulation is in the same line as the general outward motion of the wave.

A longitudinal wave.

Sound waves are longitudinal waves. A sound is a disturbance of the air that spreads outward from a point source. For example think of a drum. When struck, the drumhead vibrates in and out, each outward vibration striking the air on its surface and driving it rapidly forward. In the next instant, the drumhead vibrates inwardly and the surface air is not struck, until the next moment, and so on. The air molecules that have been struck are propelled rapidly forward and knock into more molecules, which are then set in motion and hit those a little further away. When these wave fronts of compressed and then not compressed air reach our ears they cause our eardrums to vibrate in the same pattern as the drumhead. This is how sound works.

Young and Fresnel had imagined that light would behave in a similar manner. However, light has another feature which puzzled them, that of *polarization*. As Newton and others had noted, light that is passed through certain kinds of crystals (Iceland spar) produces a double image. When that light was passed through another crystal, the images did not recombine, in fact they might be made to disappear altogether. Newton concluded that light, like magnets, had poles, and the crystal had separated light into separate poles. Newton was thinking of particles of light, of course.

But giving this feature of light an interpretation in terms of waves proved problematical. That is, it did until Fresnel, in 1821, decided that light waves were longitudinal waves, as sound waves were, but were *transverse waves*, rather like waves on the surface of water.

In a transverse wave, the wave undulation is at right angles to the way the wave moves, just as the surface of water goes up and down as the wave moves outward. If light consisted of transverse waves, polarization would be accounted for; also the way that light waves can reinforce or cancel each other makes sense.

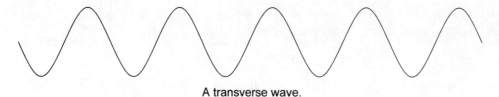

A transverse wave.

Following this idea that light comes in transverse waves led to other consequences. To begin with, the medium that light waves traveled in, called the *luminiferous æther*, could not be, as they had thought, a fluid. Light waves travel very quickly. The speed of light had been measured in the mid-18th century and a figure approximating the currently accepted figure of 300,000 km/s was calculated. A transverse wave cannot travel through a fluid at this speed.

The only way that light could move at that speed is through a very, very rigid solid body. So, the æther had to be a rigid solid, but at the same time, it had to pervade all space and had to be so thin that any and everything could pass through it unimpeded. These seemed to be totally incompatible properties, but there it was.

Æther, the Ad Hoc Hypothesis

What was this æther? Had it been seen? Had it been measured? Did it serve any purpose other than give a way of physically representing the observed features of light? This was another Ad Hoc hypothesis. An Ad Hoc explanation provides a consistent way of dealing with mysterious phenomena, but it can turn out to be either correct or incorrect. Consider, once again, Copernicus' Ad Hoc explanation for why stellar parallax was not observable: the stars are too far away; the entire orbit of the Earth around the Sun is, he said, like a point compared to the distance to the stars. On that he turned out to be correct. But when faced with the fact that the phases of Venus weren't observable even though they should have been, he snatched at the Ad Hoc explanation that Venus had its own light, instead of reflecting the Sun's light. Here he was wrong. That's the trouble with Ad Hoc explanations; they are fine as working hypotheses, but they can easily turn out to be wrong if nothing else supports them.

Chapter Seventeen
Electromagnetism and the Æther

The æther was an Ad Hoc explanation, but until a better one came along, it had to do the job.

Electricity and Magnetism

There was not a lot of useful analysis in Newton's works on electricity and magnetism. As phenomena, they did not seem to fit nicely into his mechanical matter-and-motion system. Both electricity (at least static electricity) and magnetism (e.g., the lodestone) had been known for centuries, but little progress had been made in understanding them.

One thing was clear: they seemed to work over empty space. A magnet could draw up iron filings if placed near them. A charged rod could attract or repel small bits of paper or fluff. Direct contact was not necessary. Hence, there was once again the problem of *action at a distance*. Descartes found this so abhorrent that he banished the idea from all explanations in *res extensa*, considering them an appeal to the occult. Newton was willing to tolerate the idea of action at a distance, since that is what the force of gravity is, but he refused to "frame hypotheses" to explain it. This was an area surrounded with mystery.

In the late 18th and early 19th centuries, electricity and, to a lesser extent, magnetism became all the rage among the scientific hobbyists. Showing curious electrical phenomena at social events was a popular diversion. Some serious work was done as well. The American all-round wizard Benjamin Franklin opined that since there seemed to be two kinds of electrical charges that could cancel each other out, electricity must have poles, just as magnetism does. The electrical contribution for which Franklin is best known is showing that lightning in the sky is some form of electricity. For this he flew a kite in a thunderstorm, directing the wet kite string to an electrical conductor to show the same effects that one could get with electricity from a generator or a battery.

The Frenchman Charles Coulomb devised techniques for measuring electric charges and discovered that their intensity varies inversely with the square of the distance between two charged bodies. Magnetism had already been shown to vary inversely with the square of the distance. So, of course, did gravitational attraction. For that matter, so did the intensity of light.

Was there a common thread in all of these that was being missed?

Naturphilosophie

Enter the German *philosophy of nature*, a movement in the early 19th century starting in Germany (hence the German name, *Naturphilosophie*) that sought to find unity in nature. The object was to simplify the scientific understanding of the world even more by showing how everything in nature is a manifestation of some single unifying force. *Naturphilosophie* was very interested in the issues of electricity and magnetism since they exhibited this mysterious action at a distance, like gravity, and perhaps served as some sort of "glue" for the universe.

The conviction that there was an underlying unity became much stronger when certain experiments showed that electricity and magnetism were actually related to each other. In Denmark, a physicist named Hans Christian Oersted labored for many years and finally succeeded in showing that an electric current could move a magnet. In Britain, the versatile and virtually self-taught experimental scientist Michael Faraday did the converse: he showed that moving a magnet could cause an electric current to flow.

Maxwell's Wave Equations

James Clerk Maxwell.

Implied connections such as these are fine, but they don't lead to great scientific breakthroughs until someone can put them together in a formal system that can then be calculated and tested, and then fit into a larger theoretical framework. The person who did that was the Scottish mathematical physicist James Clerk Maxwell. (This is the same person who posed the thought experiment with the demon violating the second law of thermodynamics discussed in Chapter 16.) Maxwell examined the experimental results of Faraday and Oersted and sought to put them into a mathematical form. He found that he could account for the effects of electricity and magnetism by conceiving of them as transverse waves that move at the speed of light.

Chapter Seventeen
Electromagnetism and the Æther

The coincidences were enough. Maxwell, under the influence of the principles of *Naturphilosophie*, concluded that all of these phenomena were part of the same whole:

> The velocity of transverse undulations in our hypothetical medium, calculated from the electromagnetic experiments of MM. Kohlrausch and Weber, agree so exactly with the velocity of light calculated from the optical experiments of M. Fizeau, that we can scarcely avoid the inference that light consists in the transverse undulations of the same medium which is the cause of electric and magnetic phenomena.

Maxwell leapt at the conclusions that (1) electricity, magnetism, and light were all just different aspects of the same thing, which he called *electromagnetism* and (2) that electromagnetism is a form of pure *energy*.

Energy, that totally abstract concept that provided a way of doing the accounting between different sources of power, had now shown itself more directly by effects that were familiar, but had not previously been categorized. Energy had taken a greater step in being an essential component of nature.

Moreover, the æther, which had been a convenient Ad Hoc hypothesis to account for the propagation of light now had a much more important role to play as the medium that carried all of electromagnetism. It was difficult to conceive of electromagnetism without imagining the medium in which it operated, and Maxwell had done such a good job of defining and calculating those operations. The æther seemed now to be a necessary concept in physics—as necessary as matter and energy.

For More Information

Alioto, Anthony M. *A History of Western Science,* 2nd ed. Englewood Cliffs, NJ: Prentice-Hall, 1987. Chapter 22.

Asimov, Isaac. *The History of Physics.* New York: Walker & Co., 1984. Chapters 20-21.

MacLachlan, James. *Children of Prometheus: A History of Science and Technology,* 2nd ed. Toronto: Wall & Emerson, Inc., 2002. Chapter 17.

Motz, Lloyd, and Jefferson Hane Weaver. *The Story of Physics.* New York: Avon, 1989. Chapters 9-10.

Ronan, Colin A. *Science: Its History and Development Among the World's Cultures.* New York: Facts on File, 1985. Chapter 9.

Spielberg, Nathan, and Bryon D. Anderson. *Seven Ideas that Shook the Universe,* 2nd ed. New York: Wiley, 1995. Chapter 6.

Albert Einstein.

Chapter Eighteen

Invariance and Relativity Trade Places

Just as every discussion of nature from late antiquity to the end of the Scientific Revolution had to begin with an examination of Aristotle's views on the subject, every discussion of the physical world in the 18th and 19th centuries took Newton's *Principia* as its original context. Even physical theories that had gone beyond Newton or found part of his conceptions wanting still began there, because Newton's views were the standard. And this was especially so the more general and abstract the issues were.

Absolute Space and Time

Consider concepts as fundamental to an understanding of the physical world as space and time. Right at the beginning of *The Principia*, Newton lays out what these terms mean and what we can understand about them. Prior to stating his Axioms, or Laws of Motion, Newton defines some terms that will be important throughout *The Principia*, such as matter, motion, force, impressed force, centripetal force, etc. He does not offer *definitions* of time and space, on the grounds that these terms are well known, but then, in a Scholium following the definitions, he defines them anyway, on the grounds that the "vulgar" misuse the terms. These "clarifications" turn out to be very important indeed, because they form the basis of the model of the world that is presented in *The Principia*. The major distinction that Newton makes on these "well-known" terms is between absolute and relative time, space, and place. Making these explicit is every bit as important as laying out his other assumptions. The following are definitions from Newton's Scholium.

> SCHOLIUM.
>
> Hitherto I have laid down the definitions of such words as are less known, and explained the sense in which I would have them to be understood in the following discourse. I do not define time, space, place and motion, as being well known to all. Only I must observe, that the vulgar conceive those quantities under no other notions but from the relation they bear to sensible objects. And thence arise certain

prejudices, for the removing of which, it will be convenient to distinguish them into absolute and relative, true and apparent, mathematical and common.

I. Absolute, true, and mathematical time, of itself, and from its own nature flows equably without regard to anything external, and by another name is called duration: relative, apparent, and common time, is some sensible and external (whether accurate or unequable) measure of duration by the means of motion, which is commonly used instead of true time; such as an hour, a day, a month, a year.

II. Absolute space, in its own nature, without regard to anything external, remains always similar and immovable. Relative space is some movable dimension or measure of the absolute spaces; which our senses determine by its position to bodies; and which is vulgarly taken for immovable space; such is the dimension of a subterraneous, an æreal, or celestial space, determined by its position in respect of the earth....

III. Place is a space which a body takes up, and is according to the space, either absolute or relative....

IV. Absolute motion is the translation of a body from one absolute place into another; and relative motion, the translation from one relative place into another. Thus in a ship under sail, the relative place of a body is that part of the ship which the body possesses; or that part of its cavity which the body fills, and which therefore moves together with the ship: and relative rest is the continuance of the body in the same part of the ship, or of its cavity. But real, absolute rest, is the continuance of the body in the same part of that immovable space, in which the ship itself, its cavity, and all that it contains, is moved. Wherefore if the earth is really at rest, the body, which relatively rests in the ship, will really and absolutely move with the same velocity which the ship has on the earth. But if the earth also moves, the true and absolute motion of the body will arise, partly from the true motion of the earth, in immovable space; partly from the relative motion of the ship on the earth; and if the body moves also relatively in the ship; its true motion will arise, partly from the true motion of the earth, in immovable space, and partly from the relative motions as well of the ship on the earth, as of the body in the ship; and from these relative motions will arise the relative motion of the body on the earth. As if that part of the earth, where the ship is, was truly moved toward the east, with a velocity of 10010 parts; while the ship itself, with fresh gale, and full sails, is carried towards the west, with a velocity expressed by 10 of those parts; but a sailor walks in the ship towards the east, with 1 part of the said velocity; then the sailor will be moved truly in immovable space towards the east, with a velocity of 10001 parts, and relatively on the earth towards the west, with a velocity of 9 of those parts....

Chapter Eighteen
Invariance and Relativity Trade Places

One thing that Newton makes clear is that our frames of reference are always *relative*. We have no fixed points that we can determine, so we are always measuring motion and distance relative to some reference points that may or may not be moving in "absolute" space. We may, for example, calculate the motion of the Earth relative to the Sun and consider the Sun as fixed in space, but there is really nothing observable that can determine if the Sun is still, or, if not, how fast it is moving in an absolute sense.

The Michelson-Morley Experiment

There was no way to determine absolute motion because there was nothing in the universe that we could count upon being "absolutely" still. Or so it seemed to Newton and to everyone else until the idea of the luminiferous æther became established as a concept in physics in the late 19th century. The luminiferous æther (see Chapter 17) was conceived as filling all space and being a very rigid solid. Therefore, it must be totally without motion, since there is nowhere for it to go. If there were some way of detecting motion through the æther, it would be the same as detecting absolute motion.

In the 1880s, the American physicist Albert A. Michelson thought that he had found a way. If light was some kind of undulation of the æther, and that undulation progressed at a measurable speed, then he could figure how fast he, on the Earth, was moving with or against the æther by seeing how much slower or faster light seemed to travel in one direction than in another. (In later years, Michelson developed a technique for making extremely accurate measurements of the speed of light.)

Ever since Copernicus became generally accepted, science has taken it as a given that the Earth is in motion in space. Whether the Sun is in motion was an open question, but it seemed clear that any place on the surface of the Earth, spinning on its axis daily and revolving around the Sun annually, must be moving in some direction relative to absolute space. Michelson reasoned that if he simultaneously sent a beam of light in two different directions at right angles to each other and was able to measure differences in the time each beam took to reach its destination, then he could determine which direction the Earth was moving through space.

His technique was to measure interference patterns, the same phenomenon that caused the bright and dark bands of light in Thomas Young's two-slit experiment. Michelson made several attempts to take good measurements from 1881 to 1885, but could not eliminate other factors that made his results useless. In 1886, he teamed up with a chemist, Edward Morley, and did the experiment under satisfactory conditions. Their collaborative effort is universally referred to as the *Michelson-Morley Experiment*.

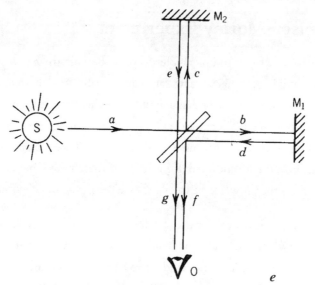

The Michelson-Morley Interferometer.

They made an experimental device that he called an *Interferometer* (see diagram); it was a combination of a light source, S, two ordinary mirrors, M_1 and M_2, and a half-silvered mirror (i.e., a mirror that lets some light through and reflects some light) set at a 45° angle. To eliminate vibrations, they dug down to bedrock, anchored the Interferometer to it, and balanced it with great precision.

The experiment worked as follows: a beam of light was emitted from the light source, S, and traveled along path *a* until it reached the half-silvered mirror. There some of the light continued through the mirror along path *b* and some of the light was reflected along path *c*. The light that was traveling on path *b* struck the mirror M_1 and was reflected back on path *d*. The light on path *c* struck the mirror M_2 and was reflected back on path *e*. The light returning on path *d* hit the half-silvered mirror and some of it was re-

Chapter Eighteen | 313
Invariance and Relativity Trade Places

flected onto path *f*. (The part of the light that passed through the light heading back to S is of no interest.) Likewise, part of the light returning on path *e* went through the half-silvered mirror and continued down path *g*. As a result, light that started from S traveled on two orthogonal paths (i.e., paths at right angles to each other), and then were reunited at the same place, shown as the eye on the figure, actually just a screen. On the receiving screen would be an interference pattern of bright and dark bands. One of the mirrors was adjustable so that once a reading was obtained, the mirror could be moved slightly closer or farther to bring the light back in phase and eliminate the interference pattern.

The idea was that if the instrument was pointed in such a way that one of the paths (e.g., *b*) was in the direction of the Earth's motion through the æther, then the other path (e.g., *c*) would be at right angles to it, and there would be a difference in the time light took to travel one path over the other, because the path in the direction of the Earth's motion would have the velocity of the Earth added to the velocity of light while the other way would not.[1] Michelson and Morley planned to aim the Interferometer in many different directions, taking readings until they found the one with the greatest difference in times. They were looking for changes in the interference patterns. When they found the biggest change, they would use the widths of the interference bands to calculate the difference in speeds of the two paths of light and thereby determine the velocity of the Earth through absolute space.

It was an ingenious plan, well conceived, carefully set up, and meticulously executed over months of time and thousands of readings. Alas, they did not find what they were looking for. No matter which way they pointed the Interferometer, no matter what time of day they did the experiment, they always got the same result, which was: no change in interference patterns.

---

1. The calculation is not difficult, but it is more complicated than it is worth getting into here. Suffice it to say that it turns out that the path that goes *with* the Earth's motion through the æther would take slightly longer than the path that goes against it. The path with the æther (e.g., *b*) has the velocity of the Earth added to it, while on the return journey (*d*) it has it subtracted. They don't cancel each other out because the lengths of the paths (in absolute space, of course) are not the same in both directions.

Modus Tollens

How one interprets and makes use of the results of scientific experiments is very much structured by the rules of logic, the system of formal reasoning that Aristotle founded and made essential for complex thinking ever since. Euclid's *Elements* owes its power to being able to deduce results from intermediate steps using the rules of logic, and Newton, in like manner, established that this is the method of building up a store of knowledge in science. Experimental results are important only because they have logical implications for understanding more than just the experiment. This is true whether the results are as expected or not.

In the Middle Ages, scholars continued the work begun by Aristotle to clarify all the forms of the logical syllogism. They did so, and gave them all names. One of those syllogistic forms is of more importance than all of the others for the development of science. It is called *modus tollens,* and it is the syllogism that applies to the results of the Michelson-Morley experiment.

The structure of modus tollens is as follows. Suppose we know this: if a statement (call it H, for hypothesis) is true, then some implication of it (call it T, for test) must also be true. Now, suppose that on investigation, the implication, T, is false. Then, modus tollens tells us, the original statement must be false.

> If H, then T.
>
> T is false.
>
> Then, H is false.

This is a valid argument regardless of the content of H and T.

In the case of the Michelson-Morley experiment, modus tollens is applied this way:

> H = "The æther is motionless in the universe and the Earth moves through it."
>
> T = "Light will appear to travel at different speeds when measured by instruments that are moving in different directions through the æther."
>
> H implies T. That is, if H is true than so is T.

Chapter Eighteen | 315
Invariance and Relativity Trade Places

But the Michelson-Morley experiment shows that T is false. Light always appeared to travel at the same speed in all directions. Therefore,

H is false.

This is modus tollens.

What is False?

Okay, so logic says that the negative results of Michelson and Morley imply that the hypothesis, H, is false. But H is itself a complicated statement, so while the whole statement must be false, what part of it makes it false?

Consider the possibilities:
- Maybe the Earth is really motionless and it is the heavens that move after all.
- Maybe the Earth drags part of the æther around with it, so near the Earth the æther is not motionless.
- Maybe the problem is with T, not with H. That is, maybe Michelson and Morley did not set up the experiment properly in order to record the correct results.
- Or, maybe the premise that H implies T is what is wrong. Maybe light won't appear to travel at different speeds for some structural reason that has not been foreseen. For example, maybe all the measurements in the direction of the motion through the æther are also affected in such a way that even though the relative speeds are affected, they can't be measured.
- And last, to cover all possibilities, maybe the whole premise of the æther is what is wrong. It was an Ad Hoc hypothesis, after all.

In fact, every one of these possible explanations was put forward and argued. The first one, that the Earth is actually stationary, was so outrageous that it was not taken very seriously inside scientific circles—but it couldn't be ruled out! The idea of dragging the æther around with the Earth might have seemed an attractive solution, but it was then hard to see how the æther could have the extreme rigidity necessary for it to transmit transverse waves at the speed of light. A great deal of attention was given to the next option, that Michelson and Morley didn't know what they were doing and set the experiment up badly. The standard solution to this, regardless of the outcome of an experiment, is for others to repeat the experiment and be sure they all get the same results. However, this was a pretty delicate experiment to begin with. It had taken Michelson five years to work out the complications so he would not get false data. Sure enough, when others tried to repeat the experiment, taking less care, they got results that Michelson and Morley did not, but their gloating proved short-lived when it turned out that they were the ones who had not done the experiment correctly.

The Lorentz-FitzGerald Contraction

Then there was the bizarre sounding explanation that the experiment failed to show altered speeds because the measuring equipment was also affected. The Irish physicist George FitzGerald proposed an explanation that seemed like Ad Hoc gone mad. He suggested that the difference in the times that should have been measurable for light traveling on the different arms of the Interferometer was not seen because the arm of the Interferometer pointed in the direction of the æther drift actually shrank in length by precisely the amount necessary for the speed of light on the two paths to appear the same.

On the face of it, this seems like as silly an explanation as Copernicus saying that Venus showed no phases because it had its own light. But then, that is the trouble with Ad Hoc explanations; since they are created to fix a particular problem, there is little to help you decide whether they are right or wrong except the extent to which they seem sensible. And one thing that science has shown dramatically since the time of Copernicus is that the sensible-sounding explanations are often the wrong ones.

FitzGerald's suggestion was taken up later by the Dutch physicist Hendrick Antoon Lorentz, who gave it a useful mathematical formulation and showed how it applied to other situations. Accordingly, it has continued to have a life and a usefulness, despite seeming like desperately grasping at straws. In its existing form it is known as the *Lorentz-FitzGerald contraction*.

Ernst Mach and Positivism

The last of the possibilities enumerated above was that the whole idea of the æther may be the problem. The æther came (back) into science as a way of explaining the polarization of light—a solid æther could carry a transverse wave, which would account for polarization. Then, the æther was useful also for giving a medium for all electromagnetic phenomena, and Maxwell's equations seemed to confirm that it was a necessary concept.

Chapter Eighteen
Invariance and Relativity Trade Places

But was it necessary, or was it just that we were unable to conceive of another explanation? Was there any evidence for the existence of the æther other than that it made a nice tidy explanation?

Just when the Michelson-Morley negative result was being discussed in scientific circles, a new approach to thinking about scientific concepts was also getting attention. This was the philosophical outlook that has come to be called *Positivism*. Its leader was the Austrian physicist Ernst Mach.

Mach argued that scientific entities that could not be independently verified did not belong in science. Positivism was used to argue against psychological theories that used unverifiable concepts such as thoughts and feelings, instead of observable behaviors. In physics, Mach took aim at any theories that made blanket statements about things that could not be directly detected. The æther was certainly such an entity. So, for that matter, was absolute time and space.

Albert Einstein

Just as Newton was considered the incomparable man of science in the 17th and 18th centuries, our own time has picked Albert Einstein for a similar honor. When I open my word-processing program, an icon of Einstein appears in the corner of the page which I can click on for help whenever I need it. He is the symbol of the person who understands everything—at least everything scientific. We revere him because he found a new and much broader interpretation of Newton's grand synthesis. He made many other important findings in physics also, and he was personally an interesting person with lovable eccentricities. Even so, the extent to which the public has latched on to Einstein as the symbol for genius itself is surprising.

Einstein was born in 1879 in Ulm, Germany, and died in Princeton, New Jersey, in the United States, where he was a professor at the Institute for Advanced Study, in 1955. As was the case with Newton, there are stories about his childhood and youth that have been made into legend. It is a curious commentary on the times that, while the stories about Newton seemed determined to confer a near god-like status to him, those about Einstein almost denigrate the man or confer an *idiot savante* status on him, suggesting that he was lacking in some ordinary abilities. As a child, Einstein was said to have learned to talk at a later than usual age—the age getting older at each telling. Then there is the often cited evidence that he did not "do well" in school. He was certainly not a docile student. When a subject did not interest him or when he thought poorly of the teacher, he ignored it, and as a result was given low marks. When a subject interested him, he worked at it readily enough, but the overall result was a less than stellar school record. His family moved from place to place, and Einstein was shunted around to several schools, not all of them to his liking. Among the subjects that did not particularly interest Einstein, oddly enough, was mathematics. He never bothered to master all branches of mathematics, but for whatever he needed in his work, he became very competent, and some of that is very difficult and abstruse indeed. Einstein believed it was important not to clutter the mind and divert it from the most important matters.

He was a life-long pacifist and hated regimentation of any kind. This rebelliousness probably affected his normal entry into the profession of science professor in the German university system. He was also a Jew, at a time when anti-Semitism was most definitely a factor.

Albert Einstein at the patent office in Bern around 1905.

Though failing to obtain a diploma from the German gymnasium where he was attending, he did get admitted to the Polytechnic Institute in Zurich, Switzerland, in 1896 after passing an entrance examination (on the second try). In four years he obtained a degree which would qualify him to be a teacher, but he was unable to obtain a position as an assistant at the Institute. In 1902, he did get a job as a technical expert in the Swiss patent office in Bern. His job there was to read over patent applications and give a technical opinion on their merits. It was steady work and not too demanding, so he had time to think about other things, which he did.

Einstein's Annus Mirabilis, 1905

Einstein's time at the patent office has similarities to Newton's time back at Woolsthorpe Manor during the plague years. He was cut off from the demands and the distractions of academic physics, he had time to think about issues that interested him, and he

Chapter Eighteen
Invariance and Relativity Trade Places

was in the prime of his youth and mental vigor. What he accomplished during his time at the patent office was of fundamental importance for physical science. All of this work came to fruition in the year 1905, which, to continue the parallel with Newton, has come to be called Einstein's *Annus Mirabilis*.

During that year, Einstein published five research papers in the German journal *Annalen der Physik*. They all represent major accomplishments. The first, and the least important of all of them, was his doctoral dissertation, which he submitted to the University of Zurich, where it was accepted. The second was a theoretical explanation of what is called Brownian motion. In 1827, the botanist Robert Brown had discovered a curious phenomenon in many of the specimens he studied with the new and greatly improved microscopes that were just coming in then. Small bits of things were constantly moving in sudden random spurts on the slide. He could not identify what they were nor what was making them move. Brown reported his findings and it became one of the unexplained mysteries of science. Since Brown was a botanist and he was reporting on an examination of living matter, this was considered a biological matter, but it had come to Einstein's attention anyway. Einstein's paper showed that the sudden motions could be caused by the collisions of molecules moving randomly in fluids, striking the visible particles from all sides, and when the collective random contacts happened to be more on one side than another, the particle went flying, even though it was much larger than the molecules themselves. This was interesting enough on its own, but it had a wider importance for physical science. Chemistry had settled on the model of atoms and molecules to explain the structure of matter, and it seemed to fit the observable facts well. However, atoms and molecules still were Ad Hoc entities that had been thought up to fit the data of chemistry. No one had directly seen an atom or a molecule, nor even seen an effect that would be explained by atoms and molecules other than chemical structure. Einstein had become very interested in the positivist philosophy of Ernst Mach, and knew the importance of confirming the existence of theoretical entities with some sort of direct measurement. His explanation of Brownian motion provided just such a confirmation.

The next paper was an explanation of what is called the *photoelectric effect*. It had been noted that under certain conditions, shining a bright light on a strip of metal will cause an electric current to start flowing across it. Einstein thought about this, and came up with an explanation. The context and importance of this paper is better explained along with the subject matter of Chapter 19, so we will return to it there. Here, to show its importance, it's worth mentioning that this work earned him the Nobel Prize in physics.

The fourth paper was an unassuming thirty-page technical analysis that seemed to be about electricity and motion. It bore the title, "On the Electrodynamics of Moving Bodies." It was, in a sense, the development of an idea that Einstein had had since he was sixteen years old. It concerned time, space, and the speed of light. This paper was the foundation of the special theory of relativity, which ultimately overthrew Newton's concepts of absolute time and space.

The fifth paper, that appeared a few months after the fourth, was really a continuation of the fourth—an afterthought, really. Thinking further about what he had said in the paper on the electrodynamics of moving bodies, he realized that it had certain consequences about the relationship between matter and energy. In particular, he expressed this relationship in his famous formula, $E=mc^2$, which has become the most famous equation in all of science.

Einstein stayed at the patent office for seven years. He had always viewed it as a temporary position, until he could get established in physics. His publications did give him the recognition he needed, and in 1909, he left the patent office for a job at the University of Zurich and the beginning of a more traditional academic career. He always looked back on his years at the patent office as especially important, because he had the time to think and no one was expecting him to come up with anything. In later years, at the height of his fame, he was consulted by the U.S. government and asked what would be the best way to foster the development of science in the United States. His reply was that the best thing would be to build a series of lighthouses all up and down the Atlantic coast and staff them with young physicists—and then leave them alone to think.

The Special Theory of Relativity

There seemed to Einstein something odd about the laws of physics as they applied to things in motion. Newton had proclaimed that though all our measurements of time, space, and motion are relative to some frame of reference, there was an absolute time,

Chapter Eighteen
Invariance and Relativity Trade Places

space, and motion in the universe. Moreover, the electromagnetic theory of light implied that there was a fixed medium, the æther, that carried light and which did not move in space. Of course, Michelson and Morley had not been able to detect it.

Using a method for which he became famous, Einstein examined the consistency of these viewpoints by imagining certain situations in which one could, theoretically, make measurements and then consider what those measurements implied. These *thought experiments* (in German, *Gedanken Experimenten*) were similar to the paradoxes that Galileo had Salviati present to Simplicio in the *Dialogue on the Two Chief World Systems* and in *Two New Sciences* (see Chapters 11 and 12) to show the internal contradictions in the Aristotelian worldview. Only Einstein was turning his mind to contradictions in Newton.

The Train Station Experiment

Suppose there is a perfectly straight railroad line running through a small town. This railway carries express trains that do not stop in this small town, but pass through the railway station without slowing. Let there be two light fixtures at either end of the station, A and B, and let there be a person standing on the platform in the station at point M, midway between A and B, holding a pair of mirrors that are joined at a right angle. With these mirrors, the person at M can hold them so he can see both A and B at once. Suppose now the person at M, holding his mirror, sees that the lights at A and B both come on at the same time. He can say that the lights came on *simultaneously*.

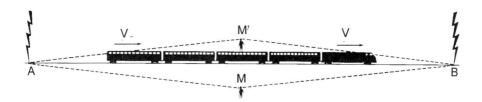

Now, suppose that while this was going on, the express train came roaring through and on that express train was a passenger with another set of mirrors. At the moment when the passenger is directly across from the person on the platform, at a place on the track designated as M', she leans out of the train, holding her mirror up and also sees the lights at A and B come on. What will she see? If the train is moving in the direction from

A toward *B*, then the passenger is moving away from *A* and toward *B*, so she will see the light at *B* come on first. The passenger can then say that the two events were not simultaneous, and the light at *B* did come on first. Who is right? Both are reporting the situation based on the same kind of information.

Suppose you say that, well, the lights are on the platform, which was stationary while the train was moving, so therefore the person on the platform was right. But Newton's physics make it clear that we have no way of telling what is moving or what is stationary other than relatively. Therefore it would be just as valid for the passenger to say she was stationary and the train station was moving. In that case she could claim to be right and the person on the platform wrong. The person on the platform orients himself with respect to fixed positions on the platform. The passenger orients herself from her position in the train. These are their respective *frames of reference*. When it comes to their measurements, they are both right—for their frames of reference.

Einstein called the issue of who was moving and who was still *Galilean relativity*, because Galileo discussed a similar situation with a ship loosing anchor and floating away from the dock. A person on the ship looking up might think that the ship was stationary and the dock was moving, Galileo's point being that the observations would be the same in either case.

Einstein takes this a step farther and shows that the observations would *not* be the same when it came to sighting distant events and indeed neither one could claim to be more correct than the other.

The issue in this example is *simultaneity*. To one observer, the events happened at the same time. To the other, they did not. If you still think that one of them is right and the other wrong, try this: take away the information given at the beginning that *M* and *M'* were midway between *A* and *B*. Suppose all you know is that the lights came from the general direction of *A* and *B*, but you don't really know how far either of them was. Maybe *A* is 100 meters away and *B* is several kilometers away. If the person on the platform sees them flash at the same time, was he right that they were simultaneous? Or what if the lights were actually attached to the front and back of the train?

Chapter Eighteen
Invariance and Relativity Trade Places

All we know about when things happen and where they are is what we get from our perceptions. If we determine that two events are simultaneous using our best equipment, then for our frame of reference, they are simultaneous. What sense does it make to try to figure out what "time" these events occurred in some absolute time at some absolute place that we can never perceive anyway?

Ernst Mach's positivism would have us discard the notions of absolute time and space entirely since they cannot be detected. The measurement of time is relative to a particular frame of reference. Positivism also dictates that the measurements that we do make—the world as we in fact observe it—are the only ones we really have, so we should take them literally and stop trying to fit them into something we cannot ever see.

Consider once again the Michelson-Morley experiment. The one thing that it did determine was that the speed of light appears to be a constant no matter how one is moving. Einstein decided that since this is the constant, the *invariant* that he had, he should build the rest of science from it.[2] But now, taking the speed of light as a constant, all manner of other quantities become relative to it. Consider length for example. Suppose the person on the platform makes a measurement of some length along the platform, say from A to B. Let's say he does this by walking out to the track and standing at the point M' (when the express train is no longer there!) and using some triangulation method to find the distance by sight from that point. Now, suppose when the express train is coming through, the passenger does the same when she reaches M'. Because the train is moving, she will get different sightings, just as she saw the lights come on at different times. Taking a figure for the velocity of the train as it goes through the station and a figure for the speed of light, an adjustment can be made to the figures she gets to make them compatible with those that the person on the platform got.

And now comes the surprise. The adjustment that must be made is precisely the same one that FitzGerald proposed for the Michelson-Morley experiment. These are the figures that Lorentz took and generalized into a formula which we call the Lorentz-FitzGerald contraction. The measurements made by a person moving with respect to an-

---

2. Einstein was aware of the Michelson-Morley experiment but it was but one of the pieces of experimental information that he was taking into account; he likely would have reached the same conclusions without the help of Michelson and Morley.

other object will be smaller by an amount that can be calculated knowing the velocity of the frames of reference of the two objects.

Once again, Einstein goes along with the measurements. If the measurements of length are different in different frames of reference, then he says the lengths really are different. The length of any object is a function of where it is seen from. In fact all of the measurements that we make of objects or events in time and space are relative to the frame of reference from which they are made.

The mass of an object increases with its relative speed. Its length diminishes. Time slows down. Throughout this, the speed of light remains the same in each frame of reference, and it becomes the upper limit of speed, that cannot be reached by anything other than light.

The Twin Paradox

These principles, which we call *special relativity*, give rise to all sorts of implications which take us by surprise. They are not what we expect, but they apply to situations which we have not really experienced, only imagined. We are surprised by them because we have, without realizing it, extrapolated from our own common-sense experience beyond where it is applicable.

One of the most puzzling of these, involving the relativity of time, is called the *twin paradox*. Suppose there are two twins, Alex and Bernie. Alex trains as an astronaut (I take the liberty of updating the story). At the age of, say, 25, Alex gets selected for a long voyage deep into space while Bernie remains behind. Alex heads off into space in a powerful ship that accelerates constantly at several meters per second, reaching enormous speeds quickly (these numbers have no particular significance, they are for illustration only). Alex settles into the voyage while Bernie marries and has children. On the space ship, time seems to progress in normal fashion. Days turn into weeks, then months, then years. Alex ages appropriately. After 20 years in space, the ship returns to Earth. Alex is then 45 years old and is anxious to meet Bernie, and Bernie's wife and children. Maybe Bernie will even be a grandfather. To Alex's shock and dismay, Bernie is dead, so is his wife, so are their children. And, they may all have had long and productive lives. In fact,

Chapter Eighteen
Invariance and Relativity Trade Places

many generations could have gone by while Alex was away. But Alex has aged only 20 years and experienced only 20 years of life while he was away.

How can this be? It is because time itself moves more slowly for the space ship, due to its acceleration and deceleration. Time is a measure of change, and all change takes place more slowly on the spaceship than it does on Earth, if it could be measured from Earth. Alex's metabolism would slow, his heartbeat would slow, the clocks on the spaceship would slow down, every process that takes place over time will slow down. Hence time itself slows down.

These sort of implications of special relativity seem far fetched to us and since they are based on conclusions reached by Einstein with pencil and paper, we may rightly ask whether we have any direct evidence that this is all true.

Indeed we do. Einstein was a theoretician, not an experimentalist, but he recognized the need for verification of his theories, so when he could do so, he suggested actual experiments that could be done to see if his predictions were verified. And if he did not, other physicists also tried to come up with empirical tests. In the case of the twin paradox, a test that has been performed is this: two atomic clocks—the most accurate kind of clocks we have—were mounted on a large revolving turntable that could rotate very quickly. One of the clocks was placed in the middle of the turntable, the other at the edge. They were synchronized with each other, and then the turntable was set spinning for a considerable time. The clock on the edge of the turntable was subjected to constant acceleration as it was forced into circular motion all the time. Therefore, time should slow down for it. When the experiment was concluded, the results were as expected, the clock on the edge showed an earlier time than the clock in the middle.

Mass and Energy

Newton's world had two variables, matter and motion. In the 19th century, his concepts were broadened and generalized to become mass and energy, where mass meant pretty much the same as Newton's term, matter, while energy was a broader and more abstract notion, but one that included motion.

Einstein's special relativity theory noted a relationship between mass and energy. As an object moved faster and faster, it gained mass, and simultaneously it took more energy to change its speed and make it go faster. The inertia of a body is its resistance to change in motion. Einstein realized that there is a relationship between the inertia of a body and its energy. Some months after his initial paper on special relativity, Einstein returned with another paper that explored this idea. He showed that there is in principle a continuum between mass and energy, and they are interconvertible. From this paper came his equation $E=mc^2$, stating a definite quantitative equivalence of mass and energy.

This idea has far-reaching consequences for two reasons: First, it essentially reduces the number of basic kinds of entities in the physical world from two, mass and energy, to one kind which takes different forms. It's the goal of *Naturphilosophie* reached at last. Second, it opened up the possibility of new and virtually limitless sources of energy. The equation shows that a tiny amount of mass, *m*, is equal to an enormous amount of energy, *E*. The speed of light, *c* is a huge number to begin with. Its square is unimaginably larger, and that multiplied by the mass *m* is the amount of energy that one would get if you could convert the mass entirely to energy. This relationship is the conceptual foundation for nuclear energy, which was developed in the decades following Einstein's paper.

General Relativity

Astounding as these ideas were, that was not what brought the general public to identify Einstein as the model for scientific genius. That came suddenly fourteen years later with a dramatic announcement that his predictions had proved correct.

We call the theory that Einstein developed in 1905 the "special" theory of relativity, because it concerns quantities that vary under special circumstances that require that two frames of reference be moving inertially with respect to each other, in other words, in straight lines at constant speeds. This is "special" because more generally, things moving around in the universe are speeding up and slowing down or traveling on curved paths with respect to each other; in other words, accelerated motions. Those are much more complicated relationships.

Chapter Eighteen
Invariance and Relativity Trade Places

Einstein continued to ponder these fundamental relationships between mass, energy, time, and space over the next several years. In 1916, he was ready to publish again. This time it was a far more complex and farther reaching theory that we call *general relativity*.

Here we will only be able to get a small sense of the scope of general relativity, and we will concentrate on the relationship between general relativity and gravity.

Acceleration and Gravity

It is a curiosity, noticed by Newton himself, that mass has two distinct ways of being measured. Newton's second law of motion in its modern form states that a force is to be measured as the mass of a body multiplied by the acceleration that the force causes. This can be turned around as a measure of mass. The mass of a body is the force necessary to change its motion, divided by the amount of change of motion, or acceleration. This can be called the *inertial mass* of a body because it concerns what is required to alter the inertial motion of the body. Take for example, causing a body to decelerate from a certain speed to a complete stop. Physics instructors are fond of examples from baseball to illustrate Newtonian mechanics, so here is one: A major-league pitcher can throw a fastball to the catcher at speeds that range around 90 miles per hour. The catcher, with an appropriately padded mitt and the skill to know what he is doing can catch the ball with no difficulty. The force with which the ball strikes the catcher's mitt is in effect a measure of the mass of the baseball, given its speed. Now consider another object traveling at the exact same speed toward the same catcher, who had to try to stop it with his mitt. For example, consider an 18-wheel tractor-trailer truck. The catcher would not have a chance. That is because the truck has a much larger mass. More generally, this relationship can be and is used by scientists to measure the mass of certain bodies that it would be difficult to measure any other way. For example, the mass of electrons can be measured by determining the strength of an electromagnetic field that is required to deflect the electrons from their paths through a vacuum tube, say.

The other way that mass is measured is more familiar to us. It is by weight. This makes use of the relationship discovered by Newton that all bodies attract each other by a force that is equal to a constant multiplied by the mass of both bodies and divided by the square of the distance between them. Therefore, we can consider the mass of the Earth and the distance from the surface of the Earth to its center as unchanging and measure the mass of individual objects (such as ourselves) by seeing how much we are drawn to the ground. This we call weight, or *gravitational mass*. It's a different kind of measure altogether from inertial mass. Consider a balance scale—the sort seen in doctor's offices, with a place to stand and a pivot arm on which one can move weights back and forth until the pivot arm is balanced. The weight of the person standing on the scale is calculated by seeing how much weight is necessary to place at what distance from the pivot to balance the two. What is measured are the gravitational forces that are exerted between the body and the Earth. This *gravitational mass* is a different idea from the *inertial mass* that is measured by stopping motion. But the numbers turn out the same.

To Newton, this may have been due to some Divine plan, but to a positivist thinker like Einstein, it meant something else. If the measurements always came out to be the same thing, then they *were* the same. That meant that, in some as yet not understood way, *acceleration and gravity were the same*. Remember that no one had an adequate explanation of what gravity was. Newton had calculated the effect of it and put a formula to it, but not found its cause. This was one of those great unanswered questions that was so basic that people tended to forget that it was unanswered.

Einstein's Elevator

Once more, Einstein helps us understand the issue with one of his thought experiments. This time he chooses an elevator. What is an elevator? It's a small room that moves up and down, and in most of today's elevators, once the doors close, you cannot see the world outside the elevator at all. It is its own frame of reference, but unlike the frame of reference of the moving express train that was taken to be moving inertially, this one does start up and come to a stop, changing speeds several times.

What do you know when you are standing in an elevator? Our natural assumption is that the elevator is located somewhere near the surface of the Earth and the floor of the elevator is the side closest to the Earth. We believe that in part because we are standing in the elevator and can feel our weight pressing against the floor, that is, we perceive our *gravitational mass*. If we were floating around in the elevator like an astronaut at a space station, we would have a different view, but we aren't. But what would it be like if the elevator was out in space away from any strong gravitational field and was accelerating up-

Chapter Eighteen | 329
Invariance and Relativity Trade Places

ward at a rate of 9.8 m/s? That's the same rate that a falling body accelerates near the surface of the Earth. Then we would not be floating around the cabin of the elevator, but instead would be pushed against the floor with exactly the same force as we would feel standing still on the ground. But what we would be perceiving is our *inertial mass,* measured by the force it takes to cause our acceleration. With the doors of the elevator closed, we cannot tell the difference.

Einstein wondered, then, if there is any difference. He tried to think of situations where Newtonian physics would be able to tell the difference between the two situations, and what he came up with is a beam of light. Go back to the closed elevator and imagine a person standing with his back to one wall of the elevator, armed with a pistol and a flashlight. On the opposite wall there is a target painted with the bull's-eye at the man's shoulder height. Let's say that the elevator is still and on the ground, that is, in the Earth's gravitational field.

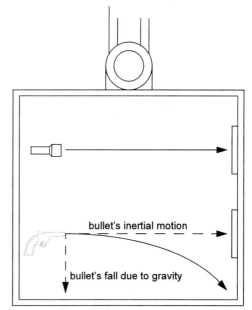

The man raises one arm perfectly level, aims the flashlight at the bull's-eye and turns it on. The flashlight will illuminate the exact center of the bull's-eye, because light travels in straight lines. He then raises the other arm, aims the pistol straight across and fires. The bullet *will not* strike the center of the target because a bullet is a projectile and its gravitational mass will cause it to travel in a parabola from the moment it leaves the gun. It will therefore strike below the center. Never mind that the distances are so short that this cannot be noticed; this is a "thought experiment"—we are concerned with the principles involved.

According to Newtonian physics, a flashlight, aimed straight at a target in an elevator resting on the ground, will illuminate the center of the target, while a bullet fired staight at a target will fall in a parabola due to the combination of its inertial motion and the downward acceleration of gravity (both represented by dotted lines), missing the center of the target as shown (in a wildly exaggerated form) above.

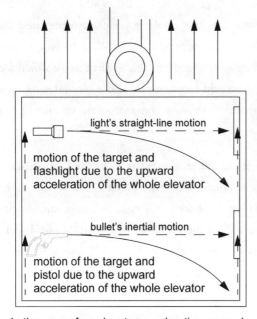

In the case of an elevator accelerating upwards at 9.8 m/s² outside of the earth's gravitational field, Newtonian physics predicts that both a bullet fired from a pistol and light shone from a flashlight will travel in a parabolic arc, missing the target at which they were aimed. Again, the bullet is expected to travel in a parabola, but this time the curve is caused not by gravity forcing the bullet to deviate from the straight path it would follow if allowed to move by inertia alone. Instead, it is the motion of the elevator that causes the bullet to miss its target. The elevator shoots upwards, carrying pistol, target, and observor with it, with exactly the same rate of acceleration as was caused by gravitation pull of the earth in the first example. This motion results in the bullet traveling along the exact same path whether the elevator is resting on the ground or rising upwards in space. However, according to Newton's system, the same is not true of a beam of light aimed at a target. There will be a difference between a flashlight aimed at target in an elevator resting on the ground and one in an elevator that is rising upwards in space. In the former case, the flashlight, sending out a beam of light which moves in a straight line, will illuminate the center of the target. But now, when the flashlight and target are accelerating upwards, the beam of light will appear, to an observer in the elevator, to be parabolic, like the path of the bullet, because the whole elevator will have moved upwards during the time it takes the light to cross the elevator.

But now, put the elevator out in space, away from gravitational fields, and accelerating at 9.8 m/s² upward. Try the experiment again. The pistol shot will again miss the target, but this time not because of the gravitational pull. It will miss the target because from the moment the bullet leaves the barrel of the gun, it will travel inertially, that is, in a straight line, to the other side of the elevator, but the elevator itself is accelerating upward. The target will have moved up farther than the bullet because the elevator as a whole has a force pushing it while the bullet is on its own once it leaves the gun. So, from the point of view of the elevator, the bullet travels in the same parabolic path it did on Earth.

What about the flashlight? If the light also travels straight across while the elevator is accelerating upward, then it too will travel in an apparent parabolic path and miss the target. How can it be that gravitational mass and inertial mass are the same numbers, that acceleration and gravity behave the same way from the viewpoint of the measurable experience, and yet light behaves one way in one situation and a different way in another?

Chapter Eighteen
Invariance and Relativity Trade Places

Einstein, pondering this and keeping that positivist viewpoint in mind, decided that we had no right to claim dogmatically that light travels in straight lines. If light appears to bend where the environment (e.g., the elevator) is subject to acceleration, it must also bend where the environment is subject to gravitational pull.

The Bending of Starlight

We think that light travels in straight lines because that is what our experience tells us. Even the most precise measurements of light confirmed straight-line motion. But what if the deviation from straight lines was very, very small. Was there any way to test this? Einstein tried to imagine the most extreme situation that he could: a long path of light with a huge gravitational pull on it. For the long path for light to travel, he could choose the light from any star. For the gravitational pull he could consider the situation where light passed near another star. If he was right, then the light going by would be slightly deflected from a straight-line path as it passed the other star.

This would only be good evidence if there was a control situation to compare with. So, Einstein's idea was to take sightings of distant stars on a clear night and determine their apparent angular distance from each other. Then take sightings of the same stars during the day when the light of the stars would have to pass close to the Sun before reaching the observer. Of course, the stars would not be visible during the day normally, but during a solar eclipse when the direct light from the Sun is blocked by the Moon, the sky gets dark enough that some stars are visible in the sky. If he was right, the starlight would be bent toward the sun from each star. If one star was to the left of the Sun and the other to the right, then the apparent angle between them during an eclipse would be slightly larger than they were at night.

The calculations required to figure this out were incredibly difficult. The easiest part was adjusting for the change in parallax, since the Earth would be in different positions for the two sightings. Einstein worked months and months on this and finally came up with a number representing the tiny shift in apparent position that would show up during the eclipse. He specified the expected amount, and an expedition was sent out to measure the angle during the next eclipse.

Now, when a scientific theory can predict something that was previously unknown it is a lot more convincing than if the theoretical explanation is announced after the phenomenon is known. The latter looks like an Ad Hoc explanation, rigged to fit the phenomenon. Getting the theory right at the outset is strong evidence that it has some fundamental truth to it.

So it was fortunate that when the scientists set up their telescopes to make the important observation, they were prevented from doing so due to cloudy weather. This was fortunate because actually Einstein had made a mistake in his calculations and specified an amount that was twice the shift that should have been seen. Before the next suitable solar eclipse came along, Einstein had found his error and produced the correct figure.

Einstein = Genius

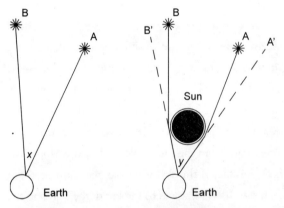

Eddington's observations confirmed Einstein's prediction that starlight would be visibly bent by the gravitation effect of the Sun. The apparent angle between stars A and B as seen from the Earth is usually *x*. During a solar eclipse, however, the sun bends the paths of their light, increasing their apparent angular separation to *y* (1.7 seconds of arc greater than *x*), and causing the stars to appear at A' and B'.

In 1919, Sir Arthur Eddington, a prominent British astronomer, set out for the equatorial island of Principe off the west coast of Africa, armed with the correct figure for starlight displacement. When the expected solar eclipse occurred, he made observations of the chosen stars and compared their angular separation during the eclipse with their normal separation and found the difference to be 1.7 seconds of arc, exactly the figure specified by Einstein.

Formal announcement was put off for six months while Eddington analyzed the photographic plates from the eclipse. Finally, his results were announced at a meeting of the Royal Society of London. This gave time for the public to be alerted to the significance of this and for much fanfare to be built up around the meeting. The public was told

Chapter Eighteen | 333
Invariance and Relativity Trade Places

that the world system of Newton was now proven to be wrong, and the system of Albert Einstein would take its place.

Suddenly Einstein was a world-renowned figure, held up for all as the smartest man in the world. In the public eye, he retained this distinction all his life, which bemused him. In 1933, when Hitler came to power, Einstein left the position he had held for the past twenty years as Director of the Kaiser Wilhelm Institute in Berlin and accepted a post at the Institute for Advanced Study in Princeton, New Jersey, where he remained for the rest of his life.

The Perihelion of Mercury

There were other experimental confirmations of general relativity than just the bending of starlight. One of these took the form of a *postdiction* rather than a prediction. This was just as good because it showed that general relativity could account for some things that Newtonian mechanics could not. A case in point was the perihelion of Mercury.

As Kepler had shown, all the planets travel round the Sun in elliptical orbits with the Sun at one focus of the ellipse. The point where the planet is closest to the Sun is called the perihelion. Another feature of the planetary orbits is that the ellipses themselves revolve around the Sun. Therefore, the perihelion is not always in the same place, it traces a path around the Sun of its own. This would not have troubled Newton too much because he was prepared to attribute differences between his physical model and the real world to the hand of God, including such "choices" as that the planets all revolve in the same direction.

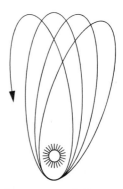

An exagerated diagram of the orbit of Mercury showing the advance of the perihelion.

But, just as Laplace was eager to show that Newton's "choices" could be accounted for by natural means, Einstein also tried to find causes for any anomaly in the predicted results. One of the problems that Newton could not solve was the advance of the perihelion of Mercury, the closest planet to the Sun. The observed rate of revolution of Mercury's perihelion was 43.03 seconds of arc per century greater than what it should have been by Newtonian mechanics. This tiny difference might have failed to get anyone's attention, except that just such information had led to the discovery of the planet Neptune in the mid-19th century. Einstein found that general relativity predicted a small difference in the gravitational pull on Mercury that would account for this discrepancy.

Einstein's general relativity could predict previously unobserved phenomena and it could postdict observations that had remained inexplicable. It was doing everything that a new theory needed to do to become established, except that it was so complex that it was intuitively impossible to grasp. That was a severe failing because it was an article of faith that ultimately nature's laws have a compelling simplicity.

The Curvature of Space

It was difficult for thinking people in the 17th century to comprehend the Copernican system because deep within them was the common-sense notion that the Earth was the center of the cosmos and immobile. Only when the public reached a point where it could reasonable contemplate the Earth in motion could it begin to see the greater simplicity of the heliocentric worldview. Likewise, part of what makes general relativity nonintuitive is that we want to look at the world through the eyes of Euclid. We think of space having three dimensions, all perpendicular to each other and going out in straight lines. We think of space as a large box in which objects are situated as though on a stage. We don't think of space itself as having properties, only the things in it.

To grasp general relativity, we have to broaden our notions a bit. We have to think of the space of the cosmos as something that is defined by the things that reside in it. Light travels in the shortest possible path through space, but that path is not a straight line because space itself curves in the presence of very massive objects, such as stars.

To speak more precisely, Einstein's relativity operates not in three-dimensional space, but in a four-dimensional space-time continuum. Where does this leave gravity, the mystery effect described by Newton? In general relativity, gravity is not a force but a curvature of space caused by the presence of matter. It is no longer a mystery; it only appears so when viewed in a three-dimensional model of the world.

Matter causes space to curve. What is matter? How about this, to continue the analogy: matter may be nothing but space with very great curvature. Space curves around matter because matter is space that is all twisted up and pulls nearby space into it.

Back to Plato and Pythagoras

For most of us, our general outlook on the world is the common-sense one that Aristotle espoused. I don't mean the idea of the Earth being in the center of things, but the idea that what is real in the world are the objects in it: things composed of matter. These can be studied with ideas and concepts, but reality lies with things. On this Aristotle had sharply disagreed with his teacher Plato, for whom the ultimate reality was the world of Forms, and Forms had no materiality. Plato himself was broadening the view of the Pythagoreans for whom the ultimate reality was number, or we might say, more generally, mathematics.

General relativity takes us back in the direction of Plato and Pythagoras. We can talk about things in the world, but when we try to understand the world fully, we are led to the position that all that really *is* is space-time and its properties. The mechanist viewpoint has lost ground to the mathematical, formal view of reality of the Platonists and Pythagoreans.

For More Information

Alioto, Anthony M. *A History of Western Science,* 2nd ed. Englewood Cliffs, NJ: Prentice-Hall, 1987. Chapter 23.

Asimov, Isaac. *The History of Physics.* New York: Walker & Co., 1984. Chapters 21-22.

MacLachlan, James. *Children of Prometheus: A History of Science and Technology,* 2nd ed. Toronto: Wall & Emerson, Inc., 2002. Chapter 19.

Motz, Lloyd, and Jefferson Hane Weaver. *The Story of Physics.* New York: Avon, 1989. Chapters 15.

Newman, James R. "Einstein." In Byron E. Wall, ed. *The Nature of Science: Classical and Contemporary Readings.* Toronto: Wall & Emerson, 1989.

Ronan, Colin A. *Science: Its History and Development Among the World's Cultures.* New York: Facts on File, 1985. Chapter 10.

Speyer, Edward. *Six Roads from Newton: Great Discoveries in Physics.* New York: John Wiley, 1994. Chapters 5 & 7.

Spielberg, Nathan, and Bryon D. Anderson. *Seven Ideas that Shook the Universe,* 2nd ed. New York: Wiley, 1995. Chapter 6.

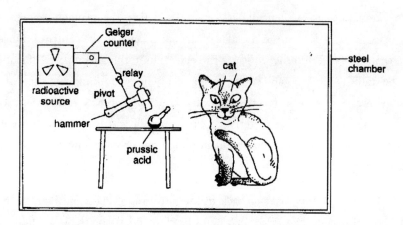

Schrödinger's Cat.

CHAPTER NINETEEN

Science Isn't So Sure

Anyone who is not shocked by Quantum Theory has not understood it!
Niels Bohr

I think I can safely say that Nobody understands Quantum Mechanics!
Richard Feynman

The Lord God does not play dice.
Albert Einstein

I don't like it and I'm sorry I ever had anything to do with it.
Erwin Schrödinger

Is the fabric of Nature smooth and infinitely divisible, or is it ultimately granular? To put it in terms of ordinary experience, is Nature a bucket of water or a bucket of sand? This is a dispute that has been argued since the time of the pre-Socratics. For a time, one answer seems to have won out, then the tide will turn and the other alternative gains ascendancy.

Among the ancient Greeks, Parmenides was the leader of the "smooth" or *continuous* viewpoint. He argued that existence was a unity. Change was a transformation of what already existed and no abrupt boundaries could exist between the things that are, because that would imply the existence of nothings, which was a logical absurdity. The "granular" or *discrete* viewpoint was represented in antiquity by Democritus, among others, who argued for an "atomic" theory of matter. Matter took the form of tiny indivisible (being the meaning of *a*-tomic) particles that existed in a great empty space. All different properties of matter were to be explained by the different form of these atoms, and the way they cohered to each other. In ancient times, the continuous viewpoint was dominant. Parmenides' dictum that "nothing" could not exist was convincing. The atomists represented a definite minority viewpoint.

Each viewpoint does some things better than the other. The discrete viewpoint explains change well. Everything is accounted for by the reconfiguration of the ultimate particles. Change of place and change of alignments are easy when there is empty space to move around in. The mechanist philosophy that became dominant in the 17th century with Descartes and then Newton sought to explain all phenomena of nature as mechanical motion of particles. Chemical change was to be explained as atoms of matter that joined together in different configurations and thereby produced what we took to be qualitative changes, all due to structural differences. Aspects of matter that we would experience with the senses, such as taste, color, softness or hardness, and odor would be accounted for by different arrangements of hard, indivisible matter.

The continuous viewpoint was better at explaining stability. And, it did not have that problem that was so troubling to the mechanists, action at a distance. How could forces such as gravity, electricity, and magnetism work over empty space? Not a problem with the continuous viewpoint because there was no empty space. Matter, from the continuous viewpoint, doesn't really occupy a definite space; it just looks like it does. It's just more concentrated in certain spots.

Near the end of the 19th century, science was wavering between the two views. The mechanist model was still strong and had great explanatory value. On the other hand, electromagnetism was best described by Maxwell's field equations, which were continuous functions based on the trigonometric sine function. The luminiferous æther was postulated in order to provide a medium for electromagnetic waves, but what it did was return to the continuous model that denied that there was any empty space. Physics was balanced between two basically incompatible models, each of which was better at explaining some aspects of nature.

Radiation

Into this morass, came evidence of some previously unrecognized forms of nature that didn't seem to fit either model all that well. These were in the first instance, forms of *radiation*. We tend to use the word radiation in a restrictive sense now, but, generally speaking, radiation is anything that is transmitted outward in all directions from some center. Sunlight is radiation; sound waves radiate; drop a stone in the middle of a still pool of water and a wave radiates outward. In the continuous viewpoint, radiations are waves; in the discrete viewpoint, they are rays.

Chapter Nineteen
Science Isn't So Sure

Cathode Rays

Physicists began studying new forms of radiation in the late 19th century, with interesting results. In the 1870s, British physicist William Crookes made use of the recently invented vacuum tube to help study electricity. He found that if he pumped an electric charge into a metal plate at one end of the tube, the *cathode*, it caused a glow down the tube at the other end where there was another metal plate, the *anode*. Moreover, he discovered that if he brought a magnet near the tube, it made the glow move. In fact, if he turned the magnet one way, the glow moved in one direction, but if he turned it in another way, bringing the other pole of the magnet forward, the glow moved in the opposite direction. Since the glow seemed to move outward in straight lines and did not appear affected by gravity, Crookes concluded that the glow must be an electromagnetic wave form, like light itself. But since the glow could be moved by magnets at right angles to it, he thought that the glow must be caused by charged particles, "atoms" of electricity. Note the flip-flopping between a continuous and a discrete perspective.

A German researcher working on the same phenomenon of the glow caused by an electric charge at the cathode called these *cathode rays*. The name stuck, and these devices are still called cathode ray tubes. Modern versions of them are around us all the time in the form of television picture tubes and computer displays. At about the same time, an Irish physicist, adopting the particle view, called these "atoms" of electricity *electrons*. We freely use both names, depending on context. A cathode ray is therefore (taking the discrete viewpoint) a stream of electrons.

X-Rays

In the last two decades of the 19th century, many experimental physicists focused their research on these mystery radiations. In 1895 the German physicist Wilhelm Röntgen, working with a cathode ray tube, was studying the luminescence that these rays caused when they struck certain chemicals. He wrapped the cathode ray tube in black cardboard, darkened the room, and turned on the electricity. He noticed that there was a faint flash of light from across the room well away from the cathode rays. What was there was a sheet of paper that had been coated with a chemical compound. When he turned off the electricity to the tube, the paper darkened. When he turned it on again, it glowed once more. When he took the paper into the next room and closed the door, it still glowed when the cathode ray tube was in action. Röntgen concluded that radiation was being emitted from the solid materials being struck by the cathode rays, which then released a new, unknown form of radiation that could penetrate layers of paper, wood, and even thin sheets of metal.

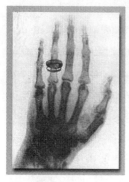

X-ray taken by Röntgen of his wife's hand wearing jewelry.

Officially these rays are called, in his honor, *Röntgen rays*, but more often they are called by the name that Röntgen himself gave them, *X-rays*, after the mathematical symbol x for unknown quantities. The property of X-rays of being able to penetrate soft material and leave a visible trace on photographic paper was quickly noticed by scientists in many fields. X-rays quickly became a research tool in the sciences and a diagnostic tool in medicine. It was also all the rage in public. People flocked to get "pictures" taken of their insides. Entrepreneurs set up stands on downtown streets with X-ray machines with strong emissions and no safety features. It was quite some time before science discovered that these miracle rays were dangerous.

Radioactivity

The very next year, 1896, a French physicist, Henri Becquerel made a startling discovery when he was investigating *fluorescence*. Flourescence is a property of certain materials wherein after being exposed to strong light (e.g., sunlight) for a time, they will continue to glow on their own for a certain period. It was a phenomenon that had been known for a time and had been investigated by a number of physicists earlier in the century, including Becquerel's father.

Becquerel, the younger, working with a compound of uranium that his father had also studied, exposed a sample to sunlight, then took it out of the light and noted that it emitted radiation that would penetrate black paper and then darken a photographic plate. He suspected that the flourescence was partly X-rays, and was trying to determine this precisely.

However, accidentally he discovered that the materials that he had been using for its fluorescent properties would also emit light even though they had not been in the sunlight at all. Not only did they radiate light, they were *radioactive*. The term *radioactivity* was coined by the Polish-French physicist Marie Curie in 1898. Madame Curie showed that all uranium compounds were radioactive, and moreover, the intensity of radioactivity was proportional to the amount of uranium in the compound. It was

Marie and Pierre Curie in a Paris laboratory, 1903.

not difficult to conclude that it must be the uranium itself that was giving off the strange radiation. Another element that exhibited the same properties was thorium. These were the heaviest elements known in the late 19th century. Madame Curie and her husband Pierre Curie together began a determined search for radioactivity wherever they could find it. They found it in two previously unknown elements, which they named *polonium*, after Poland, Mme. Curie's native country, and *radium*, because of its intense radioactivity.

Atoms are not Atomic

Back in England, physicist J. J. Thomson, working at the Cavendish Laboratories at Cambridge University, tried to measure the effects in cathode ray tubes. He found that cathode rays could be generated using any element whatsoever. Since they behaved like a stream of particles, he believed that those particles had to come from

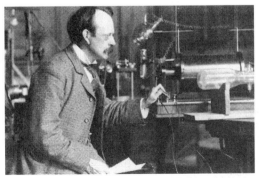

J. J. Thomson.

somewhere—since, of course, he believed in the conservation of matter. They came, he believed, out of the atoms of the materials he was using to generate them. Taking the "discrete" name for cathode rays, *electrons*, he concluded that those

Thomson's watermelon-seed model distributes electrons throughout the body of the atom.

electrons must somehow be imbedded in, and then spit out of, atoms. If so, the *atom*, the indivisible ultimate particle of matter, was not so indivisible. It had parts and it could be made to throw off those parts. Thomson's model envisaged an atom as being a hard sphere in which were imbedded much smaller electrons, spread throughout the atom, like seeds in a watermelon.

The Rutherford Experiment

Thomson had a student named Ernest Rutherford, originally from New Zealand. He studied with Thomson at Cambridge, then taught for a few years in Canada at McGill University before returning to England to head up a laboratory of physical research at the University of Manchester. Rutherford had been studying the curious radioactivity observed by Becquerel and the Curies. He found that there were actually three different emissions. With a magnet, one could deflect part of the radiation in one direction, part in another, and part could not be deflected. He decided that these were different entities and gave them the names of the three first letters of the Greek alphabet, *alpha*, *beta*, and *gamma* rays. Since each could be sent in different directions (or not) by a magnet, he could study each one separately. He found that the first two rays showed the characteristics of particles, and alpha particles were considerably more massive than beta particles. Gamma rays appeared to be simply electromagnetic radiation, rather like X-rays.

Ernest Rutherford.

Taking alpha particles as a tool, he decided to use them to see what he could find out about the atom itself. If alpha rays were streams of fairly heavy particles, he could aim them at a thin sheet of metal, say, gold foil, with a fluorescent screen behind the foil and see what showed up on the screen. It was rather like trying to figure out the shape of something from the shadow it cast on a distant wall. Rutherford expected that the alpha particles would all penetrate the atoms and he would get some outline showing up on the screen. Indeed, most of the alpha particles did just that, but to his amazement, some of the particles bounced off something and were deflected back in the direction they came from. He is said to have remarked that it was like shooting a cannon at a piece of tissue paper and seeing the cannon shell come back.

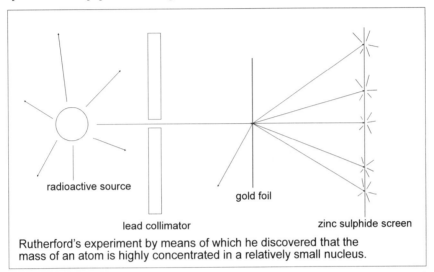

Rutherford's experiment by means of which he discovered that the mass of an atom is highly concentrated in a relatively small nucleus.

The implication of this, following Newton's third law, is that the alpha particles must have hit something as massive or larger than themselves. That must mean that the atom was not some uniformly distributed unit of mass, but mostly empty space, with nearly all the mass concentrated in very small spots. Rutherford reasoned that the electrons that Thomson had thought imbedded in the atom must be out in that empty space. He therefore proposed a model of the atom that was rather like a solar system, with the electrons circling around a nucleus like planets around the sun. Rutherford's conception is still the most popular image of the atom, and is familiar to anyone who has seen an icon representing atomic energy.

The Ultraviolet Catastrophe

While this work on sorting out the structure of the atom was going on, radiant energy was getting a lot of attention from those trying to work out all the interrelations of electromagnetism and thermodynamics. Consider a piece of metal, an iron rod, for example. As it is heated, it changes color. There are a series of colors that bodies move through as they change temperatures, and "red hot" is certainly hot enough to give a severe burn, but it is relatively cool compared to the hotter alternatives. Scientists wondered why. Why do bodies emit light of different colors, or as they had come to think of that, different frequencies, when their energy content, in the form of heat is raised? This was going to be a very difficult problem to study because there were so many complicating factors. But, experimental science since Galileo has learned that to study a feature of nature effectively, it must be isolated as much as possible.

Black Body Radiation

In this case, it is the relationship between absorbing energy (e.g., heat) and emitting energy (e.g., radiant light) that is the problem to be separated from all else. In 1859, the German physicist Gustav Kirchhoff, founder of the science of spectroscopy, came up with the idea of a *black body*. A black body, to simplify the concept slightly, is one that does not reflect light. Shine a light on any ordinary body and it will reflect some light, enabling you to see it. A white body reflects a lot of light. The darker a body is, the more it absorbs light and the harder it is to see. For physics, a "black body" is the idealized end of the road on this, a body that absorbs all the light and other forms of energy put to it. This is a theoretical concept. There are no real bodies in nature with this property. But that is what the physicists would like to have to study the phenomenon.

Now, while a true "black body" does not exist, physicists set out to make a piece of laboratory equipment that would get as close to the ideal as possible. When they had done so, a device called a cavity radiator, they began to take measurements on the relation between energy in and energy out. A black body (the ideal one) would not reflect light, but if it absorbed energy in any form, it would begin to radiate energy back, by beginning to glow.

Is the First Law of Thermodynamics Wrong?

While the experimentalists were measuring how the cavity radiator was behaving, the theoreticians calculating how the ideal black body would behave. The wave equations of Maxwell indicated that bodies would emit radiation in all possible wavelengths. But the shorter the wavelength, the more energy should be emitted, and in the case of very short wave lengths—those in the ultraviolet range—the amount of energy that should be emitted is *infinite*.

Since the first law of thermodynamics stated that the amount of energy in any closed system is a constant, this was clearly in violation of this notion. What was wrong? The whole fundamental basis of thermodynamics or the all-encompassing electromagnetic analysis of Maxwell? If there were basic faults in either of these, all of 19th century physics could come crashing down. This was the *ultraviolet catastrophe*. The catastrophe was what was in store for physics if this could not be solved.

The Quantum of Energy

Meanwhile, back in the laboratory, a different set of numbers was emerging. Work with the cavity radiator indicated that the amount of energy radiated did increase as the frequency of light rose, but then at some point it reached a maximum and tailed off. Just where the theoretical calculations showed infinite amounts of energy emitted, the measured results showed the radiated energy falling off to zero. Theory and practice did not agree.

Max Planck.

Among those puzzling over these anomalies was the German physicist Max Planck. Planck agonized over the calculations for months and could not find any way out of the contradictory results. The conceptual models that physicists were working with were curious mixtures of the continuous and the discrete. Matter was assumed to exist in discrete atomic form, but energy had been analyzed with continuous functions and was thought to radiate in waves. In the fall of 1900, Planck tried something totally unjustified, which he later described as an act of desperation. He began thinking of energy as existing in discrete bundles. This provided a threshold below which there was no energy released. Planck calculated this threshold on the basis of the experimental data and put it into his calculations. That did the trick. The calculations made the theoretical result match the experimental measurements.

But what did it mean? Planck calculated a number, which he called the *quantum of action* and gave it the symbol *h*. The number is unimaginably tiny, 6.63 x 10^{-27} erg/sec. What it implied was that energy was not some continuous fluid, but a discrete collection of tiny grains, so tiny that we cannot perceive their granularity in any measurements. From the point of view of observable results, energy is a continuum, except for this solution to the ultraviolet catastrophe. Was this yet another desperate Ad Hoc assumption that allowed one to push on with an unsatisfactory theory or was this the first indication of something real? Planck himself leaned toward the unsatisfactory trick point of view. What was needed was some other natural phenomenon that was explained by these *quanta*.

The Photoelectric Effect

Planck's brainstorm of the quantum was in 1900. For the next five years, no one knew what to make of it. Then Einstein, working at the Bern patent office, thought about it and decided to treat *h* as something real. This was a characteristic positivist move. Planck had detected that this tiny number had significance and made theory match the experiment, so why not treat it as a feature of nature and see if it can explain anything else. Einstein had been thinking about another problem that mystified science toward the end of the 19th century, called the *photoelectric effect*. It had been noted that some metals produce an electric current when a light is shining on them. Physicists were coming around to the idea that electricity was carried by these charged particles that they had called electrons, but those particles were supposed to be bound up in the atoms of the metal. (Remember the Thomson model of the atom.) How could shining a weak electromagnetic wave onto a firmly constructed piece of metal dislodge electrons and make an electric current?

Einstein's explanation was that light itself is not just emitted in streams of energy; it exists in bursts, in packets of a definite discrete size. These packets are sizeable enough that when they strike the electrons in atoms they may have enough kinetic energy to send them flying. He worked out equations to track this process and showed that it accounted for the observed phenomenon well. Planck's quantum was thereby given a physical form and shown to be useful for explaining something new. It therefore entered the catalogue of physical concepts that were considered to be reasonable descriptions of nature.

Interestingly, it was this work on the photoelectric effect that was cited by the Nobel committee as the reason for giving Einstein the Nobel prize in physics in 1921, not his work in special or general relativity.

Light: Discrete and Continuous

But what about light? Einstein had proposed that light travels not in some continuous wave, but in packets of energy, which he called *light quanta*. Later the term for this was changed to *photons*. These are discrete bits, particles, "atoms" of light. That would imply that light was a stream of particles after all, as Newton had maintained, though a hundred years of wave theory had seemed to establish its wave nature again and again.

The answer to this dilemma is a clue to the solution to the discrete-vs.-continuous controversy. Light has wave properties and it has particle properties. Sometimes it helps us to think of it as a wave; sometimes we are better off thinking of it as particles. The important point to remember is that particles and waves are just metaphors expressed in terms of familiar common-sense notions that we can relate to. We sometimes call the mechanical model of the world the "billiard ball" universe. We can imagine billiard balls racing around a billiard table and crashing into each other. By analogy we can imagine what a particle universe would be like. Similarly we are familiar with waves on the surface of water, we can see how they move up and down as they spread outward—the defining characteristic of transverse waves. From that we can imagine the wave propagation of light, wavelengths, wave frequencies, and so on. But these are analogies. Light shows wave-like and particle-like features. Depending on what phenomenon we are trying to understand, we may be better off with one or the other metaphor.

And thus it is in general with the discrete and the continuous view of the world. Sometimes one model serves better than the other, but both are really just analogies and neither is sufficient on its own. This is hard for us to swallow. It means that ultimately the universe is more than we can imagine with common-sense notions.

Part III
From Certainty to Uncertainty

The Bohr Atom

Niels Bohr.

The Danish physicist Niels Bohr had been studying with Rutherford in Manchester trying to further Rutherford's understanding of the structure of the atom. Rutherford's model had made the atom a "solar system" with electrons circling a nucleus. Bohr was struck by an issue parallel to that which captivated Kepler about the real solar system: where are the electrons located and what explains their positions around the nucleus?

Electrons had been found to have a very small, but nevertheless measurable, mass, and the nucleus of course had mass. Therefore, Newton's law of universal gravitation applied to them. The force of gravity was proportional to the product of the masses, which would be small, divided by the square of the distance between them. Since the distance would be incredibly small, the force of gravity would be large. How, then, was it possible for the electrons to orbit the nucleus without spiraling into it immediately? Moreover, according to electromagnetic theory, if the electrons did crash into the nucleus, they would radiate energy constantly, not just from time to time as had been observed.

Bohr came up with this ingenious idea: the electrons orbited the nucleus in fixed orbits that were spaced a determined and unalterable distance apart. (*Not*, as in Kepler, separated by the five Platonic solids!) Electrons could, however, hop from one orbit to another, and when they did, a certain quantity of energy was either absorbed or emitted. The emitted energy would be detected as light of a particular wavelength, which would vary from element to element.

How are these electron orbits determined? The key to understanding that is the idea that each orbit represented a level of energy. For an electron to jump out to farther orbits or fall in to closer ones meant gaining or losing a certain amount of energy, and as Planck's constant, h, demonstrated, that energy came in indivisible units. The orbits were spaced by an amount determined by the quantum, h.

Chapter Nineteen | 349
Science Isn't So Sure

This was yet another use for the quantum and helped to establish its existence as something more than a mathematical trick, but after all, this was merely an imagined model of the atom. Only when this model was shown to be able to account for unexplained phenomena would it be taken seriously and further support the quantum theory.

The Spectral Lines of Hydrogen

The first phenomena that the Bohr model helped account for was the characteristic colors produced by different elements. This was the subject matter of spectroscopy, that field founded by Kirchhoff in the mid-19th century. Spectroscopists had found that each element produced a characteristic color when heated, and that color was very precise, an exact frequency of light. Sodium, for example, produces a yellow color when burned in a bunsen burner. (Robert Bunsen, indeed, was one of the early researchers in spectroscopy.) But elements can produce more than one color, depending on how hot they are. This is why a rod of metal glows differently when heated to different temperatures.("White hot" is hotter than "red hot," etc.)

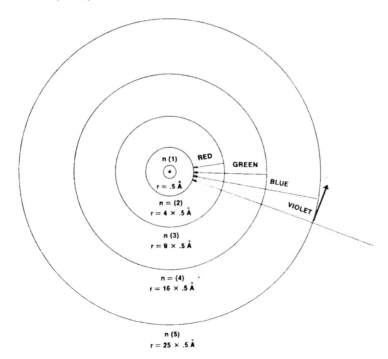

The spectral lines of hydrogen.

The simplest element, hydrogen, had been shown to exhibit several spectral lines, which had been identified and analyzed in the late 19th century and found to be predicted by a simple formula. However, there were no known causes of these spectral lines. Bohr found that the spacing of the fixed orbits for electrons in his model nicely corresponded with the spectral line spacing of hydrogen. Each color emitted by hydrogen atoms exactly fitted what he had calculated as the energy level of different possible orbits for the single electron of a hydrogen atom.

The Periodic Table Explained

Bohr published his findings in 1913 and started a number of people on a search for other properties of matter that could be explained with this model. The most dramatic confirmation of the model came with its explanation of the periodicity of the Periodic Table of Elements.

Chemistry had come a long way since the Scientific Revolution. Most of the elements had been identified and their basic properties identified. The big breakthrough of the mid-19th century was the discovery of natural groupings of the elements. The Periodic Table, a nearly simultaneous independent discovery by Dmitri Mendeleev and Julius Meyer, arranged all the known elements in a rectangular chart where the elements of one column all combined with the elements in another column in similar ways. Gaps in the table represented elements that had not yet been discovered. Just on the basis of the pattern established in the Periodic Table, Mendeleev was able to predict what the physical properties of these unknown elements were and what they would combine with. When elements were discovered that fit these detailed descriptions, it was clear that the Periodic Table had detected some feature of the structure of the elements, but what that was remained a mystery.

Bohr's model revealed the reason for the pattern. Bohr thought of the electron orbits as concentric shells. (This is more like Eudoxus and Aristotle than like Copernicus and Kepler!) Each "shell" had room for a certain number of electrons. The innermost shell could accommodate 2 electrons; the next could take up to 8; the third could fit 18; the fourth, 32; etc. But each element had a fixed number of electrons, and they were arranged in their shells according to specifiable rules. Suddenly the pattern in the periodicity became clear. Atoms formed compounds by exchanging and binding with electrons from other atoms. The "face" that an atom presented to another atom was its outermost shell. Elements that had the same number of electrons in their outermost shells would tend to form compounds in similar ways with other elements. For example, take elements from the group now known as 1A, the alkali metals, Lithium, Sodium, and Potassium. Each of these forms compounds with the elements in the group known as 7A, the halogens, Fluorine, Chlorine, Bromine, among others. Each of the elements in group 1A has a single electron in their outer shells; each of the elements in group 7A has seven electrons in their outer shells. They all have a strong tendency to form compounds with each other—one element from each group. The single electron from 1A elements forms a shared pool with the seven electrons of the 7A group, making a "shell" of eight electrons, which is very stable. All of the compounds formed from one atom of each group are salts, the most common being sodium chloride, table salt. And the other combinations are possible, such as sodium fluoride and potassium chloride. All of these have the same feature of forming a stable outer shared shell of eight electrons.

The Bohr model has been very useful in accounting for known phenomena and finding the keys to unknown phenomena in physics and chemistry. It is still very useful in explaining chemical structure in an elementary way. But as a true description of nature, it had many faults. It persisted with the solar system idea of an atom, portraying electrons as tiny hard balls racing around a slightly less tiny hard nucleus. It mixed notions from classical physics with the emerging notions of quantum physics. But it was a model that could be pictured, even if it left as many questions unanswered as it did answer.

Matter Waves

Louis-Victor-Pierre-Raymond prince de Broglie was a French nobleman, son of Victor, Duc de Broglie, and brother of Maurice de Broglie, seventeen years older, who was a respected physicist, one of the scientific secretaries at the first Solvay conference on physics in Brussels attended by almost all important European physicists of the time. Louis grew up in the ancestral manor house and was sent to the University of Paris to study the classics and history, with an intended career in the civil service, taking a degree in 1910. When his brother returned from the Solvay conference, he gave Louis a full report of the proceedings, especially the discussions about Einstein's theory of the particle nature of light. Louis was so fascinated that he decided to switch careers and went back to the university to study physics, obtaining a degree in 1913.

Louis de Broglie.

Then World War I broke out. De Broglie was conscripted and posted to the wireless section of the army, stationed at the Eiffel Tower. As soon as the war was over, de Broglie resumed his studies, his research centering on the Einstein photon theory and how it could possibly explain wave phenomena. In time, de Broglie decided that light must be both a wave and a particle at the same time, a notion that seemed like nonsense at the time. In 1924 he submitted a doctoral thesis to the University of Paris which put forth this idea, and the more general principle that matter does have a wave form.

The university was not impressed. Since de Broglie's work was in the area for which Einstein had recently received the Nobel Prize, someone at the university thought they could put an end to this assault on science by asking Einstein to be on the examining committee for de Broglie's thesis. They wrote to Einstein describing the thesis. He was interested and agreed to read it. They sent it to him. To the amazement of the university, Einstein wrote back that this was really important ground-breaking work. De Broglie got his doctorate, and five years later, in 1929, he was awarded the Nobel Prize for his work.

The essence of de Broglie's thesis is that if waves (e.g., light) can behave like particles (e.g., photons), then maybe particles can behave like waves. Take electrons for example. De Broglie proposed that electrons are waves of matter. Bohr had worked out how many electrons could fit in each shell, but he had no compelling reason why the number should be what it was. Why did two electrons fill in the first shell and eight electrons fill in the second?

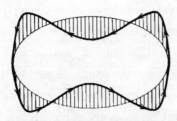
Bohr's first shell with two electrons shown as de Broglie matter waves.

De Broglie had a reason: the number was fixed at what it was because electrons were waves of a given length and they had to fit around the circumference of the shell.

Einstein did more than help de Broglie get his Ph.D.; Einstein's interest in de Broglie's ideas led him to mention it to others, who began to do experimental work to try to discover direct evidence for matter waves. Einstein also mentioned de Broglie's idea in a published paper in 1925 that was read by the Austrian physicist Erwin Schrödinger, then Professor of physics at the University of Zürich. Schrödinger was a firm believer in the adequacy of wave mechanics to account for all physical phenomena, and he saw in de Broglie's matter waves the possibility of restoring physics to a *continuous* viewpoint, somehow incorporating the apparent discreteness of quantum explanations in a wave model. In 1926, Schrödinger published a general theory of matter waves that put all of reality back in a continuum with no hard boundaries.

Erwin Schrödinger.

The Uncertainty Principle

Schrödinger intended his wave equations to be a completely deterministic description of subatomic nature of the form that had served science well since it began formulating natural laws. But the mathematical structure of his wave equations admitted of another interpretation. Maxwell's wave equations for electromagnetism used regular sine wave functions. Schrödinger's wave mechanics made use of the bell-shaped curves of probability functions. It was therefore possible to view his equations for matter waves of electrons as a mathematical statement of the *probability* that a particle electron was located at a particular place at a given time. If so, then his equations are not complete descriptions of nature, but only the best we can do given what we know. This was not what Schrödinger had intended. In this interpretation, Schrödinger's equations cannot tell us where a single particle in the universe is located. Any electron *could* be anywhere at all; it's just that its probability of being in a particular very small space is very, very high. But not certain.

354 | Part III
From Certainty to Uncertainty

Werner Heisenberg.

Around the same time that Schrödinger was working out his wave equations for quantum mechanics, another very different approach was being pursued by a brash young German physicist named Werner Heisenberg. Heisenberg, preferring a discrete view of matter, imagined the electron to be a bit of matter with a specific location and momentum—a classical mechanical conception. Given what was known about how the electron was related to the nucleus of an atom, Heisenberg sought ways to determine its position and momentum with accuracy. The way to do this would be to bounce light off of the electron and interpret the resulting energy emissions from the electron (loosely speaking, the "reflection") to determine position and momentum.

Alas, there was a quandary. Position and momentum cannot be determined at the same time, because to get a good fix on position, you have to use enough light energy to disturb the momentum, and to get a fix on momentum you lose the accuracy of position. In fact Heisenberg worked out just how accurate a picture you could get of the position and momentum of an electron and found that there is a range of indeterminacy that just cannot be reduced. That range is defined by Plank's constant (actually by $h/2\pi$). The process of observing alters the position of what is observed. At the subatomic level, this, according to Heisenberg, is unavoidable.

What does the Uncertainty Principle mean? Does it mean simply that we do not have the means to get close enough to nature to figure it out and we have to make do with incomplete knowledge? This way of thinking is dubbed an *ignorance interpretation*. It does not suggest that the world lacks deterministic laws, only that we can't know them. This interpretation would fit well with the viewpoint expressed by Pierre Laplace in the 18th century when he said that if he knew the position and momentum of every particle in the universe, and the laws that governed their motions, he could predict the future with absolute certainty. He was not suggesting that there was any practical possibility of having all this information, but he was asserting that it did exist and the universe was proceeding with certainty down this utterly determined path.

Heisenberg and Bohr spent some time together trying to sort out what it all meant and came up with a different view, often called the *Copenhagen interpretation*. In this view, there is no deterministic description of nature to be had because the wave (*continuous*) and the particle (*discrete*) descriptions of nature each measure different things. Quantum mechanics as a set of principles is a complete description of nature, but not one with a definite outcome until we observe it and choose an outcome.

If this sounds to you like these physicists are going around in circles, it would not be the first time that conclusion was reached. One of Heisenberg's attempts to explain this understanding of nature to a group of scientists was reported in the New York Times as follows:

> LEEDS, England, Sept. 1. -- Of thirty addresses delivered today before the various sections of the British Association for the Advancement of Science, one of the most important was that of a young German, Dr. W. Heisenberg. Fully 200 mathematical physicists listened to his brief exposition of a conception which will make it necessary to modify belief in what we are pleased to call "common sense" and "reality."
>
> The layman without knowledge of higher mathematics, listening to Dr. Heisenberg and those who discussed his conclusions, would have decided that this particular section of the British Association is composed of quiet and polite but determined lunatics, who have created a wholly illusory mathematical world of their own...[1]

Schrödinger's Cat

The idea that the world has a built-in randomness to it was highly offensive to many scientists who were raised on the firm belief that though we may never understand it, everything in nature is governed by fixed laws with fixed outcomes. Einstein could not stomach the Copenhagen interpretation and retorted, "The Lord God does not play dice." Einstein and others tried to imagine situations, in effect, thought experiments, that would show that the Copenhagen interpretation was wrong.

The difficulties of the interpretation were put most colorfully by Erwin Schrödinger who also found them unacceptable. Quantum mechanics concerns itself with events at the subatomic level, where we cannot observe them directly anyway. The Copenhagen interpretation asserts that reality is described fully by the mathematical probability function that spreads out the chances of incompatible events across a spectrum of possibilities.

---

1. *The New York Times,* September 2, 1927.

To Schrödinger, this made no sense. Either an event happened or it did not. It did not depend on an observer looking to see. He illustrated this with his famous *Cat Paradox*.

First, consider a radioactive substance that has a known rate of decay. That means that, following quantum mechanical principles, the probability of any atom of the substance decaying and releasing a radioactive particle can be specified, and therefore the probability of a fixed quantity of the substance decaying in a given time can also be specified. Note that it cannot be determined whether a given atom will or will not decay, only what the chance is that it will do so in a certain period.

Now, Schrödinger says, take a small sample of this substance, the amount chosen so that the probability that some of it will decay and release a radioactive particle within the next hour is exactly fifty-fifty. Put this sample in a container next to a Geiger counter. If an atom of the sample decays, the Geiger counter will register it and that will cause a latch to be released which will allow a hammer to drop upon a small flask of a poison, breaking the flask. (He suggests hydrocyanic acid.) All of this is set to happen inside a closed steel chamber in which you have also placed one live cat, who is prevented from interfering with this diabolical mechanism. If the flask of poison is broken, it will release a gas that will kill the cat. But because the chamber is sealed shut, no one will know whether the atom has decayed until the hour has passed and they open the chamber to see. At that point there will either be a live cat or a dead cat, with 50% probability of each.

But what about before the chamber is opened? According to quantum mechanics, the decay of the atoms of the sample is completely described by the probability function. In other words, there is a sense in which both events occur with equal probability. Schrödinger wanted to emphasize that even if we are willing to accept subatomic events as having mixed states, if those events are tied to events in the macro world, it is absurd to say that they don't happen until we look. The cat would have to be both alive and dead—half of each—until we opened the chamber and forced one of the events to happen by looking at it.

Chapter Nineteen
Science Isn't So Sure

The Many Universes Interpretation

It is not easy to find a way out of this logical minefield without setting off some other unforeseen bombshell. There is however, one interpretation of all this that seems to violate no logical principles. This idea was put forward by a graduate student at Princeton University, Hugh Everett, in 1957 as his Ph.D. thesis. In this he was encouraged by John Wheeler, his supervisor, who was one of the most important physicists of the mid-20th century.

Everett's interpretation has a very complex mathematical formulation, but the basic idea is simple enough. He says that whatever can happen does happen. If there is a chance that a radioactive atom will decay or not, it will do both. If there is more than one choice for the momentum and position of an electron, all the possibilities happen. The situation of the cat that is dead and alive at the same time is easily resolved. There is a cat that is dead and a cat that lives. This is all possi-

Hugh Everett's many universes interpretation.

ble because, according to Everett, whenever there is more than one possible outcome of anything whatsoever, the universe splits into separate pieces and the possibilities each happen in different worlds. There is a world in which you just read that last sentence and another world in which you decided enough was enough and gave up before you got to it. The universe is constantly splitting into more and more complete daughter worlds that go off into their separate ways.

Bizarre as this interpretation seems, it has appealing logical consistency. As for its violation of all precepts of common sense, Everett's supervisor, John Wheeler compared the feelings of dislocation that the many universes interpretation gives us to similar jolts experienced from scientific revolutions of the past:

> One's initial unhappiness at this step can be matched but a few times in history: when Newton described gravity by anything so preposterous as action at a distance; when Maxwell described anything as natural as action at a distance in terms as unnatural as field theory; when Einstein denied a privileged character to any coordinate system...[2]

We may not be able to accept the many universes interpretation, or any of the interpretations that try to make sense of quantum mechanics for us, but while we are trying to grasp the implications of the theory, it is serving us quite well, whether we understand it or not, in guiding the design of nuclear power stations, computer chips, lasers, and all of modern chemistry. It has helped us understand more of nature, and made us aware of the greater mystery of it all.

For More Information

Alioto, Anthony M. *A History of Western Science,* 2nd ed. Englewood Cliffs, NJ: Prentice-Hall, 1987. Chapter 23.

Asimov, Isaac. *The History of Physics.* New York: Walker & Co., 1984. Chapters 21-22, & 31.

Gribbin, John. *In Search of Schrödinger's Cat: Quantum Physics and Reality.* New York: Bantam Books, 1984.

MacLachlan, James. *Children of Prometheus: A History of Science and Technology,* 2nd ed. Toronto: Wall & Emerson, Inc., 2002. Chapter 19.

Motz, Lloyd, and Jefferson Hane Weaver. *The Story of Physics.* New York: Avon, 1989. Chapters 15.

Newman, James R. "Einstein." In Byron E. Wall, ed. *The Nature of Science: Classical and Contemporary Readings.* Toronto: Wall & Emerson, 1989.

Ronan, Colin A. *Science: Its History and Development Among the World's Cultures.* New York: Facts on File, 1985. Chapter 10.

Speyer, Edward. *Six Roads from Newton: Great Discoveries in Physics.* New York: John Wiley, 1994. Chapters 5 & 7.

Spielberg, Nathan, and Bryon D. Anderson. *Seven Ideas that Shook the Universe,* 2nd ed. New York: Wiley, 1995. Chapter 6.

---

2. John Wheeler in *Reviews of Modern Physics,* Vol. 29 (1957), p. 464.

Chapter Nineteen | 359
Science Isn't So Sure

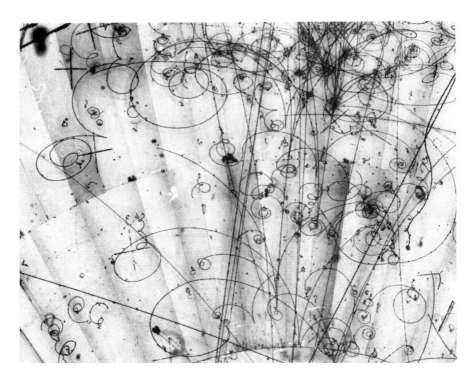

The trails of charged particles in a bubble chamber, revealing the existence of the positron, the antimatter particle corresponding to the electron. Such trails can be used to calculate the mass of the antimatter particles. From the Fermi National Accelerator Laboratory, Batavia, Illinois, 1998.

Edwin Hubble beside the 48-inch telescope at Mt. Palomar.

CHAPTER TWENTY

The Universe Will Go On Forever—Or It Won't

Meanwhile, what about the universe? How big is it? What shape is it? How old is it? Will it come to an end someday? How were physics and astronomy answering these questions?

The ancients, you will recall, generally thought of the universe as a large sphere with the Earth in the middle, the stars all attached to the outside and the planets somewhere between. The planets were the Moon, the Sun, Mercury, Venus, Mars, Jupiter, and Saturn, the counting beginning with the closest to the Earth. Each planet had an "orb," which was an invisible spherical shell concentric with the Earth. These orbs filled the spaces between the Earth and the orb of the stars. Saturn, the farthest known planet, was sometimes thought of as being pretty near the stars, but later this was revised to the view that after the orb of Saturn there was about as much distance from it to the stars as there was from the Earth to Saturn. The earliest philosophers thought of the Earth and the Heavens as about on an equal scale, but the astronomers of late antiquity, such as Ptolemy realized that the Earth must be very small in relation to the Heavens.

Stellar Parallax

Copernicus revised this picture by putting the Sun in the center, making the Moon a satellite of the Earth, and making the Earth one of the planets, all of which now orbited the Sun instead of the Earth. And he changed very little else. The universe was still a sphere; the stars were still stuck on the inside of the outer edge of the universe (or near to it), though they no longer spun around every day, and the planets still occupied spaces defined by "orbs."

The one major change with Copernicus' theory was its size. This is an implication of the pathetic rejoinder that Copernicus came up with to parry the objection about stellar parallax. Since the Earth was in motion around the Sun, it would be getting closer to stars on one side of the heavens at one time of year than it would be six months later. Therefore the visual angle between two stars should change, getting larger as the Earth got closer to them. This change of visual angle is parallax, and the problem for Copernicus was that though it was implied by his theory, it was not visible. The reason why it cannot be seen, said Copernicus, was that the entire orb of the Earth around the Sun is but a point in relation to the size of the heavens.

That made stellar parallax impossible to detect, for all practical purposes. For Copernicus, that was the end of the matter. But this was before the invention of the telescope. With telescopes much greater detail could be seen and smaller differences in angles could be detected. Astronomers began looking again for parallax. It was 300 years before it was detected.

The German astronomer Friedrich Bessel finally detected stellar parallax after four decades of patient looking for slight angular irregularities. In 1838, he found that the star with the totally unmemorable catalog name 61 Cygnus A had a parallatic angle of 0.2 arcseconds. (An arcsecond is actually a unit in the Babylonian sexagesimal number system. Just as a second is 1/60 of a degree, an arcsecond is 1/60 of a second.) This was a tiny shift indeed. No wonder it could not have been detected by the naked eye in Copernicus day.

With the parallatic angle of a star, one could use triangulation and get a figure for the distance of the star. When Bessel did this he found that 61 Cygnus A had to be over 100,000 times more distant than Saturn. The universe now had a size, and it was far larger than anyone had imagined. Bessel had found a way to measure the distances to the stars. He was able to measure the parallatic angle of 61 Cygnus A because it was just large enough for him to detect it. That implied that it had to be one of the nearer stars. Anything farther out would have still smaller parallatic angles.

With the invention of photography in the mid 19th century, parallatic angles could be determined with much greater accuracy. But even so, there were limitations with this method, not the least of which is the distortion produced by having to look through the atmosphere which turns every point of light into a disk, and a twinkling one at that.

Chapter Twenty | 363
The Universe Will Go On Forever—Or It Won't

Stellar parallax was an effective method for locating star distances, and really the only method in the last half of the 19th century. But by then it had reached its limit as a tool, finding stars up to 30 times more distant than 61 Cygnus A.

Cepheid Variables

Photography made it possible for astronomers to lay images of the stars taken at different times side by side and look for minute differences that would have escaped notice if all that was recorded from an observation was what the astronomer noted down. One such phenomenon was the *variable star*. A variable star is one that changes in its brightness over time. These were interesting to astronomers because they indicated that a process was going on in the star that perhaps could be analyzed, and from that conclusions could be drawn about the size, the mass, and the composition of the stars.

One particular kind of variable star was of special interest for those trying to determine distances. These were the *Cepheid variables*, named after the constellation Cepheus, where the first such star, Delta Cephei, was found. A Cepheid variable is a star that varies greatly in brightness on a regular cycle, generally five to ten days, or ten to thirty days. The astrophysical understanding of why these stars have these varying cycles is that there is some sort of constant tug of war between the energy of the light radiating outward and the gravitational pull of the stars. For any star to have a gravitational pull sufficient to draw light back, it must be massive indeed. The pressure of light would push the star outward in every direction, but then, just as on Earth, gravity would overtake it, drawing the outer edge of the star back. Then the light would build up again and the process would repeat. The result is that from a great distance, the star would become brighter and then dimmer in a predictable cycle.

The amount of time between these pulsations is a measure of the strength of the light pressure, which itself is a measure of the *absolute brightness* of the star. The absolute brightness is just how much light is being produced. This has to be related to the *relative brightness*, which is how bright the star appears to be from where you see it. As Newton knew well, the intensity of light decreases with the square of the distance. This can all be put into a formula with the three variables, absolute brightness, relative brightness, and distance. Know any two, and you can calculate the other. (Actually the formula uses the *magnitude* of the brightness, which is a logarithmic measure. The calculation is different, but the principle is the same.) Once astronomers (astrophysicists, really) worked out the absolute brightness of a Cepheid variable from the length of its pulsation period and took a reading of the relative brightness off their photographic plates, they could figure the distance.

In 1912, the American astronomer Henrietta Leavitt, who discovered the relationship between the period of pulsation and the magnitude of the star, used this as a tool to determine the distance to Cepheid variables that appeared in different parts of the sky. She traveled to the southern hemisphere and studied the Large and Small Magellanic clouds, so named because they were discovered by the explorer Magellan on his voyages around the world. These "clouds" were nebulae, that is, clusters of huge numbers of stars that appear as a faint lightness in the sky to the naked eye. Leavitt found that these nebulae contained many Cepheid variables with very faint magnitudes. Applying the formula she had worked out, she calculated that these "clouds" must be at least a thousand times more distant than 61 Cygnus A.

Given the apparent size of the "clouds," astronomers could infer that they were billions of kilometers across and contained millions of stars. Hence, the universe was seen to contain not just a vast number of stars but also large conglomerates of stars that appeared to be very far away indeed.

Is the Universe Finite or Infinite?

Copernicus, like the ancients, thought of the universe as a closed, finite sphere. So did Tycho Brahe, Kepler, Galileo, and Descartes. But if the stars were no longer required to move in a complete circle every day and whatever motion they appeared to have was shown to be an optical illusion due to the Earth's rotation, there was no special reason to think of the universe as having any particular shape, nor any particular size. Even though Newton believed there was such a thing as absolute space and absolute motion in it, he realized that there was no way to determine one's place in space, and therefore no real reason to assume that we know where we are in the universe. It could be that the universe goes on forever.

Chapter Twenty
The Universe Will Go On Forever—Or It Won't

General thinking after Newton tended to favor the infinite universe idea. It certainly disposed of the awkward problem of what could be on the other side of the edge of the universe. But an infinite universe had its own problems.

Olbers' Paradox

In 1826, the Swiss astronomer Heinrich Olbers posed a question of disarming simplicity and profound implications. He asked: why does it get dark at night? Here's the logic behind the question. If the universe is infinitely large and infinitely old, then every possible line of sight in the sky would sooner or later run into a star. There would be no place in the sky at all that was not lit by stars, in fact, by an infinite number of stars.

Remember the syllogism *modus tollens*. Let H be the hypothesis that the universe is infinitely large and infinitely old (and the relative configuration of stars more or less the same throughout). The test case, T, is that the sky is bright at night. H implies T. But, as anyone who goes outdoors at night can determine, T is false. Therefore, according to *modus tollens*, H is false. This is the most powerful and compelling logic that exists in science. So, until equally strong evidence to the contrary is presented, the presumption must be that the universe is finite, or at the very least, has a finite number of stars.

The Milky Way Universe

From the earliest days of studying the sky at night, a prominent whitish smear in the heavens had been noted and, had been labeled. Since the sky was seen as the province of gods and goddesses, whatever was visible in the sky over a long period of time was interpreted to be some trace of the divine. Recognizable patterns of stars were given names suggesting that they were just the bright points of light that reflected off some gods going about their business. The hunter Orion, the chained lady Andromeda, the swan Cygnus, and so on. Ptolemy listed 48 such "constellations" of stars, and for the most part, the mythological names have stuck. There was also a bright smear in the sky that did not look like stars so much as spilled milk. It was given the name *Milky Way*. The word for milk in ancient Greek was γαλα ("gala"). The adjective "milky" was γαλακτινος ("galactinos"). From this we get the word, *galaxy*. The point of this is that, originally, the terms Milky Way and galaxy meant exactly the same thing. There was only one, and it referred to that whitish spot in the sky at night. The notion that our solar system is part of the Milky Way came later, when we could understand that if we were on the edge of something very large and relatively flat, it could look like it was far away and had nothing to do with us.

At about the same time that Henrietta Leavitt was using Cepheid variables to find the distance to the Magellanic clouds, another American astronomer, Harlow Shapley, was concentrating his studies on globular clusters. These were more or less spherically shaped clusters of huge numbers of stars, as many as 100,000 in a single cluster. Trying to discover their distance from the Earth, he had planned to use Leavitt's technique with Cepheid variables. Though globular clusters contain relatively few Cepheids, he found enough to determine that all these clusters were far away, beyond the edge of the Milky Way. Shapley's view was that the universe *was* basically what we call the Milky Way, bordered by globular clusters all around its edge. This view was consistent with the concept of a finite universe—in fact it bore some resemblance to the ancient notion of the spherical universe with the Earth in the center. Here, it was the Milky Way in the center and the globular clusters all around the perimeter, where the ancients had put the "fixed stars."

More Than One Galaxy

Another topic of research interest in the first two decades of the twentieth century was the spiral nebulae. It was Pierre Laplace's interest in a spiral nebula that led him to his theory that the solar system had evolved naturally out of matter in space being drawn into a vortex by gravity and slowly coalescing into the planets and the Sun. Two hundred years later, with greatly improved telescopes, astronomers had found that there were many spiral nebulae out there and they all seemed to be very far away.

A dispute arose between Shapley and the astronomers who were studying the nebulae. Shapley took the view that the universe was finite and bounded by the surrounding globular clusters. The nebulae, he claimed were actually part of the Milky Way. This was based on some studies that had reported finding stellar parallax within some of the nebulae, which would imply that they had to be relatively close.

His opponents believed that the spiral nebulae were completely different systems, on a par with the Milky Way. In other words they were full-fledged *galaxies* themselves. If these galaxies had characteristics like our own Milky Way, then they were totally inde-

Chapter Twenty
The Universe Will Go On Forever—Or It Won't

pendent worlds, so far away that we could never have contact with them. They were really island universes. There was no telling how big the whole cosmos was if it was populated with a large number of complete galaxies.

The National Academy of Sciences in the United States sponsored a public forum in 1920 to discuss these issues, with American astronomer Heber Curtis representing the multiple galaxy viewpoint and Harlow Shapley representing the view that the Milky Way was the entire universe. It came to be called the Great Debate on the Nature of the Universe.

Both sides made very good arguments, but the sum total was inconclusive. What the debate did more than anything was show that new and better tools of measurement were needed for such vast distances.

Edwin Hubble

The key person in 20th century astronomy, upon whose work every theory of the cosmos now depends, is Edwin Hubble, the 20th century's counterpart of the 16th century's Tycho Brahe. Hubble was born in 1889 in Marshfield, Missouri, and attended the University of Chicago where he juggled interests in science with the humanities. He first took a two-year Associate in Science degree in science and mathematics, and then stayed on to complete a Bachelor's degree in 1910 with studies in French, the classics, and political economy. His hope was to win a prestigious Rhodes Scholarship to Oxford, which he did. Hubble went to Oxford, and there decided to study law. Hubble returned to the United States, where his father had just died, and spent some time settling family affairs. He then taught high school for a year. From then on, his career was in astronomy.

When he was at the University of Chicago as an undergraduate, he had come to know the astronomer George Hale, who was the person responsible for organizing the funding and construction of most of the major observatories built in the United States in the first half of the 20th century. Hale had successfully planted in Hubble the seed of irrepressible fascination with the stars. Hubble enrolled as a graduate student in astronomy at the Yerkes Observatory at the University of Chicago, which, of course, had been founded by George Hale, though he was no longer there. As a student, Hubble spent his time photographing and analyzing faint nebulae, working out a classification system for them, which got him his Ph.D. in 1917. Then World War I intervened.

Hubble had been offered a position at the Mount Wilson observatory in California where he would be able to work on the new 60-inch reflecting telescope that was there already and soon after on the 100-inch telescope then under construction. But instead Hubble volunteered for the army where, during the course of the war, he was promoted to the rank of Major. When the war was over, the job at Mount Wilson was still waiting for him.

At Mount Wilson, Hubble's first big breakthrough with the 100-inch telescope was finding Cepheid variables in one of the spiral arms of the Andromeda Nebula and using the data from them to determine that the Nebula was almost one million light years away. This was three times what Shapley had come up with as the size of the entire universe. (Later measurements corrected the distance to over two million light years.) This work, followed up with more confirming results, established the existence of a universe filled with many galaxies of which the Milky Way was only one, and not a very big one at that.

Redshift

Just as stellar parallax could no longer be used to measure the distance to stars beyond a certain point, the Cepheid variable trick was running out of usefulness too, as farther and farther galaxies were discovered. Yet another way of estimating really vast distances needed to be developed.

While pure astronomers were determining the sizes of stars and galaxies and the distances to them, their colleagues, the astrophysicists were busy determining the composition of the stars by examining the spectral lines in the light coming from them. Spectroscopists, who study the properties of light under refraction, could identify which elements were present in any sample of material that was radiating energy by examining the precise wave lengths of the light radiated. Each element had its own set of spectral lines that could be used to identify it. Astrophysicists then could use this information to determine the composition of the stars. It was a well-known and much used technique. In the 19th century, the English astronomer Norman Lockyer (founder of the journal

Chapter Twenty
The Universe Will Go On Forever—Or It Won't

Nature) discovered spectral lines in sunlight that did not correspond to any known element on earth. He asserted that these must represent a new element, which he called *helium* from the Greek *helios,* the Sun. In 1895, that element was finally discovered on Earth, which lent confirmation to the method. The strength of spectral analysis came from its precision. Spectral lines were very precise. As Niels Bohr showed, they were a product of the structure of the electron shells in the atoms themselves (see Chapter 19) and could not vary.

That made it even more puzzling when light from the distant galaxies was analyzed and found to have consistently longer wavelengths than what was expected. The meaning of this was clear: the galaxies were moving away from the Earth. The phenomenon of shifting wavelengths was already known. It was discovered in 1842 by the Austrian physicist Christian Doppler and has been named, in his honor, the *Doppler effect.* Doppler had discovered it in light from stars and used it as a tool to determine the motion of individual stars relative to us. We are more familiar with the effect as it applies to sound waves. The whine of an approaching railroad train or the siren of, say, a fire truck is at a higher pitch when it is approaching us than it is when it is going away from us. The sound waves have shortened or lengthened accordingly, making the frequency higher or lower.

Applied to light, a longer wave is closer to the red end of the spectrum while a higher wave moves toward the blue end. Hence these are called *redshifts* and *blueshifts.* Doppler was familiar with both kinds. Stars that showed redshifted light were moving away from us while stars that showed blueshifted light were moving toward us.

What was surprising to the 20th century astronomers was that the light from other galaxies was always red-shifted, meaning they were always moving away from us. The universe, or what we could see of it, was expanding.

At about the same time, Hubble and the American physicist Howard Robertson both compared the distances that Hubble had determined for some galaxies using Cepheids with the amount of redshift of the light from those galaxies. They both concluded that there was a close relation: the farther a galaxy was from us, the more its light had shifted, and therefore the faster it was moving away from us.

Hubble's Law

Hubble plotted his known distances and redshifts as points on a graph and put a line through the points that fit them as close as possible. This line can be expressed as a linear equation (thanks to Descartes' analytic geometry, of course); that equation we now know as Hubble's Law, and is generally written as

$$V = H_0 D$$

where V is the velocity of redshift, D is the distance to the galaxy, and H_0 is a constant of proportionality—it's the slope of the graph. If the relationship represented by this equation holds for more distant galaxies, where the distance is not known from other means, it can now be calculated from the equation. The amount of redshift therefore, was a measure of the distance to the galaxies.

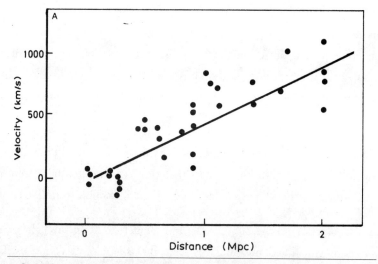

One of Hubble's earliest redshift vs. distance plots. From 1929.

In the next few years, Hubble had applied this formula to galaxies with fainter and fainter light, presumably farther away, and, using their redshift as his observational data, calculated their distance. If what he was finding was true, the universe was much more vast than anyone had expected, and it was getting bigger all the time.

Chapter Twenty | 371
The Universe Will Go On Forever—Or It Won't

The Big Bang

If the universe is expanding at a terrific rate, it must have been smaller in the past. How far can one go with this idea? If Hubble's Law holds for all times and all places, it must be that the universe had a beginning when it was as small as possible. Questions about the creation of the universe sound more like they belong to the province of religion than of science. Maybe they do. In any case the scientist who worked through these implications of Hubble's Law was a Jesuit priest, Father George LeMaître. LeMaître was a Belgian professor of astronomy at the Catholic University of Louvain. He had studied with Arthur Eddington at Cambridge and then for a time at Harvard and at the Massachusetts Institute of Technology (M.I.T.) before taking up his position at Louvain in 1927, where he remained for the rest of his career.

The high point in Eddington's career was when he reported the results of the observations made of the solar eclipse in 1919, which confirmed Einstein's theory of general relativity (see Chapter 18). Eddington went on to become one of the most prominent supporters of relativity theory as it applied to astronomy. Not surprisingly, LeMaître was well versed in relativity and thought easily from that perspective.

LeMaître realized that not only did Hubble's Law imply that the universe was smaller in the past, it also fit in with Einstein's theory of general relativity. In other words, just as James Clerk Maxwell had put a theoretical framework on the experimental work in electromagnetism and Erwin Schrödinger had come up with a theoretical model for quantum mechanics, George LeMaître had placed Hubble's Law into the framework of general relativity. Once part of a theoretical framework, its other implications could be examined.

George LeMaître, left, with Albert Einstein.

LeMaître's model described a universe that began with a primeval atom of super dense matter all packed into a sphere about thirty times the size of the Sun. This "primeval atom" then exploded, breaking into fragments that ultimately became galaxies, stars, planets, atoms, and every other kind of material that exists.

The religious aspects of LeMaître's model did not escape the notice of other astronomers. It was, after all, a creation story, an updated version of the book of *Genesis* that passed scientific muster. And the fact that LeMaître was a priest kept this idea in people's minds. To some scientists, this was going beyond science into wishful thinking. Their view was that there must be other ways to account for redshift that would not imply a creation.

Among those who opposed LeMaître was British astronomer Fred Hoyle, who wrote disparagingly about LeMaître's model, calling it the *Big Bang*. This was intended to underscore its ludicrous nature, but it had the opposite effect. It made the idea accessible to the general public by giving them something they could imagine and think about. Never mind that the images that it brought to mind were often contrary to what the model specified—it made astronomy important and interesting to a great many people. And as these things happen, the intended insult stuck as the name for the theory, and it is now called that by all, advocates as well as opponents.

The Cosmic Background Radiation

Was this yet another Ad Hoc hypothesis being forced upon the empirical data behind Hubble's Law? Sure, the Big Bang was one way of accounting for redshift, but were there others? Was a whole theory of the universe to be connected to observable reality only by the thread of a correlation between distance and the lengths of light waves? It would help support the Big Bang hypothesis a great deal if the theory could predict some phenomenon that had not yet been seen which could then be sought and found. This is the kind of confirmation that Eddington's observation of the bending of light waves did for Einstein's general relativity itself or that the discovery of elements that had been predicted by Mendeleev did for the Periodic Table or that Galileo's observation of the phases of Venus did for Copernicus' heliocentric system.

It was not LeMaître himself who came up with a prediction of his theory, but several others who had been considering that if the universe was expanding rapidly from some

Chapter Twenty
The Universe Will Go On Forever—Or It Won't

earlier more compact state, then in addition to the energy contained in the stars, there must be some general diffuse background radiation all over the universe. The idea was put forth by several people in letters, conversations, even some published papers from 1946 onward. However, these were theoretical speculations; no serious efforts had been made to detect this radiation. One of the earlier speculators was the Russian-born American physicist George Gamow, who was one of the major developers of the Big Bang theory. Another was Robert Dicke, who later became the chairman of the department of physics at Princeton University. It was Dicke who, in the 1960s, began thinking about background radiation again and suggested to one of his young researchers at Princeton, P. J. E. Peebles, that he try to work out if the background temperature of the universe would change as the universe evolved and expanded. Peebles did calculations and determined that if the universe had started out as a Big Bang, then there should be a dim background radiation everywhere in the sky that would correspond to a temperature of something around 10 degrees Kelvin. Dicke set two other researchers onto the task of trying to detect this radiation, and they set up a small antenna on the roof of the Princeton physics laboratory to look for the radiation.

No one knows how long it would have taken them to find what they were looking for, because the radiation was found by accident in 1964 at the Bell Research Laboratories just 30 minutes down the road by two radio astronomers who were trying to adjust an antenna designed for receiving signals from early versions of communications satellites. These two, Arno Penzias and Robert Wilson, had been given the go-

Robert Wilson (left) and Arno Penzias standing before their microwave antenna.

ahead by Bell Labs to conduct some basic radio astronomy research. They set to work redesigning the antenna to be able to receive very weak signals from space. They had to eliminate as far as possible all the static that was inherent in electrical systems. The engineers working with the antenna in its earlier role had narrowed down the various causes of static and found that there was remaining a radiation equivalent to about three degrees Kelvin that could not be accounted for. That was not important for the communications satellite, but it would be fatal for radio astronomy, so it had to be found and eliminated.

They failed. Penzias and Wilson tried everything they could think of, including sweeping out the accumulated pigeon droppings in the parabolic dish antenna. They were getting nowhere. The breakthrough came in a chance conversation that Penzias had with a colleague at the Massachusetts Institute of Technology (M.I.T.), Bernard Burke, after Penzias had attended a conference in Montreal. Penzias told Burke of his difficulties with the background noise; Burke knew someone who had heard a talk by Peebles about their work at Princeton, so he suggested that Penzias get in touch with Dicke at Princeton, who may be able to help him.

The Princeton people realized that this was the data they were looking for. Now, Penzias and Wilson had an explanation, but alas no solution to their problem. Both the Princeton group and Penzias and Wilson published papers alongside each other in the *Astrophysical Journal*. The Princeton paper explained the significance of the detected *cosmic background radiation* and that it was a confirmation of the Big Bang explanation. Penzias and Wilson, still more concerned with the static than the cosmic significance, entitled their paper "A Measurement of Excess Antenna Temperature at 4,080 Mc/s."

In the end, this did provide an independent confirmation of the Big Bang theory and helped it become established over other possible interpretations of Hubble's results. Penzias and Wilson received the Nobel Prize for their discovery in 1978. Neither the Princeton team nor any of the others who had predicted that there should be a cosmic background radiation were honored.

Is the Universe Open or Closed?

There will never be full agreement on any of the issues concerning the origin, size, shape, and destiny of the universe, but there are theories that can be said to have reached a consensus. The Big Bang theory is now the generally accepted view of the origin of the universe. Since the discovery of the cosmic background radiation, the Big Bang had been

Chapter Twenty
The Universe Will Go On Forever—Or It Won't

the standard theory. But having said that, there are a great many other issues that come up within the Big Bang theory about which there is raging controversy.

One issue is simply whether the universe will go on forever or come to a natural end. Hubble's observations showed the farther that galaxies are from Earth, the faster they seem to be receding from us. The implication of that is that the rate of expansion of the universe is decreasing. When astronomers look at distant stars or galaxies, the light they see has been eons coming to them. The closer a star or a galaxy is, the more recently it was that that light started coming our way. For the really distant galaxies, that light took hundreds of millions of years to reach us. If the farther away galaxies appear to be receding from us faster than the closer ones, it means that when the light from those stars first headed our way, the universe was expanding faster than it is now. So, it is expanding and continues to do so, but at a rate that is slowing.

Then there is the issue of why it is slowing. General relativity would demand that gravity itself is putting the brakes on the expansion. How much it is slowing it down depends entirely on the amount of matter in the universe, all of which is attracted to every other bit by gravity.

Dark Matter

How do we know how much matter there is in the universe? There are various tests that astrophysicists can use to estimate the amount of matter in a star or a galaxy. Gravitational attraction is the main tool. A galaxy is a system in which the gravitational attractions among its stars are strong enough to keep them together. Galaxies rotate slowly around a center and that rate of rotation can be used to figure out the masses of the stars within it.

But here there was a surprise. Take the Milky Way for example. All the stars of the Milky Way rotate around a core group of stars, taking about 100 million years to make a complete revolution. The principles are similar to those operating on the Sun and planets in our solar system. In the solar system, Kepler's third law specified the relative rates at which the planets completed one revolution around the Sun. His formula accounted for the fact that the inner planets, such as Mercury and Venus, have much shorter periods than the farther planets, Jupiter and Saturn. Newton showed that Kepler's laws were a direct consequence of his physical system. The same principles would dictate that in the Milky Way, the stars closer to the core would complete their revolutions in a much shorter time than those near the periphery.

They don't. The stars near the edge of the galaxy go around in almost the same amount of time as those near the core. And the same is true of other galaxies that can be seen clearly enough. One explanation for this is that there is much more matter in the galaxies than appears on the telescopes. Instead of matter all being concentrated in the small spaces of the stars with vast empty space in between, if matter were spread out more evenly through the galaxy, then it would rotate more like a disc.

How much more? Here the estimates vary, but nine or ten times as much as there is of visible matter is a common estimate. In other words, astronomers, to their considerable embarrassment face the possibility—some might say the almost certain probability—that they have spent their careers looking at 10% of the universe, thinking they were seeing almost all of it.

The Critical Mass

Why does it matter? It matters because if there is not enough matter to stop the expansion, the universe will continue to get larger and larger, though at a decreasing rate. Eventually all heat producing bodies, i.e., stars, will be so far apart from each other that the universe will become incredibly cold and all processes will come to a stop. (I don't say all life processes because they would have stopped long before.) This is called an *open* universe.

On the other hand if there is enough matter for gravitational attraction to halt expansion, the universe will begin to contract. When that happens, an inexorable process gets started that reverses all the expansion and will pull the universe back to the primordial atom. That situation is called a *closed* universe, and the end result has been nicknames the *Big Crunch*.

At this point, no definitive evidence has been produced that can decide the issue one way or the other, but it is clear that the amount of mass out there is so close to what is called the *critical mass* that it is too hard to call.

Chapter Twenty | 377
The Universe Will Go On Forever—Or It Won't

The Age of the Universe

What does all this mean for the age of the universe so far? How much time has there been since the Big Bang? Here, like everywhere else, there are sharp differences of opinion, and they go back to different stances on issues, such as the amount of dark matter, the correct value of the Hubble constant H_0, whether there was a period of hyper-expansion soon after the Big Bang that is not reflected by Hubble's Law, and many other issues too complex to even begin to discuss here.

However, there are basically two schools of thought that define the range of possible ages. One says the universe is about 20 billion years old. The other says it could be as young as 10 billion years. In either case, it is of an age that humankind cannot even begin to grasp.

For More Information

Alioto, Anthony M. *A History of Western Science,* 2nd ed. Englewood Cliffs, NJ: Prentice-Hall, 1987. Chapter 23.

Asimov, Isaac. *The History of Physics.* New York: Walker & Co., 1984. Chapters 21-22.

Cohen, Nathan. *Gravity's Lens: Views of the New Cosmology.* New York: John Wiley, 1989.

Davies, Paul, and John Gribbin. *The Matter Myth: Dramatic Discoveries That Challenge Our Understanding of Physical Reality.* New York: Simon & Schuster, 1992.

Gribbin, John. *The Birth of Time: How we Measured the Age of the Universe.* London: Weidenfeld & Nicolson, 1999.

_____. *In Search of the Big Bang: Quantum Physics and Cosmology.* New York, Bantam Books, 1986.

MacLachlan, James. *Children of Prometheus: A History of Science and Technology,* 2nd ed. Toronto: Wall & Emerson, Inc., 2002. Chapter 19.

Motz, Lloyd, and Jefferson Hane Weaver. *The Story of Physics.* New York: Avon, 1989. Chapters 15.

Newman, James R. "Einstein." In Byron E. Wall, ed. *The Nature of Science: Classical and Contemporary Readings.* Toronto: Wall & Emerson, 1989.

Ronan, Colin A. *Science: Its History and Development Among the World's Cultures.* New York: Facts on File, 1985. Chapter 10.

Schatzman, Evry. *Our Expanding Universe.* Translated by Isabel A. Leonard. New York: McGraw-Hill, 1992.

Speyer, Edward. *Six Roads from Newton: Great Discoveries in Physics.* New York: John Wiley, 1994. Chapters 5 & 7.

Spielberg, Nathan, and Bryon D. Anderson. *Seven Ideas that Shook the Universe,* 2nd ed. New York: Wiley, 1995. Chapter 6.

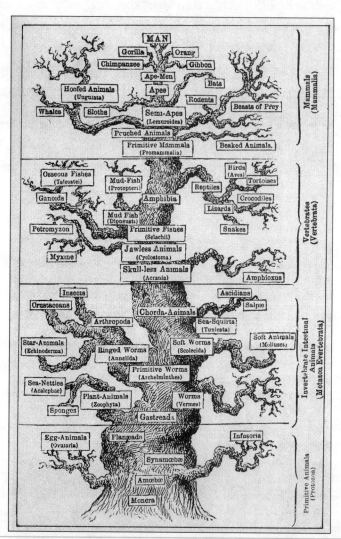

The tree of life from Ernst Haeckel's *Evolution of Man*, 1879.

PART IV

What is Life?

The first three parts of this book trace where physical science has taken us, from the earliest speculations about the stuff of the world, through elaborate theories to calculate the mysteries of motions in the heavens and familiar events on Earth, through a period of great confidence that the world was really some kind of big clock or other machine and that we soon would have it all figured out, and then back to the mysteries again with concepts that we cannot imagine and a scale both so large and so small that there seems nothing human-sized left at all.

That has been the path of the physical sciences. Meanwhile, we have also been trying to understand life. If the physical, i.e., non-living, world is too complex to understand completely, it nevertheless has some simple principles that give us a place to start. The living world is not so kind. Right from the beginning everything about life is incredibly complicated.

For starters, compared to the physical world, life has inherent complexity. Life comes in distinct forms, the species, of which there are millions. Then there is the really bizarre matter of replication. Life begets life. Every individual plant or animal is a self-contained entity with a life span of its own. Yet these individuals somehow manage to create successors to themselves. Worst of all for our having a hope of understanding life, we are life forms ourselves. We have a tendency to view the world in human terms. When it comes to understanding life, we think of our own lives and extrapolate from that, or think of all life as somehow existing for the sole purpose of serving ours. (For example, carrots and apples are food, mosquitoes and rats are pests, snakes and sharks are threats.)

Understanding life has been a goal of thinking people for as long as there has been philosophical speculation about the world in general. Aristotle's philosophy of nature was primarily an attempt to understand what we might call life processes. For Aristotle, the sublunar world, that is, the world all around us, is to be understood in terms of the cycle of generation and corruption that is exhibited by every living thing. Extrapolating from that, he thought of all of nature as being somehow in a life course. Even rocks and mountains were analyzed as though they were alive. The key feature of Aristotle's analysis was the emphasis on the fourth cause, the Final Cause, or purpose, of something. To Aristotle, you knew nothing if you could not state the purpose of a thing. The comparison to living matter is clear. What are eyes? Organs that exist for the purpose of sight. Any discussion of eyes that neglects to mention their function of vision totally misses the point. What are acorns? Little nuts, to be sure, but their role as potential oak trees is the key to understanding them.

When Aristotle applied that kind of analysis to the non-living world, he inevitably concluded that much of nature existed for a premeditated purpose. This was not very useful. To say that a heavy object falls because the natural place of heavy things is at the center of the world is not going to go very far toward an understanding of nature's laws.

The Scientific Revolution of the 16th and 17th centuries rejected Aristotle's analysis of nature and in its place put a mechanical model of the world. Proper science was to show how the machinery of nature worked and to stay away from all questions of purpose. This rather put the life sciences in an untenable position. Virtually the only progress that had been made in understanding life was to discover what the purpose of some organ was or what role some species played in what we would now call an ecological environment. Aristotle's own work on dissecting, analyzing, and classifying animals was prodigious. So were the works of others in antiquity who performed a similar service for botany and human anatomy. Medicine itself was largely an attempt to understand how the body works as a whole and what parts of it serve what function.

After the Scientific Revolution it was hard to know how to regard all these studies. Were they science? Was the science of life so different from the science of inanimate nature that completely different criteria applied to it?

As it turns out, the life sciences did climb out of this quandary and did so ultimately by adopting the methodology and the criteria of the physical sciences. Ironically, the life sciences now are more committed to the goal of a thoroughgoing mechanistic viewpoint than the physical sciences are. Finding the blind mechanisms that make living things work has been just as fruitful for biology as finding the laws of motion have been for physics. But to get there a vast change had to take place in the objects of biological study, the form of admissible explanations, the tools of research, and the usage of what was known already in other fields of science.

This discussion is divided into four chapters. The first reviews theories about the structure of the earth and the classification of species found upon it up to the early 19th century. This might be called the approach of looking at life from the outside. That looking led to the recognition of patterns that could be explained a number of ways, including regarding life as something that had changed over time. The second chapter is a closer look at one of the conceptions of how life has changed, Darwin's theory of evolution. The third chapter follows another course of development which in time came to define the nature of biological research, looking at life from the inside, made possible by the invention of the microscope, which led to a better understanding of the entire question of inheritance, certainly one of the most difficult and important topics in any understanding of life. Finally, the fourth chapter is a look at the key to our present understanding of how life comes to be, controls its own development, and makes new copies of itself: DNA.

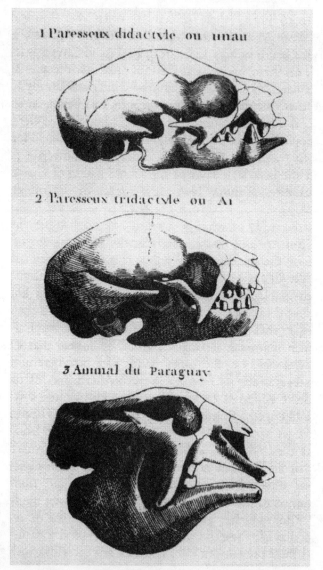

The skulls of two species of modern tree-sloth compared with the huge fossil of the ground-sloth, *Megatherium*. An early example of Cuvier's comparative anatomy.

Chapter Twenty One

The Earth and Its Inhabitants Classified

For Europe in the Middle Ages and early Renaissance, intellectual life was dominated by the Catholic Church and its teachings. The dogma of the Church that went beyond theological questions was filled in with philosophical analysis provided by the Scholastic scholars. That meant that the views of the non-Christian Aristotle were of prime importance in understanding and interpreting the world. The Scientific Revolution changed this, and Aristotle lost his position as the backup answer for everything. Without Aristotle to rely upon, there was a huge vacuum when it came to answers to all sorts of questions that the new science had not yet investigated.

People like to have answers. Aristotle was good at giving answers to most any question about the world that might occur to people. If they could not look to Aristotle for authority, where could they look? Curiously, the answer seems to have been the Bible. Where science was given sway over discovered truths, religion still claimed authority over revealed truth. So, if science was silent on a subject, maybe the answers could be found in the Scriptures.

Take for example, the origin of life. Science was hardly ready to venture any useful theories on that subject, but no matter, it was all explained in the Book of *Genesis* in the creation story. How about the age of the Earth? That could be figured out from the Bible too. In the 17th century, James Ussher, an Irish Archbishop took it upon himself to find out how old the world really was. He did this by going backwards through the Bible and reckoning the number of years accounted for by every biblical story all the way back to the creation and the Garden of Eden. When he had it all figured out, he announced his result: the world was created on October 23, 4004 B.C. Some say he even gave the time of day as 9:00 a.m. And what became of this nonsense? It was adopted by all and sundry as *the* date of creation. It was printed in the English Bible; it was repeated again and again. Before long how this date had been calculated was no longer known, but still it was the answer to the question.

Theories of the Earth

When it came to the issues of the nature of life on Earth, how it got to be, and how it is organized, the new sciences of Newton *et al.* were not a great deal of help. Most of the informed work of careful observation and classification had been done by Aristotle and those working within his intellectual tradition. Since Aristotle could not be relied upon, that only left Scripture.

When Aristotle's physics and cosmology were discarded because they were found faulty, along with them went a tradition of careful observation and cataloguing. On the one hand, it was a huge step backwards, but on the other hand, it wiped the slate clean for a fresh look with the more sophisticated eyes of the post-Newtonian scientific mind.

Consider the issues of the structure and composition of the Earth itself. Here scientists really began with a clean slate since there was almost nothing said about it in the Bible except in the creation story. But these were issues of more than just academic interest. After the Scientific Revolution came the Industrial Revolution and with that came an enormous demand for materials that can be found in the ground, such as metals and minerals. Mining, though a very ancient activity, gained a new importance when industry could not progress without its help. And, with the Industrial Revolution came construction of canals and railroads that necessitated doing a lot of digging into rocks and soils and draining swampland. There was a great need to know what lay ahead in an excavation project to know where best to dig and build. And when the digging commenced, a wealth of new and sometimes surprising objects and formations turned up to think about and make sense of.

Neptunism

A key problem that needed understanding was the curious feature of *stratification*, that seemed nearly universal on the surface of the Earth. The typical experience of a miner or an excavator for a building project was to find that the ground came in distinct layers. After digging down for a period through one kind of soil or rock, the texture changes abruptly, continues with that for a time, then changes again and so forth. These *strata* are found almost everywhere, though not always of the same composition. Also, they are not always horizontal. Sometimes the strata are laying on their side as though the

Chapter Twenty One
The Earth and Its Inhabitants Classified

Illustration of stratification from A. G. Lehmann's *History of Stratified Mountains*, 1756.

ground had experienced some great upheaval in the distant past. Finding the pattern in these strata and explaining their formation was a timely endeavor.

Abraham Werner made his name by explaining all this. Werner was a professor at the mining school in Freiberg, Saxony, where he was a famous and popular teacher. Werner's approach to the stratification issue was to assume, to begin with, that the strata found at the bottom were formed first, and all other strata were laid on top of them one by one. Note that this assumed a sequence of events over a considerable period of time. It, therefore, right from the start was incompatible with the story in *Genesis*. His theory has been called *Neptunism* after Neptune, the Roman god of the sea, because of the prime importance he gave to water in his explanations.

Werner's idea was that originally all of the surface of the Earth was submerged in water, holding vast amounts of material in suspension. As the sea began to recede, those rocky materials crystallized and formed a layer at the bottom. This would typically consist of granite or quartz. This layer need not be flat since it was not assumed that the Earth below the waters formed a smooth sphere. As the seas receded further, some promontories of land appeared and from there on the formation of layers was unevenly distributed because some parts were already out of water. Then a transitional layer formed out of materials that had partly crystallized and partly formed as sediment. This layer had such materials as limestone. As the seas subsided even further, erosion began to take place and bits of the first two layers got washed into the sea where they mixed and regrouped and got laid down in a third layer, which he called *Flötz*, or "swept away." These were minerals such as salt, coal, and basalt. The next layer was similar, but had been subjected to greater mechanical decomposition. This included sand and clay. These are the basic layers that are found, in that order, in most places on Earth, but not always all present. Mountains, for example, would have been some of the first land to appear above water and therefore would consist entirely of the first layer. And these layers need not be horizontal at all because they formed on top of whatever surface was already there. Finally, there was a fifth kind of material, volcanic rock, that was not formed by this layering process but instead was thrown up from deep within the Earth by volcanic eruption.

Werner's theory was very popular. Students came from all over Europe to hear his lectures and then returned to their home countries to apply his theories. This bare outline of Neptunism does not convey how well Werner's classification system accounted for a great variety of different minerals. It provided a useful framework where none had existed before.

The obvious criticism of Neptunism is, where did all this water come from and where did it go? Werner, interestingly enough, was not concerned about this. He claimed it was not a question for scientific geology and therefore he had no need to try to answer it. For Werner, it *must* be true, because he could find no other explanation.

Plutonism

Werner's theory de-emphasized the role of volcanic rock, and more generally, of the idea that great eruptions from the deep did more than lay down some lava and other volcanic deposits here and there. He did not see the heat of the Earth as an important factor in the formation of the Earth's crust, giving that role entirely over to water.

An alternative view to Neptunism was to attribute the formation of strata to churnings that were caused by great heat within the Earth, such heat as was seen in volcanic eruptions, and having great power which was also exhibited by earthquakes. Several simplified theories based on these notions had been put together in the generations immedi-

Chapter Twenty One
The Earth and Its Inhabitants Classified

ately preceding Werner's, but their shortcomings were such that they did not develop large followings. These are collectively known as *Vulcanism*, named after Vulcan, the Roman god of fire.

A much more sophisticated theory that bore a resemblance to those of the Vulcanists was the theory of James Hutton, which, to distinguish it from the others, has earned the nickname of *Plutonism*, after Pluto, Roman god of the underworld.

Hutton was a Scottish geologist who pursued a number of other career interests as well. After making some money from a chemical process that he invented, he bought a farm and retired to it for several years, which is where he was when the basic ideas of his geological theories came to him.

Hutton noticed that wind and rain and all the normal features of the weather will, over time, cause considerable erosion, carrying soils out to sea. Riverbeds will change direction; rocks found in riverbeds will have no rough edges. Earthquakes can separate landmasses, produce chasms, and cause one side to be raised from the other, forming cliffs and breaking layers of strata apart. This, he thought was all brought on by the normal churning of the heat of the Earth.

Hutton asserted that all the features of the terrain that people were so anxious to explain were just the natural result of natural processes happening over vast periods of time. This, he believed, was God's way, to work through natural laws, rather than to intervene. An important aspect of his theory is that it depended entirely on natural processes that were still in effect. This is called the principle of *actualism*. That is, he was not attributing any of the phenomena seen on the Earth to cataclysmic events of the past nor to laws of nature that somehow were no longer in effect. To do so would have seemed unscientific to him.

He wrote up his theory first in a scientific paper published in the *Transactions of the Royal Society of Edinburgh* in 1788, and then in much greater detail with a great number of illustrative examples in a book *The Theory of the Earth*, published in two volumes in 1795. He was working on a third volume when he died. He was not a particularly felicitous writer, so his book attracted little attention when it came out. One sentence of it was memorable though and is often quoted, as it sums up his conclusions. Here it is with the paragraph that precedes it:

> For having, in the natural history of this earth, seen a succession of worlds, we may from this conclude that there is a system in nature; in like manner as, from seeing revolutions of the planets, it is concluded, that there is a system by which they are intended to continue those revolutions. But if the succession of worlds is established in the system of nature, it is in vain to look for anything higher in the origin of the earth. The result, therefore, of this physical inquiry is, that we find no vestige of a beginning,—no prospect of an end.[1]

With those words he separated himself from any of the attempts some people were making to find traces of the origin of the Earth in its geology or to see impending apocalypse, and his view demanded that the Earth had been around for a vastly longer time than Bishop Ussher's 4004 B.C.

Fortunately, while Hutton could not write, he had a disciple who could. In 1802, John Playfair published a much more readable summary of Hutton's Plutonism in his *Illustrations of the Huttonian Theory*.

Catastrophism

In the minds of many, Hutton's theory suffered from being just too tidy. It was fine to say that mountains and rivers and gorges and waterfalls and all manner of purely geological formations had come to be over vast time by natural processes. You could believe this or not. But there was a particular class of natural objects that troubled a lot of people and seemed hard to fit into any of these theories.

These were *fossils*. The word itself originally meant something "dug up"—from Latin *fossilus*, "dug out", or "dug up." Fossils were mysterious objects that appeared magically in the stratified layers of rock. When we speak of a fossil now, the word carries with it our concept that it is the remains, or the imprint, of a plant or animal long dead. But that is not the only interpretation possible. When fossils were first found, some people entertained the idea that they were sort of molds for making the plant or animal, and what they saw was a remnant of the tools in God's workshop.

---

1. James Hutton, *Theory of the Earth* (Edinburgh, 1795), Vol I, p. 200.

Chapter Twenty One
The Earth and Its Inhabitants Classified

Whatever they were, fossils needed explaining, and since they were to be found in the strata that geologists were accounting for in their theories, theories about fossils and theories about the structure of the Earth answered similar questions.

A person whose theory of the Earth was guided by his theory of fossils was Georges Cuvier, who spent years studying the wealth of fossilized remains in the Paris Basin. Because of the impetus of the Industrial Revolution, Paris, like many urban areas was growing quickly. The area around Paris contained a large amount of swampland that was being drained and put to other use. No sooner had the draining begun than a huge number of fossils were uncovered. Cuvier, who was working at the *Muséum d'histoire naturelle* in Paris, began a study of these remains. As he soon discovered, the fossils varied along with the strata in which they were found—and, many of the fossils could only be the remains of creatures that were no longer to be found on earth.

Cuvier had a different explanation for the stratification he found, which also explained why fossils of extinct creatures appeared in different layers. For Cuvier, each stratum marked an era of some sort that was terminated by some great cataclysm. These catastrophic events of the past were due to causes that are unlike those which are at work in the world today. This is the very opposite of Hutton's "actualist" view that normal processes of wind and water account for all the geological formations. Cuvier is the prime representative of the geological theories called *Catastrophism*, in which major structures of the Earth are attributed to unknown causes in the past.

Cuvier's importance in geological theories is minor compared to his importance for identifying and classifying the fossilized remains of animals from all over the Earth. Cuvier can be considered the founder of the science of *comparative anatomy*, the study of the structures of different animals in comparison to each other, and also he was the inventor of the technique of *correlation of parts*, with which he could reconstruct an animal's skeleton on the basis of only a few bones, because any one bone had implications for the body in which it must occur. In some of the more dramatic examples of comparative anatomy, Cuvier was able to classify the remains of woolly mammoths from Siberia and mastadons from the U.S. and show how they were related to, but distinct from modern elephants. In other cases, he reconstructed the skeletons of totally unknown creatures from a handful of scattered bones, and gave his contemporaries images of huge monsters from the past.

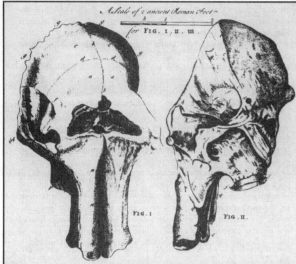

The cranium of a woolly mammoth found in a riverbank in Siberia in the late 18th century. This drawing was later used by Cuvier to establish the mammoth as an extinct species of the elephant genus.

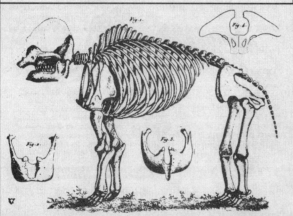

Skeleton of an American mastodon. This drawing was used by Georges Cuvier to distinguish the mastodon from the woolly mammoth and from modern elephants.

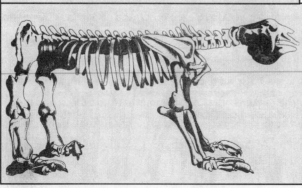

A Megatherium, drawn by Georges Cuvier, 1796.

Chapter Twenty One
The Earth and Its Inhabitants Classified

Diluvialism

These theories, Neptunism, Plutonism, and Catastrophism, were all proposed on their scientific merits, without a conscious attempt to fit in with any religious interpretation. Doubtless, each scientist was guided by their own beliefs and outlook and tended to see things a certain way, but by and large they left theology out of their science.

Not so everyone. There were also those who saw geology as visible proof of the Biblical creation story. The most colorful of these was William Buckland, Reader in Geology at Oxford University in the early 19th century. Buckland was a very popular lecturer and a true eccentric. One of his goals was to eat his way through the entire animal kingdom, with a special fondness for animals that most people would rather live without entirely, such as mice and rats.

Buckland lecturing at Oxford.

Buckland was a Catastrophist to the extent that he attributed many geological formations to a catastrophic event. However, in his case there was really only one catastrophic event that was of importance and that was the flood of the Bible, survived only by Noah and those with him on the Ark. Accordingly, Buckland's brand of Catastrophism has its own name, *Diluvialism*, because so much was due to the great deluge. In his book *Relics of the Flood (Reliquiae Diluvianae)*, Buckland cited a cave in Yorkshire that had been recently uncovered containing bones of hyenas who were not native to England. Buckland said that the hyenas had all drowned in the flood, which was God's way to make England suitable for its present inhabitants.

Taking particular aim at Hutton's claim that there were no vestiges of a beginning nor prospect of an end, Buckland claimed that the fossil record contradicted this, showing that there was a time prior to the appearance of the species that exist today, and proof of the extinction of other species:

> In the course of our enquiry, we have found abundant proofs, both of the Beginning and the End of several successive systems of animal and vegetable life; each compelling us to refer its origin to the direct agency of Creative Interference; "We conceive it undeniable, that we see, in the transition from an Earth peopled by one set of animals to the same Earth swarming with entirely new forms of organic life, a distinct manifestation of creative power transcending the operation of known laws of nature: and, it appears to us, that Geology has thus lighted a new lamp along the path of Natural Theology."[2]

Uniformitarianism

Buckland's real importance for the history of geology was the result of his—to all accounts—provocative and interesting lectures. That is because among those to whom he lectured was a young Oxford student named Charles Lyell. Lyell, from a wealthy Scottish family, went to Oxford to study law, and might have continued with that were it not for Buckland. Lyell was fascinated by Buckland, but not convinced by the diluvian explanations.

Charles Lyell.

When Lyell finished at Oxford, he wanted to study some geological formations that would best show the processes of nature and reveal the vestiges of the flood if there were any to find. He had read a work by George Scrope on volcanoes and became convinced that a close study of volcanoes would give him the experience he needed to make up his own mind. He chose to go to Sicily and study the major, and still active, volcano Etna. What he saw on Mount Etna convinced him not only that Buckland had been wrong, but that he, Lyell, was the man to set it all straight.

One of the features of Etna that Lyell had gone looking for and found readily enough were parasitic cones. These are mini-volcanoes that erupt on the side of the main volcano when pressures build up that are not relieved through the main crater. Of these there were about eighty cones visible around the periphery of Etna, and, he could see, there were

---

2. From William Buckland, *Geology and Mineralogy Considered with Reference to Natural Theology* (London, 1836). The citation within is Buckland quoting himself from the *British Critic*, XVII, Jan. 1831, p. 194.

likely hundreds more that had become covered over. He made inquiries among the local population and found out that only one of the eighty cones had erupted during all of the period of written history in the area. This would suggest that Etna had been forming for a very long time—200,000 years he figured, at the very least. Also he noted that the entire volcano sits over a stratum of sedimentary rocks that contained fossils of mollusks that were virtually identical with those found in the nearby Mediterranean. Elsewhere on the island he found that the sedimentary stratum sat high up on hills. Since this stratum contained fossils of shellfish throughout, Lyell concluded that it must have been pushed upward by earthquakes over a long period of time—of which there was plenty since Etna had to be very old indeed. Therefore these formations were not necessarily caused by violent and sudden catastrophes, but could have arisen from natural causes.

Lyell's brand of geology bears the name *Uniformitarianism*, because Lyell attributes all the physical characteristics of the Earth to natural causes operating in a uniform manner over long periods of time. In general viewpoint, Lyell is quite similar to his fellow Scotsman James Hutton, but their theories can be distinguished by the role that Lyell allows for violent eruptions on a local level. Lyell's work also had immensely greater impact than Hutton's, because Lyell also took on the task of reforming geology and putting it on a solid scientific footing.

Lyell returned from Sicily and set to work on a massive opus that would be called *Principles of Geology*. It was published in three volumes between 1830 and 1833. Lyell emphasized the *actualist* philosophy that underlay his explanations by stating that in his subtitle: "Being an Attempt to Explain the Former Changes of the Earth's Surface by Reference to Causes Now in Operation." Volume One attacks the attempts to reconcile geology with Scripture by ascribing *different causes* to events in the past, in the manner of Buckland. Lyell also reviewed every existing geological theory before either refuting it or agreeing with its points. Since the geological terrain is populated with plants and animals, Lyell also reviewed all theories of how these creatures have existed through the ages, whether they have changed or not, whether any forms are extinct, and what the causes might be for any of this. If nothing else, Lyell's work was unique in its thoroughness. Anyone who knew nothing about geology before picking up his three volumes could be quite expert after digesting all of them. And this is precisely what did happen to one of the people who eagerly awaited each volume as it was published and devoured every word of them. That was a young man taking a voyage around the world, named Charles Darwin. But we will return to that in Chapter 22.

Taxonomy

The geologists who were classifying and analyzing the strata of the Earth were in effect catching up with those who had been classifying and analyzing plants and animals for millennia. *Taxonomy, the* classification of the living world, goes back to Aristotle at least, and for Aristotle was a major part of his biological work.

The unit of classification in biology is the *species*, a term which is easier to use in an informal context than it is to define. The simplest and most useful definition of a species is all the creatures (animals or plants) that can mate and reproduce themselves. This sounds fine, but it is not without exceptions. To the ancient philosophers, a species was some sort of natural category that had a meaning of its own. The ability of its members to reproduce themselves merely underscored that there was something that unified each group. Naming and grouping species was therefore the first step toward understanding life. Since living creatures clearly fell into distinct groups (species) and since the groups bore more or less resemblance to each other, there must be some order to living things that reflects this natural order. But that does not mean that it is easy to find this natural classification. The best that the philosopher can do is to divide life into categories that put those with the greatest number of similarities together.

This basically was Aristotle's plan to bring order to a study of the living world. Aristotle proceeded to divide living nature into distinct groups and subgroups by dividing and then subdividing again and again using the principle that all members of a group that possessed a particular characteristic went into one subgroup and all those that did not when into another. For example, living creatures could either move around (were "animated") or they could not (were "planted"). Among the animals, some had backbones (the vertebrates) and the others did not (the invertebrates). This is a fail-safe system for dividing everything into separate compartments, but it may pick the wrong feature to use as a major division, as Aristotle himself pointed out. For example, he said, if, among the vertebrates, one chose to divide by the number of feet the animals possess, then birds and humans would end up in the same category, and humans would then end up being classified as featherless bipeds. This is the pitfall of an artificial classification system, such as this. These strange groupings would not be made in a *natural* classification that represented nature's true divisions, but that was not given to us to know. (Aristotle himself was such a careful observer that he was able to avoid most of the mistakes that were common among other classifiers. For example, he realized that whales and dolphins are mammals, not fish, and classified them accordingly.)

In the Middle Ages, anthropomorphism entered classification systems in a big way. Not only was the classification system to divide life into distinct species, it was also to rate every life form on a scale from the lowest to the highest, with, of course, human beings at the top. Thus, the system known as the *scale naturae* or Great Chain of Being appeared, based upon Aristotelian categories. Every living thing had a place in the pecking order, from rocks and minerals at the bottom, through plants, invertebrate animals, vertebrates, and on to humans, with angels and God above that. And, with some more elaboration and some variations, that is where things stood to the end of the Scientific Revolution.

Linnæus

Science was high in the pecking order of human activities in the 18th century. The Scientific Revolution had shown the power of scientific thinking to solve the mysteries of nature and to give the human species the ability to make nature bend to its will. There was scarcely an activity with higher social status than science during the Enlightenment.

But alas, science had become difficult, abstruse, and, even worse, highly mathematical. Except for those with mathematical aptitude and training, scientific activity was not open to the general educated public. But there was a way. The life sciences were still accessible to the layman. Ordinary people could possibly do some useful work of a biological nature and thereby participate in the glory of science.

This is the context of the life of Carl von Linné, a Swedish botanist, who later adopted the Latinized name Linnæus. Linnæus was arguably the most popular and respected scientist in the 18th century. The work he did could be understood by anyone and anyone could participate in it.

Part IV
What is Life?

Linnæus.

Linnæus took it upon himself to straighten out the ever more inadequate taxonomical systems that existed in his day. Expeditions to faraway places in the world returned with life forms never seen before in Europe. Even closer to home, many new plants and animals were constantly being discovered, especially as the Industrial Revolution caused more terrain to be explored and surveyed for possible future use. Where the public came in was by acting as Linnæus' corps of assistants in the field. Anyone who found a plant or animal that they thought might be something not known before could package it up and send it to Linnæus at the University of Uppsala, in Sweden, where he would examine it, classify it, and give it a name. If the "assistant" was really lucky, Linnæus would name the new species after the person who sent it to him, thereby immortalizing him forever.

Linnæus set out to establish a system flexible enough to accommodate every new species that turned up and organized enough that logical groupings could be preserved. He pioneered a system of naming species that has been retained to the present, the binomial (i.e., "two-named") system of genus and species. Any new plant or animal that came to him for classification was first categorized into the general group (genus) to which it best belonged. Then, a single feature of the plant or animal was chosen to distinguish it from all others in the genera. This was the specific name. For this feature, an easily identifiable visible characteristic was chosen and given a name. For example, for plants, the arrangement of pistils and stamens in the flower of the plant was often used. Every species carried a double-barreled name: the genus name followed by the species name, for example, *Homo sapiens*. The first edition of Linnæus' new taxonomic system *The System of Nature (Systema Natura)* was a mere 12 pages long. Linnæus continued to update this and publish new editions throughout his life, and also publish other works on more specific classification issues.

Linnæus' system was clearly artificial, which was the only thing practical, but his lifelong goal was to find the key to the natural classification system that God intended. Philosophically, Linnæus was committed to the idea that all species were immutable. All members of a given species were direct descendants of an original pair created by God. Linnæus was not committed to the idea that the ancestors of all living things were the survivors of the Biblical flood who were on Noah's Ark, but he seemed to believe that each species was called into being at a specific time in the past.

Le Comte de Buffon

Linnæus greatest rival was the French naturalist Georges Buffon, who was also interested in studying and cataloguing all the species. Buffon worked at the *Jardin du Roi* in Paris and undertook to describe and try to make sense out of all the diverse nature that was brought to him. Buffon was struck by the diversity of life, noting that basically any way that life could be put together exists, starting from a number of predetermined molds. Unlike Linnæus, Buffon was not committed to the idea that all living species are descended from an original created pair. Instead, he saw a process, in which some species blend together and other spread apart, a process he called "degeneration" which required vast amounts of time.

Quite independently of the geologists who were trying to find the age of the Earth from their studies, Buffon made an estimate based upon the time he figured that "degeneration" would have needed to produce the species that were known in his time. His estimate was 72,000 years at least, and probably closer to a million. And what about the creation story in *Genesis*? Buffon was not willing to write this off as untrue, but, rather, gave it a metaphorical interpretation that the six "days" of creation actually referred to six long epochs of time.

One other way that Buffon can be distinguished from his contemporary Linnæus was his style of publishing. While Linnæus was revising the *Systema Natura* again and again and reissuing it with corrections and many additions, Buffon was simply adding to his monumental work, the *Histoire Naturelle*. By his death he had published 36 volumes of it and eight more appeared posthumously.

Lamarck's Theory of Evolution

How did Buffon's theory of "degeneration" fit the accepted canon of scientific explanation that had been established since Newton? Under the new spirit of thoroughgoing mechanism, an explanation is not scientific unless it can specify how everything happens. Buffon saw and reported a slow change in the characteristics of animals as they deviated from what he took to be a basic model. But he had no mechanism to account for it.

Linnæus did not see change, once the species existed, but he was prepared to accept that species were specially created at different times through history when they were required to fill some role. To some, this was not a very scientific explanation either, because it required intervention from outside of nature to produce the effect.

These were not acceptable as scientific explanations in the way that, say, the geological theories of Hutton and Lyell were. The existence and/or the development of species did not follow from natural principles that could be stated.

A theory that could explain the variety of species on Earth, take into account the apparently extinct forms of life that resembled present species, and do so without appealing to forces other than natural ones would be leapt upon as proper explanations, explanations that could make biology scientific.

Such a theory was put forth by Jean Baptiste Pierre Antoine de Monet, chevalier de Lamarck, a man who made a career of being a naturalist. He, like Cuvier, worked at the *Muséum d'histoire naturelle* in Paris, the successor to the *Jardin du Roi* in Paris which had been Buffon's place of employment, clearly the center for serious work in classification and the study of fossils. Lamarck's theory was a long time in gestation and had many complexities to it, which have basically all been forgotten today. In Lamarck's early work, he, like Linnæus, was committed to the idea that the species were immutable, but he changed his mind after he joined the staff at the *Muséum* in 1794 at the age of fifty. Lamarck was assigned to classifying the invertebrates, at which he excelled to the point where he can be considered a founder of invertebrate taxonomy.

It was only after the turn of the century when Lamarck was already in his late fifties that he began to think in terms of the evolution of species. While thinking about the development of species, he had also proposed a theory of matter that was rather different from the chemistry that was being developed by his countryman Lavoisier. One of the features of his chemical/biological theories was that he suggested that life arose through a

Chapter Twenty One | 399
The Earth and Its Inhabitants Classified

The botanical and zoological *Jardin du Roi* in Paris where Jean Lamarck worked after 1788.

process of spontaneous generation from inert matter. These are among the many complexities of his thought that are virtually unknown to us today.

The Inheritance of Acquired Characteristics

What Lamarck is remembered for now is a single aspect of his theories that was latched onto and held up as the key to understanding evolution, though for Lamarck himself, it was only a secondary factor. That key is his proposal for the *inheritance of acquired characteristics.*

If it can be taken for granted that there is something natural and unvarying in the reproduction of plants and animals from generation to generation, then any suggestion that a slow process of change takes place while this is going on will require a cause to make it happen. If the cause proposed for this is external to the known laws of nature, the explanation may be deemed unscientific. But if a cause can be proposed which seems quite natural and easily understood, the explanation fits the model of a scientific theory, even if the actual physical mechanism cannot be completely specified.

Lamarck's theory fit this model. The basic structure is as follows: a population of plants or animals lives in an environment and has done so for, say, many generations reproducing itself regularly. Then, some change is introduced into the environment. This might be a change in climate, a change in food supply, or the introduction of predators. The individuals who live in the environment are presented with new challenges. To survive the cold, or get more food, or stay away from predators—whatever the situation is—the individuals in the population adapt by growing more fur or stronger legs, for example. Clearly this is possible. Bodies develop protective layers to combat the cold. Any animal that runs more develops stronger legs and can better outrun prey or escape predators. What is new with Lamarck's theory is the dictum that these acquired characteristics can be inherited. The offspring of these individuals will be born with a greater propensity to meet whatever the need is. All it would require is some slight part of the acquired characteristic to be passed on, and then over time and many generations, the species can change dramatically, fitting the changed environment.

The classic example of Lamarckian adaptation is the giraffe's neck. An illustrative scenario would go like this: in ages past there were some creatures resembling ordinary horses living in an environment where their food supply lay on low lying bushes and shrubs. After a time, the food supply was no longer enough to feed the population, perhaps from overgrazing, but there was food to be had in the foliage higher up in trees. The giraffes began stretching to reach these leaves and developed slightly longer necks. In succeeding generations, giraffes were born with necks a bit longer than the necks their parents had been born with. After many generations the necks had stretched so much that the giraffes could reach to the tops of tall trees to feed.

The beauty of this explanation is that it is all quite natural. One can make a neck longer by stretching, or legs stronger by exercising. The population would do this to enable it to live in its environment. If it can pass on a portion of that acquired characteristic to its offspring, the succeeding generations will fit their environment even better. One of the extraordinary facts about nature is that living things are superbly suited to the environments in which they live. This would explain why. Also if a population of some creatures (including plants) wandered into a different environment for which it is not ideally suited, it would begin to adapt to it and after many generations, would appear to have been created to fit it.

The difficulty with this is that how that characteristic gets passed on is not spelled out. But then nothing was known about how any inheritable traits are passed on anyway,

Chapter Twenty One
The Earth and Its Inhabitants Classified

so this was nothing new. One might as well assume that acquired traits get passed on as easily as native traits.

The idea of extinct species was troubling for many people, who thought that an extinct species implied that God had made a mistake. Lamarck himself did not believe that there had been extinctions in the past. Any species found now in fossil remains that were remarkably different from their closest present-day matches were assumed to be simply earlier versions of the present ones that had lived in different environments.

Lamarck's theory was difficult and in some ways obscure because it was bound up with a number of other pet theories of his that were not in the mainstream. But the idea of species changing over long periods of time and doing so by passing on acquired characteristics was an attractive one, which gained a following.

Lamarck's thought might have slipped into obscurity by the mid 19th century were it not for summaries of his thought that were presented by others as points of discussion. Chief and foremost among the summarizers of his work was the geologist Charles Lyell, who discussed Lamarck's ideas in some detail in Volume Two of his *Principles of Geology*, where he presented them in order to disagree with them. The following is part of Lyell's exposition (and criticism!) of Lamarck's argument for evolution through the inheritance of acquired characteristics:

> The name of species, observes Lamarck, has been usually applied to "every collection of similar individuals produced by other individuals like themselves." This definition, he admits, is correct; because every living individual bears a very close resemblance to those from which it springs. But this is not all which is usually implied by the term species; for the majority of naturalists agree with Linnæus in supposing that all the individuals propagated from one stock have certain distinguishing characters in common, which will never vary, and which have remained the same since the creation of each species.
>
> In order to shake this opinion, Lamarck enters upon the following line of argument: —The more we advance in the knowledge of the different organized bodies which cover the surface of the globe, the more our embarrassment increases, to determine what ought to be regarded as a species, and still more how to limit and distinguish genera. In proportion as our collections are enriched, we see almost every void filled up, and all our lines of separation effaced! we are reduced to arbitrary determinations, and are sometimes fain to seize upon the slight differences of mere varieties, in order to form characters for what we

choose to call a species; and sometimes we are induced to pronounce individuals but slightly differing, and which others regard as true species, to be varieties.

The greater the abundance of natural objects assembled together, the more do we discover proofs that every thing passes by insensible shades into something else; that even the more remarkable differences are evanescent, and that nature has, for the most part, left us nothing at out disposal for establishing distinctions, save trifling, and, in some respects, puerile particularities…

Every considerable alteration in the local circumstances in which each race of animals exists causes a change in their wants, and these new wants excite them to new actions and habits. These actions require the more frequent employer of some parts before but slightly exercised, and then greater development follows as a consequence of their more frequent use. Other organs no longer in use are impoverished and diminished in size, nay, are sometimes entirely annihilated, while in their place new parts are insensibly produced for the discharge of new functions.

I must here interrupt the author's argument, by observing, that no positive fact is cited to exemplify the substitution of some entirely new sense, faculty, or organ, in the room of some other suppressed as useless. All the instances adduced go only to prove that the dimensions and strength of members and the perfection of certain attributes may, in a long succession of generations, be lessened and enfeebled by disuse; or, on the contrary, be matured and augmented by active exertion; just as we know that the power of scent is feeble in the greyhound, while its swiftness of pace and its acuteness of sight are remarkable—that the harrier and stag-hound, on the contrary, are comparatively slow in their movements, but excel in the sense of smelling.

It was necessary to point out to the reader this important chasm in the chain of evidence, because he might otherwise imagine that I had merely omitted the illustrations for the sake of brevity; but the plain truth is, that there were no examples to be found; and when Lamarck talks "of the efforts of internal sentiment," "the influence of subtle fluids," and "acts of organization," as causes whereby animals and plants may acquire new organs, he substitutes names for things; and, with a disregard to the strict rules of induction, resorts to fictions, as ideal as the "plastic virtue," and other phantoms of the geologists of the middle ages…

But to proceed with the system: it being assumed as an undoubted fact, that a change of external circumstances may cause one organ to become entirely obsolete, and a new one to be developed, such as never before belonged to the species, the following proposition is announced, which, however staggering and absurd it may seem, is logically deduced from the assumed premises. It is not the organs, or, in other words, the nature and form of the parts of the body of an animal, which have given rise to its habits, and its particular faculties; but, on the contrary, its habits, its manner of living, and those of its progenitors, have in the course of time determined the form of its body, the number and condition of its organs—in short, the faculties which it enjoys. Thus otters, beavers, waterfowl, turtles, and frogs, were not made web-footed in order that they might swim; but their wants having attracted them to the water in search of prey, they stretched out the toes of their feet to strike the water and move rapidly along its surface. By the repeated stretching of their toes, the skin which united them at the base acquired a habit of extension, until, in the course of time, the broad membranes which now connect their extremities were formed.

In like manner, the antelope and the gazelle were not endowed with light agile forms, in order that they might escape by flight from carnivorous animals; but, having been exposed to the danger of being devoured by lions, tigers, and other beasts of prey, they were compelled to exert themselves in running with great celerity; a habit which, in the course of many generations, gave rise to the peculiar slenderness of their legs, and the agility and elegance of their forms.[3]

Since Lamarck had written his works in French while Lyell wrote in English, it was mainly due to Lyell that the English-speaking world came to know of Lamarck's theory. And, as was mentioned before, Lyell had one reader of his *Principles* who read every word he wrote very carefully, namely Charles Darwin.

For More Information

Alioto, Anthony M. *A History of Western Science,* 2nd ed. Englewood Cliffs, NJ: Prentice-Hall, 1987. Chapters 19-20.

Appleman, Philip, ed. *Darwin.* A Norton Critical Edition. New York: Norton, 1970.

Bailey, Edward B. *James Hutton: The Founder of Modern Geology.* Amsterdam: Elsevier, 1967.

Bowler, Peter J. *Evolution: The History of an Idea*, Rev. ed. Berkeley: University of California Press, 1989.

Bowler, Peter J. *Fossils and Progress: Paleontology and the Idea of Progressive Evolution in the Nineteenth Century.* New York: Science History Publications, 1976.

Buckland, William. *Geology and Mineralogy Considered with Reference to Natural Theology.* 2 vols. London, 1836.

_____. *Reliquiae Diluvianae: Or Observations of the Organic Remains Contained in Caves, Fissures and Diluvial Gravel, and other Geological Phenomena, Attesting

---

3. From Charles Lyell, *Principles of Geology*, Vol II, Book III, Chapter 24. Originally published in 1832. This excerpt is from the ninth edition published in 1853.

Part IV
What is Life?

the Action of a Universal Deluge. London, 1823. Reprinted New York: Arno Press, 1977.

Burnet, Thomas. *The Sacred Theory of the Earth.* London, 1691. Reprinted with an introduction by Basil Willey. London: Centaur Press, 1965.

Chambers, Robert. *Vestiges of the Natural History of Creation.* London, 1844. Reprinted with an introduction by Sir Gavin De Beer. Leicester: Leicester University Press.

Cohen, I. Bernard. *Revolution in Science.* Cambridge: Harvard University Press, 1985. Chapters 18 & 21.

Coleman, William. *Biology in the 19th Century: Problems of Form, Function and Transmutation.* New York: Wiley, 1971.

Eiseley, Loren. *Darwin's Century: Evolution and the Men Who Discovered It.* New York: Doubleday, 1958.

Gillespie, Charles Coulston. *Genesis and Geology: A Study in the Relations of Scientific Thought, Natural Theology and Social Opinions in Great Britain, 1790-1850.* Reprinted New York: Harper, 1959.

Greene, John C. *The Death of Adam: Evolution and Its Impact on Western Thought.* Ames, Iowa: The Iowa State University Press, 1959.

Lovejoy, Arthur O. *The Great Chain of Being: A Study in the History of an Idea.* Reprinted New York: Harper, 1960.

MacLachlan, James. *Children of Prometheus: A History of Science and Technology,* 2nd ed. Toronto: Wall & Emerson, Inc., 2002. Chapter 16.

Ronan, Colin A. *Science: Its History and Development Among the World's Cultures.* New York: Facts on File, 1985. Chapter 9.

Rudwick, Martin. *The Meaning of Fossils: Episodes in the History of Paleontology.* 2nd ed. New York: Science History Publications, 1976.

Wall, Byron E., ed. *The Nature of Science: Classical and Contemporary Readings.* Toronto: Wall & Emerson, Inc., 1990.

Winsor, Mary P. *Starfish, Jellyfish, and the Order of Life.* New Haven: Yale University Press, 1976.

Chapter Twenty One
The Earth and Its Inhabitants Classified

Title page of Thomas Burnet's 1684 book, *The Sacred Theory of the Earth. The stages of the Earth are depicted in the spheres that surround the title. Note the depiction of the Great Flood in third sphere, counting clockwise from the top.*

Darwin's study at Down House.

CHAPTER TWENTY TWO

Evolution by Natural Selection

It is out of fashion these days to speak of the *Zeitgeist*—the spirit of the times—of an age. The concept is too broad and too much wishful thinking gets hidden under its umbrella. But sometimes the notion does seem to convey something. For the 19th century, especially for Britain and probably for the rest of the western world too, the *Zeitgeist* was very much the idea of *Progress*.

The 18th century was the time of great confidence in human reason to figure everything out—to apply the methods of the Scientific Revolution to all issues and expect to discover their fundamental principles. The Industrial Revolution, starting in the middle of the 18th century, brought with it prosperity and a general rise in the standard of living. Life was getting better all around.

It was not too much of a stretch to begin thinking that not just human conditions, but somehow nature itself was on an upward path. Lamarck's evolutionary theory was an expression of this sentiment. In England, the view of nature in perpetual improvement was explicit in another evolutionary theory that bore some resemblance to Lamarck's, though in a much less systematic presentation. This was the theory of Dr. Erasmus Darwin, who happened to be the grandfather of Charles Darwin.

Dr. Erasmus Darwin.

Erasmus Darwin's Zoonomia

Erasmus Darwin's evolutionary ideas were more characteristic of the progress *Zeitgeist* than Lamarck's in that they were not hampered

Part IV
What is Life?

so much by the effort to provide evidence or to connect the evolutionary conception with other scientific principles. To a much greater extent, they represented the free rein of imagination. The basis of Dr. Darwin's evolution was first an assumed inherent propensity for all living matter to transform itself to purposeful ends. And second was the general observation of individual development that all living things go through as they are born, grow, and mature.

A few excerpts from his writing serve well to convey this progressive, optimistic spirit:

> From the meditating of the great similarity of the structure of the warm-blooded animals, and at the same time of the great changes they undergo both before and after their nativity; and by considering in how minute a proportion of time many of the changes of animals described have been produced; would it be too bold to imagine, that in the great length of time, since the earth began to exist, perhaps millions of ages before the commencement of the history of mankind, would it be too bold to imagine, that all warm-blooded animals have arisen from one living filament, which THE GREAT FIRST CAUSE endued with animality, with the power of acquiring new parts attended with new propensities, directed by irritations, sensations, volitions, and associations; and thus possessing the faculty of continuing to improve by its own inherent activity, and of delivering down those improvements by generation to its posterity, world without end?[1]

A few pages later, Dr. Darwin gushed forth with the crux of the idea of progress in nature:

> [A]ll nature exists in a state of perpetual improvement by laws impressed on the atoms of matter by the great CAUSE OF CAUSES... This idea is analogous to the improving excellence observable in every part of the creation; such as in the progressive increase of the solid or habitable parts of the earth from water; and in the progressive increase of the wisdom and happiness of its inhabitants; and is consonant to the idea of our present situation being a state of probation, which by our exertions we may improve, and are consequently responsible for our actions.[2]

Erasmus Darwin's works appeared in the last decade of the 18th century. He died in 1802, but his works continued to be popular in the early 19th century and were published

---

1. Erasmus Darwin, *Zoonomia; Or the Laws of Organic Life*, 4th ed. (Philadelphia, 1818), Vol I, p. 397.
2. *ibid.* p. 400-401, 437.

both in Britain and abroad. He found a ready audience for his ideas among the general public.

Robert Chambers' Vestiges

The scientific community was not so well disposed to the sort of unfounded rambling enthusiasm that was characteristic of the elder Darwin. Another work of what might be called pseudo-science was the very popular *Vestiges of the Natural History of Creation* by Robert Chambers, published anonymously in 1844. (Note the reference in the title to James Hutton's *Theory of the Earth*—see Chapter 21.) Chambers' work was another imaginative synthesis, drawing together many of the ideas in science that were swirling about and looking for a great synthesis. This was the period in which the German *Naturphilosophie* movement was most influential. The idea that everything in nature could be somehow related to one or two overriding forces was destined to appeal to a wide audience.

Chambers' work, like Erasmus Darwin's, asserted that God had created a world that then ran according to a grand plan. This plan, according to Chambers, had two "principles of action": gravitation—for the inorganic world, and development—for the organic. Development referred to both the process of growth and maturation in every individual living being and to the evolution of the species. As an illustration of the principles at work in the inorganic world, Chambers mentioned the nebular hypothesis of Laplace (see Introduction to Part III) wherein gravity pulls brute matter into the organized form of the solar system.

Not afraid of sweeping generalizations, Chambers asserted that life originally arose from spontaneous generation from inert matter and had an internal driving force toward perfectibility that propelled evolution of all species *and* had produced the human species. This was dynamite, leading to wide condemnation and charges of atheism—probably why Chambers published the book anonymously. However, it did not prevent the book from selling well and further pushing the idea of a mechanist explanation.

The Design Argument

The charges of atheism that Chambers' book brought on, despite his attributing the "principles of action" to God, were given extra force in the British Isles, because they also ran counter to a tidy bit of logic that purported to prove the existence of God. The argument is basically simple: look around you; nature is complex and wonderful; everything works together so beautifully and so purposefully that we can speak of nature having a grand design. Designs do not just happen accidentally. There must be a designer of nature, and that designer is God. This is called the *Design Argument*. It has a long history and a tight logical structure. It was given a particularly British flavor by none other than Isaac Newton.

Newton's Effect of Choice

The system of the world that is modeled by Newton's *Principia* successfully accounted for much of the observable phenomena of the world by his formulas of motions and forces. But not everything. Newton found in his system no reason for the fact that the planets all circled in the same direction and in nearly the same plane. Since they do, it must be the work of God. This is the very issue that Laplace addressed with his nebular hypothesis, showing that Newtonian-like calculations *could* account for the formation of the solar system as the natural result of gravity in the same way that it would account for the motions of the spiral nebulae. Laplace concluded that he did not need this hypothesis of God to explain this.

But it was not just planetary motions that evaded Newton. The whole living world was beyond his ken, and, he thought, beyond any scientific explanation from natural laws. Wherever there was a purposeful design in nature that Newton could not explain, he attributed to the *effect of choice*.

> All material Things seem to have been composed of the hard and solid Particles...variously associated in the first Creation by the Counsel of an intelligent Agent. For it became him who created them to set them in order. And if he did so, it's unphilosophical to seek for any other Origin of the World, or to pretend that it might arise out of a Chaos by the mere Laws of Nature; though being once form'd it may continue by those Laws for many Ages. For while Comets move in very excentric Orbs in all manner of Positions, blind Fate could not make all the Planets move one

and the same way in Orbs concentrick, some inconsiderable Irregularities excepted, which may have risen from the mutual Actions of Comets and Planets upon one another, and which will be apt to increase, till this System wants a Reformation. Such a wonderful Uniformity in the Planetary System must be allowed the Effect of Choice. And so must the Uniformity in the Bodies of Animals, they having generally a right and a left side shaped alike, and on either side of their Bodies two Legs behind, and upon their Shoulders, and between their Shoulders a Neck running down into a Back-Bone, and a Head upon it; and in the Head two Ears, two Eyes, a Nose, a Mouth, and a Tongue, alike situated. Also the first Contrivance of those very artificial Parts of Animals, the Eyes, Ears, Brain, Muscles, Heart, Lungs, Midriff, Glands, Larynx, Hands, Wings, swimming Bladders, natural Spectacles, and other Organs of Sense and Motion; and the Instinct of Brutes and Insects, can be the effect of nothing else than the Wisdom and Skill of a powerful everliving Agent, who being in all Places, is more able by his Will to move the Bodies within his boundless uniform Sensorium, and thereby to form and reform the Parts of the Universe, than we are by our Will to move the Parts of our own Bodies.[3]

Natural Theology

The way Newton structured the reasoning, if a scientific (i.e., mechanistic) explanation could not be found for something that served a purpose, it could only have arisen by design (the "effect of choice"), and therefore was the work of God. Laplace had effectively shown that unexplained phenomena in the inorganic world might yet be subject to mechanical explanations, but the living world remained a testimony to divine intervention.

Thus, there arose a new kind of devotional study, *natural theology*. There were two ways to try to know the mind of God. One was the study of the Scriptures. The other was the study of God's handiwork in nature. To examine nature and report in detail on all those features that defied mechanistic explanation was an act of piety. In Britain especially, some of the best naturalists—those who examined nature, collected samples,

---

3. Isaac Newton, *Opticks*, 4th ed., Bk III, Pt. I, Query 31 (London, 1704). Newton's *Opticks* was published late in his career. The book is, as the title indicates, mostly about the phenomena of light, but Newton took the opportunity to append to this work a number of speculative musings about nature which concerned him, but for which he had no definite answers. These took the form of "queries," most of which are one or two paragraphs, many of them about alchemy. The last query, number 31, is longer, broader, and more philosophical. There, Newton discusses scientific method and the limits of science and of proper human inquiry.

Part IV
What is Life?

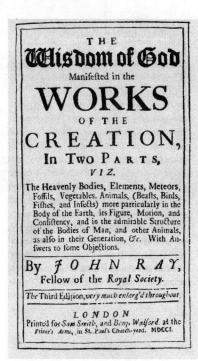

Title page of John Ray's *The Wisdom of God Manifested in the Works of Creation*, a classic Design Argument. Ray was a clergyman, naturalist, and one of the founders of natural theology. From the third edition, 1701.

helped in the taxonomic classification process, and wrote works describing what they found—were often clergymen. Nature was a second Scripture.

The Bridgewater Treatises

In 1829, the Eighth Earl of Bridgewater died, leaving in his will a commission to underwrite the publication of a series of works to illustrate the "Power, Wisdom and Goodness of God in the Works of Creation." In other words, the Earl undertook to support works of natural theology, which would, of necessity, bolster the Design Argument. There were eight *Bridgewater Treatises* published in the 1830s, written by eminent scientists of the day exploring the different aspects of nature in which they were expert and making the case for divine intervention in the world. The series was administered by the Royal Society of London and was representative of prevailing scientific opinion of the day. The work by William Buckland, *Geology and Mineralogy Considered with Reference to Natural Theology* (from which a quotation was reproduced in Chapter 21), was one of the Bridgewater Treatises.

Another was by the distinguished Scottish anatomist, Sir Charles Bell, entitled *The Hand: Its Mechanism and Vital Endowments as Evincing Design*, published in 1833. Bell considered not just the human hand, but the hands and hand-like appendages of many animals and showed how every difference in visible structure was accompanied by a multitude of internal relations that made the entire animal adapted to its function in nature. Hence, only God could have designed it so. Bell particularly took aim at the evolutionary theory of Jean Lamarck (see Chapter 21), which he considered "miserable" and sought to refute by his own work.

Chapter Twenty Two | 413
Evolution by Natural Selection

> ...In seeking assistance from the works of distinguished naturalists, we do not always find indications of that disposition of mind prevailing, which we should be apt to suppose was a necessary result of their peculiar studies. We do not discover that combination of genius with sound sense, which distinguished Cuvier, and has been the characteristic of all the great men of science. It is, above all, surprising with what perverse ingenuity some will seek to obscure the conception of a Divine Author, an intelligent, designing, and benevolent Being, and clinging to the greatest absurdities, will rather interpose the cold and inanimate influence of the mere "elements" in a manner to extinguish all feeling of dependence in our minds, and all emotions of gratitude.
>
> Some will maintain that all the varieties in animated beings are merely the result of a change of circumstances influencing the original animal; that new organs have been produced by a desire, and consequent effort, of the animal, to stretch and mould itself into a shape suitable to the condition in which it is placed, ...that, as the leaves of a plant expand to light, or turn to the sun, or as the roots shoot to the appropriate soil, so do the exterior organs of animals grow and adapt themselves. We shall presently find that an opinion has prevailed that it is the organization of animals which has determined their propensities, but the philosophers of whom we are now speaking, imagine the contrary, ...they conceive that, under the influence of new circumstances, organs have accommodated themselves, and assumed their particular forms.

At this point Bell appended a footnote in which he named the culprit whom he had been maligning as none other than Jean Lamarck, then cited an example or two of Lamarck's principle of the inheritance of acquired characteristics. Lamarck's theory had been summarized (and condemned) by Charles Lyell in volume two of his *Principles of Geology*, which appeared the year before Bell's book. Bell's examples from Lamarck were the same as those reported by Lyell (see Chapter 21) and in almost the same wording, suggesting that he was familiar with Lyell's summary. In this footnote, Bell also referred to Lamarck's concept of the evolution of man from the orangutan (also in Lyell's summary, though not reproduced here in Chapter 21). These indignities were enough to warrant the summary condemnation of Bell:

> That a man, in jest, or in mere idleness, or to provoke discussion, should have given expression to such fancies, is probable: but that any one should have published them, as a serious introduction to a system of natural history, is, indeed, surprising. It is a miserable theory, to which we can only conceive a man driven by the shame, or fear, of being thought to harbour the belief of vulgar minds.

Part IV
What is Life?

Having shown all the complexities and intricate relationships concerned with hands, Bell came finally to the inevitable conclusion of every Design Argument, that only God could have done it.

> It must now be apparent that nothing less than the Power which originally created is equal to effect those changes on animals, which adapt them to their conditions: and that their organization is predetermined; not consequent on the condition of the earth or of the surrounding elements...
>
> It has been shown, that...there is nothing like chance or irregularity in the composition of the system. In proportion indeed as we comprehend the principles of mechanics, or of hydrolics, as applicable to the animal machinery, we shall be satisfied of the perfection of the design. If anything appear disjointed or thrown in by chance, let the student mark that for contemplation and experiment, and most certainly, when it comes to be understood, other parts will receive the illumination, and the whole design stand more fully disclosed.[4]

The lines were drawn. The force of the design argument according to its own criteria depended upon the impossibility of accounting for the evident design with uniform physical laws. Anyone who claimed to have found natural laws to account for the living world was subject to ridicule by the scientific establishment and charges of heresy from established religion, often represented by the same people. Progress was in the air. Even theories of evolution were bandied about in the general public. However, the scientific establishment, not to mention the religious one, was having none of it.

Charles Darwin as a young man.

Charles Darwin

Into this environment came Charles Darwin, born in 1809 into a wealthy middle-class English family. His father was a prominent physician; his father's father was the famous Dr. Erasmus Darwin, one of the founding members of the Lunar Society of Birmingham that fostered scientific and technical interests and, of course, the author of one of those early speculative

---

4. Charles Bell, *The Hand: Its Mechanism and Vital Endowments as Evincing Design* (London, 1833). These quotations are from Chapter 6 of the 1837 edition.

theories of evolution. The grandfather on his mother's side was Josiah Wedgwood, also a founding member of the Lunar Society, whose Wedgwood china was the standard of excellence in British porcelain.

Darwin was expected to continue the tradition of his father and paternal grandfather and become a physician. He dutifully went off to the University of Edinburgh to study medicine. He did not last long. On the one hand, the tedium of medical studies bored him. On the other hand, watching surgery in the operating theatre (in an era before anæsthestics) nauseated him. He was not cut out for medicine. He returned home in some shame, and his father, seeking some other respectable profession for his son, sent him off to Cambridge University to study for the ministry.

This went rather better, especially since theology at Cambridge included natural theology. Darwin was an enthusiastic supporter of the Design Argument and was fascinated by the examples of biological adaptation that were cited in support of it. He was said to have found the logic of natural theology as convincing as Euclid. He developed a close relationship with the botanist Reverend John Henslow and the geologist Adam Sedgwick, both of whom took Darwin on field trips to study nature first-hand. Through them, Darwin gained a thorough education as a naturalist, while acquitting himself satisfactorily enough in theology to qualify for his degree.

Professor John Henslow.

The Voyage of the Beagle

Then came the turning point in Darwin's life that fixed his career path forever. The British Admiralty was planning a voyage that would go around the world and map the coastlines of South America. The captain of the voyage, Robert Fitzroy, sought to find a "gentleman-naturalist" to accompany him on the voyage and make careful notes of the flora, fauna, and geological formations they encountered along their way. Fitzroy appealed to Henslow to find a person for the job, and Henslow suggested Darwin. It was an unappealing assignment for anyone of Darwin's age, just embarking on a career. It was an unofficial add-on to the voyage, so it did not contribute to a naval career. It was going to be a long voyage of uncertain duration and destination, and not without dangers. Anyone returning from it would have lost several crucial years and be out of touch with the people who could help him get established. It could be a dead end.

But Darwin wanted to go. The alternative was to head straight for a life as a country parson. But there were complications: Fitzroy didn't like him very much; he viewed him as "unsuitable"—too young and inexperienced and perhaps too rebellious for his taste. Darwin's father was mortified. This was no proper undertaking for a son of his. In time, all these feathers were smoothed, and Darwin got the position. The voyage lasted five years, from 1831 to 1836.

When the Beagle returned, Darwin took lodgings in London and made something of a name for himself as a naturalist giving reports on his findings from the voyage. In 1839, he published his *Journal of Researches into the Geology and Natural History of the Various Countries Visited by H.M.S. Beagle*, which, despite a title that sounds deadening to our ears, sold well. Darwin was a careful observer, a good artist, and a facile writer. He had an assured career as an observational naturalist, with the abilities necessary to communicate both to the scientific community and to the general public.

He had seen a great deal on the voyage and took his time drawing conclusions. Two parts of his voyage have generated the most interest among those trying to reconstruct the development of Darwin's thought. The first is along the coast of South America. The second is at the Galápagos Islands.

As the Beagle made its way down the east coast of South America, Darwin went ashore wherever the ship stopped and observed the terrain and the animal and plant life, collecting samples when possible. As the ship continued south down the coast, the prevailing climate at each locale slowly changed the farther they got from the equator. One thing that Darwin noticed was as the climate changed, the flora and fauna changed. In any one location there would be a balanced array of plants and animals filling (as we might say) ecological niches. In the next location, the array would be similar, but slightly different, and the differences reflected the climate. Plants of similar type would be hardier to survive a harsher climate. Animals would have thicker fur, and so forth. The life forms were adapted to their environments. Were there separate creations of entire environments for each set of climatic conditions or was there a process of adaptation going on?

Chapter Twenty Two
Evolution by Natural Selection

417

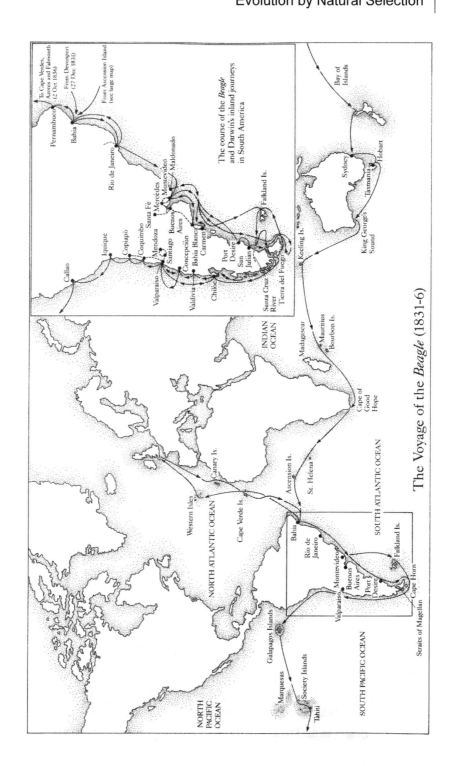

And then there was another kind of variation: over time. In the same locale where Darwin encountered new animals, distinctive to the region, he also found fossils of extinct creatures with very similar features. For example, in a region where the exotic *armadillo* with its armor covering of bony plates was native, fossils of the extinct *edentates* were also found. The *edentates* were far larger than the *armadillos* but had the same unusual anatomical structure. Why would there be changes over time of these very similar creatures?

Darwin was carrying with him Lyell's *Principles of Geology* with its summary of Lamarck's theory of evolution. Darwin was therefore familiar with both Lamarck's principle of the inheritance of acquired characteristics, which could account for both of these kinds of adaptations, and with Lyell's condemnation of Lamarck for his lack of evidence. Lyell had argued for the fixity of species and for special creations for each different kind of creature. At the same time, Lyell argued for what appeared to be just the opposite view for the terrain. His uniformitarian geological theory called for slow, imperceptible change over eons of time, which would be able to account for all the natural features that Darwin was encountering. Why was Lyell opposed to slow change due to natural causes in species while advocating it for the Earth itself? Was it because he had not seen the kind of evidence that Darwin had seen on the voyage?

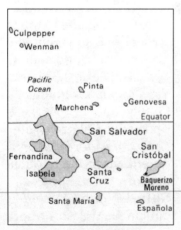

The Galápagos Islands.

The second part of the voyage of the Beagle that has garnered great interest among historians was the visit to the Galápagos Islands. After the Beagle made it down to the bottom of South America, it went through the Straits of Magellan, then up the west coast to Ecuador, where it turned west and traveled about 600 miles to the Galápagos Islands, a group of nine volcanic islands lying directly on the equator.

The abundance of exotic life here is extraordinary. By recent standards, half of the birds and plants are different from species elsewhere in the world, about a third of the shore fish and nearly all of the reptiles are unique to the Galápagos. At the outset of his visit, Darwin was so impressed by the diversity that he did not see the patterns that struck him later. What was so curious about the nine islands was that although they all had exactly the same climate and the same geological conditions, the species varied from island to island. The giant tortoises had such different markings that the local inhabitants knew at once which island they came from. (It is the tortoises that gave the islands their name. When Spanish explorers discovered the islands, they named them "Islands of the Tortoises." The Spanish word *galápago* means "tortoise.") It was not until Darwin was ready to leave the islands that he realized that this was important information, and he got the natives to tell him as much as possible about where each species that he had collected had come from.

On the islands themselves, Darwin noticed that what he at first took to be different varieties of mockingbirds must surely be different species. And this set him to thinking. Why would there be different species of mocking birds for each island when they all have the same physical characteristics? Was he to believe in a special creation for each island? Was it not more likely that a single species of mockingbird migrated to the islands and then evolved in separate ways? But then, if Lamarck was correct and individuals of a species adapted to local conditions and then passed on their acquired characteristics, why were they different?

Darwin's drawing of finches with different beaks.

On his return to England he illustrated this divergence of form with examples of finches that, with help from an ornithologist, he identified as all different, but similar, species. The finches, that were found on all the islands, had different beaks on different islands. On one island, finches existed that had strong, sturdy beaks for breaking open hard nuts and thick shells; on another, the beaks were still sturdy but less rounded, more suitable for slightly softer and smaller bits; still smaller beaks were found on finches who fed mostly on insects; and very narrow pointed beaks suited those who dug for grubs. In each case, the physical characteristics were adapted to the local situation.

Part IV
What is Life?

Darwin the Patient Collector of Evidence

Darwin had asked himself many questions on the voyage of the Beagle and had not found satisfactory answers to them. But where others would have either put the questions out of their minds or settled on answers that were more or less satisfactory, Darwin did neither. Instead, he settled into years of clarifying the issues, breaking them into smaller, easier questions, and then investigating all of them to the best of his abilities for as long as it was going to take to solve them.

Darwin's home in the village of Down, near London.

After three years in London, Darwin had had enough of city life. He married his cousin, moved to the village of Down outside London, raised a family, and remained there for the rest of his life. Once settled in at Down, Darwin began to settle into a routine that rarely varied. He saw very few people, preferring the company of his close and trusted colleagues to any contact with a wider public. He worked for only a few hours each day, went for long walks, took afternoon naps, and wrote letters. He had some sort of long-lasting illness that he may have contracted while on the Beagle. No one knows for sure.

He made definitive studies of barnacles and of orchids. He visited breeders of cattle and pigeon fanciers and they discussed their methods. All the while, he lived as a recluse and worked slowly and methodically. He continued this for twenty years. Still he had not found everything that he thought he needed to know before he could answer those questions that he first began to formulate on the Beagle.

As Darwin had seen on the coast of South America, species vary systematically from place to place and over long periods of time. How could one explain these variations? For that matter, how could you explain the similarities? Darwin realized that before one attempts to explain how species may change, it is necessary to have a theory of how they remain the same. Very little was known about inheritance and how it works. Even if he could explain how a species reproduces itself, he had to explain why the individual members of that species were then not all identical.

Darwin Reads Malthus

In 1798 the Reverend Thomas Malthus published a provocative book called *An Essay on Population*, which sent a fright through the British reading public by arguing that it was inevitable that humanity was either going to run out of food and starve or begin fighting over the remaining resources. The thrust of the argument was that the natural rate of population growth is greater than the most optimistic rate imaginable of increase in food supply. Someone recommended the book to Darwin and he read it in the year 1838—before he moved to Down House. Malthus' *Essay* started a train of thought in Darwin.

The Reverend Thomas Malthus.

Suppose it were also true in the plant and animal world that the birth rate of all populations would naturally lead to a vast increase in population, but the resources available could only sustain a fixed number. Then, there would be just what Malthus predicted for the human population; there would be competition over the resources. The competition would have winners and losers. The winners would be those who were able to get enough to eat and stay away from predators and disease. The losers would succumb one way or another. The winners would mate and produce the next generation. The losers would die out without progeny.

The result is that there would be a natural mechanism for selecting the strong and fit from any population and letting them reproduce, while weeding out the weak. If the environment changed, the attributes that best led to survival might also change, but the individuals who were best suited to the new environment would be the ones to leave offspring.

Continuous Variation and Selective Breeding

In order for this to produce a continuous evolution of attributes, it was necessary that the offspring in a given generation not be all the same. If they were, a process that weans out the best of them would have nothing to choose from. So, there must always be differences within a population, some of which are more desirable than others. Moreover, the variations have to keep varying. If each generation had just the same set of possibilities, no great changes could be accounted for. Certainly, the sort of changes that Lamarck illustrated with the giraffe would not be possible. But what causes variations?

Darwin had to be satisfied that if those individuals with the most suitable characteristics would survive and breed, then that process, repeated again and again, could add up to significant change. Darwin knew that farmers for years had been selecting which of their livestock to let breed and which not. He also knew that some remarkably different characteristics had been bred this way over many generations. To get more details on this, Darwin visited with farmers in the vicinity to learn more about their methods. Likewise, Darwin knew that people who raised pets and animals for show could produce extraordinary visible characteristics after a number of generations. Among the most extraordinary of these were the pigeon fanciers who managed to raise pigeons with truly outrageous features, by selecting which bred with which. Darwin therefore went to bird shows and talked with the breeders there. He had to know how many generations were required to change certain features and how pure the stock had to be, among other issues. Darwin's own extensive work on raising orchids sought to find the same answers in the plant kingdom.

All this is what Darwin was doing for twenty years of patient work, accumulating the evidence he needed to present *his* theory of evolution, which would be in a proper scientific form and which would have the convincing evidence that the earlier theories all lacked. Then the bombshell hit.

Alfred Russel Wallace

Darwin is the most famous naturalist of all time, but he was one of many in this odd profession that was characteristic of 19th century Britain. Naturalists came in many kinds. Darwin was what might be called a "gentleman-naturalist." He was from a wealthy family; he had a university education. Some other naturalists came from poorer social circumstances and chose their career rather as one might choose a trade, relying for income on funded expeditions and a ready market for their collections in museums back home. Such a person was Alfred Russel Wallace.

Wallace was born in 1823, fourteen years after Darwin, into a poor working-class family. Liking the outdoors, he trained to be a surveyor. (Because of the Industrial Revolution and the building of the railroads and canal systems, there was a lot of work for surveyors.) On his days off, he liked to wander in fields and collect interesting specimens of animal life. After a time, and with the encouragement of a like-minded friend, Henry Bates, he decided to try to make a profession of specimen collecting. For Wallace, like for Darwin, the critical maturing experience was a long trip to faraway places. Wallace

Alfred Russel Wallace.

made an expedition to South America to travel up the Amazon with Bates, during which he saw some of the same phenomena that puzzled Darwin. Unfortunately for Wallace, on the return voyage home, his ship caught fire and he lost all his specimens. Nevertheless, unfazed, he soon set off on another venture, this time to the Far East and the Malay Archipelago.

Wallace had taken Lyell's *Principles of Geology* with him to South America and became as familiar with it as Darwin had been. He had also carefully read Robert Chambers' *Vestiges of the Natural History of Creation* and had begun to think carefully about the transmutation of species. Like Darwin, he sought a mechanism to drive the change. Like Darwin, he thought he could perhaps solve the species question and planned to write a book on it. In 1855 he published an article in which he proposed that species "come into existence" at a time and place where there is already a very similar species existing. Darwin read the paper and thought well of it.

Part IV
What is Life?

Wallace and Darwin began to correspond with each other—Wallace using Darwin as a sounding board for his ideas, and Darwin giving encouragement, but not sharing very many of his own thoughts with Wallace. Darwin had discussed his developing ideas about evolution only with a very small number of trusted colleagues and was not ready to announce them to a wider audience. Among those was none other than Charles Lyell, who, despite being set against the idea of transmutation of species, kept an open mind and encouraged Darwin in his work. Lyell wrote to Darwin in 1856, after Wallace's paper had appeared, urging Darwin to publish at least a sketch of his ideas, otherwise someone else would come along and claim to have thought of them first. Darwin wrote back to Lyell that to give a fair sketch would be "absolutely impossible." Yet, he did worry about being upstaged:

> ... I do not know what to think; I rather hate the idea of writing for priority, yet I certainly should be vexed if any one were to publish my doctrine before me.[5]

To Wallace, Darwin began to hint that he had been working on these issues for a long time, without giving him any details. In 1857, Darwin replied to a letter from Wallace saying:

> ... I can plainly see that we have thought much alike and to a certain extent have come to similar conclusions... This summer will make the 20th year (!) since I opened my first note-book, on the question how and in what way to species and varieties differ from each other. I am now preparing my work for publication, but I find the subject so very large, that though I have written many chapters, I do not suppose I shall go to press for two years.[6]

Wallace pressed on with his ideas. Then he just happened to read that very same book by Thomas Malthus, the *Essay on Population*, which had started a chain of thought in Darwin twenty years before. Unlike Darwin, though, Wallace wasted no time incorporating these new ideas into his grand scheme and committing it all to paper. The lack of mountains of evidence that Darwin thought necessary was not going to hold up Wallace. One day, when Wallace was fighting an attack of ague (probably malarial fever) and had to stop working, he thought the whole thing through in a flash, and wrote it up in a short

---

5. Letter from Darwin to Charles Lyell, May 3, 1856. In Philip Appleman, ed. *Darwin* (New York: Norton, 1970), p. 63.
6. Letter from Darwin to Alfred R. Wallace, May 1, 1857. *Ibid.*, pp. 65-66.

paper for publication. He thought Darwin would appreciate it, so he sent it off to him asking for his opinion of it and asking him to show it to Lyell as well.

Darwin in a Dither

Darwin received Wallace's paper in June of 1858 and was aghast at what he saw. Wallace had sketched out a theory of evolution that in all essential respects was what Darwin had been struggling with for over for twenty years. Wallace had tossed it off in a few hours after his reading of Malthus. Were this to be published before Darwin got any of his work into print, Wallace would get all the credit for the theory, and Darwin's work would just be viewed as filling in the details. Even worse, Wallace had, in good faith, sent the work to Darwin asking for his opinion, so Darwin, as a gentleman, was honor bound to treat it with respect. He could hardly now rush into print with his own theory, claiming that he was unaware of Wallace's work. Darwin wrote an agonized letter to Lyell:

> [Wallace] has to-day sent me the enclosed, and asked me to forward it to you. It seems to me well worth reading. Your words have come true with a vengeance—that I should be forestalled. You said this, when I explained to you here very briefly my views of "Natural Selection" depending on the struggle for existence. I never saw a more striking coincidence; if Wallace had my MS. sketch written out in 1842, he could not have made a better short abstract! Even his terms now stand as heads of my chapters... I shall of course at once write and offer to send [it] to any journal. So all my originality, whatever it may amount to, will be smashed...[7]

A week later Darwin was beginning to think of ways in which his priority could be established over Wallace, but he was ashamed of his plan, which he thought lacked honor. Nevertheless, he sought the advice of Lyell.

> My dear Lyell,—I am very sorry to trouble you, busy as you are, in so merely a personal an affair; but if you will give me your deliberate opinion, you will do me as great a service as ever man did, for I have entire confidence in your judgment and honour...
>
> There is nothing in Wallace's sketch which is not written out much fuller in my sketch, copied out in 1844, and read by Hooker some dozen years ago. About a year ago I sent a short sketch, of which I have a copy, or my views...to Asa Gray, so that I could most truly say and prove that I take

---

7. Darwin to Lyell, June 18, 1858. *Ibid.* p. 67.

> nothing from Wallace. I should be extremely glad now to publish a sketch of my general views in about a dozen pages or so; but I cannot persuade myself that I can do so honourably... I would far rather burn my whole book, than that he or any other man should think that I behaved in a paltry spirit. Do you not think that his having sent me this sketch ties my hands?...
>
> This is a trumpery affair to trouble you with, but you cannot tell how much obliged I should be for your advice...
>
> I will never trouble you or Hooker on the subject again.[8]

Lyell and Darwin's other confidante, Joseph Hooker, agreed that Darwin should publish something that showed that he had been on the same track for much longer than Wallace, though not denigrating Wallace's having come up with the ideas himself. It all moved very quickly from then on. On July 1, 1858, less than a week after Darwin's distraught letter to Lyell quoted above, a meeting was held in London of the Linnean Society at which three papers were read: the paper that Wallace had sent to Darwin, an excerpt from one of Darwin's notebooks laying out the principle of natural selection, and a copy of a letter Darwin had written to the American botanist Asa Gray in 1857 in which Darwin had discussed the general principles of his theory—which could be verified by Professor Gray, if necessary.

The papers made very little impression on those in attendance at the Linnean Society meeting, and when published engendered no response. But now the cat was out of the bag. Darwin had to put misgivings aside and get his book in print soon or risk being labeled just another wild-eyed speculator.

The Book That Shook the World

As Darwin had written to Wallace, he was not yet ready to publish his theory in full, with all the evidence he needed to prove it, but he would hurriedly put together an "Abstract [which] will make a small volume of 400 or 500 pages." The "Abstract" turned out to be almost exactly 500 printed pages. It was published November 24, 1859 under the title: *On the Origin of Species By Means of Natural Selection, Or the Preservation of Favoured Races in the Struggle for Life*. The complete print run of 1250 was sold out on

---

8. Darwin to Lyell, June 25, 1858. *Ibid.*, pp. 67-68.

Chapter Twenty Two
Evolution by Natural Selection

the first day. The reaction to the book was immediate and violent. The impact of *On the Origin* on both the scientific community and the general public was greater than any other scientific work ever published.

Elements of Darwin's Explanation

Despite Darwin's view that years and years of work were necessary to provide the supporting data and 500 pages were needed to write up an abstract, the major features of Darwin's theory are easy enough to state.

The first essential is the concept of *continuous variation*. Darwin asserted that every generation of a species has a natural variation of characteristics that varies around the characteristics of the parents. The key point here is that whatever characteristics the parents have will be the center point, so to speak, of the variation of their offspring.

Title page of the first edition of *On the Origin of Species*, published November 24, 1859.

The second element is the principle of *selection*. Darwin had endeavored to establish through his visits to breeders that the direction of change in a species can be affected in a deliberate and predictable way by selecting which individuals will be allowed to reproduce. This is *artificial selection*. The comparable effect in nature is *natural selection*, where the conditions of the environment will favor individuals with certain characteristics and present challenges for others. Those with the favorable characteristics will be more likely to survive long enough to mate. There is a third aspect of this, a variant of natural selection, really, called *sexual selection*. Just because an individual survives the natural selection process, it still must find a mate to leave offspring. Sexual selection is the means through which individuals of a species find each other and mate. The point is that those who are most fit to survive are likely to choose to mate with others having favorable characteristics and thereby intensifying the effect of natural selection.

Finally, a necessary feature of the world for all of this to achieve the results ascribed to the theory is a vast amount of time—the amount of time that geologists were ascribing to the Earth, not the slim few thousand years that some religious leaders still espoused.

What is the Cause of Variation?

A serious sticking point for Darwin was the issue of continuous variation. It was essential in Darwin's theory that animals and plants continue to vary in each generation, but Darwin had no reasons to support this. He saw that it was so, wherever he looked, but that was not the same as specifying a "mechanical" cause. Without such a cause, his theory lacked scientific authority. Though Darwin sought the cause of this, he could not find it. This is surely one of the reasons that Darwin deemed he was not ready to publish. Ultimately, he had to rely upon many, many cases of observed variation, and then make the induction that it is a general principle.

Note that this was a major clash of scientific styles. The ancient Aristotelian style was exactly how Darwin was working. Examine nature closely, make generalizations, and state them as general principles. This was the style of doing science that was in disfavor since the Scientific Revolution. A mechanical cause had to be specified; then it had to be shown that such a cause would produce the desired result. Or, from the point of view of the ancient Greek conflict, Darwin did not have the explanation to "save the appearances," as Plato would have desired.

Pangenesis

Darwin's real problem was deeper still. Why were the offspring of cats more cats and not dogs? Or rose bushes, for that matter? Neither Darwin, nor anyone else, had a viable theory of how any inheritable traits are passed on. Nor how the mating process results in new births. Sure, it had been described and observed repeatedly, but that was the same problem. What was the mechanism?

This stands as a good example of how science does not proceed in an orderly logical fashion. It would have made much more sense to have figured out how genetics works

Chapter Twenty Two
Evolution by Natural Selection

first—i.e., how species remain the same—and *then* address the question of how species change. Instead, the theory of evolution was the first to be thought through.

It was clear to Darwin that he had better propose some account of inheritance or his theory would lose all claim to being a scientific account. So, Darwin fixed upon a particular brand of inheritance that he called *Pangenesis*. Darwin suggested that what happens in the mating process is this: from all parts of the body, small bits called "gemmules" were thrown off into the bloodstream and collected in the reproductive organs. In the conception process, gemmules from both parents mixed and a blend of each characteristic represented by the gemmules resulted. Since each conception resulted in a slightly different mix of gemmules, variations naturally occurred. And, since it was the characteristics of the parents that were mixing, the variations were centered on some kind of average of the parents' traits.

Once again in science the Ad Hoc explanation was offered. Pangenesis theory might have been ingenious, but there was no evidence to support it and the explanation raised a host of problems. How did these gemmules work? What was the mechanism through which they directed growth? Darwin could only speculate by comparison with embryological development, which was, to be sure, equally mysterious. What about the blending process? How did it happen? Were individual gemmules merged or did each inheritable trait take several gemmules and it so happened that some would come from one parent and some from the other?

And where did these gemmules come from? Did they originate in particular cells of the bodies of each parent? (Cells as the building blocks of living things had been proposed during the time that Darwin was thinking through his theory and was becoming the established view about the time that Darwin published *The Origin*; see Chapter 23.) Suppose that a man went off to war, sustained a serious injury, and had to have a limb amputated. (Amputations to stem the spread of gangrene were frequent battlefield operations in the 19th century.) Then this man came home, married, and raised a family. Where did the gemmules for his amputated leg come from? Were we to expect offspring with deformed limbs, or one limb that had all its characteristics drawn from the mother while the father contributed equally to the other? What if both parents had had the same amputations? (There were such cases, and the children were all normal.)

It may be that no other explanation occurred to Darwin, but he cannot have been satisfied with this one without a considerable amount of confirming evidence. Darwin did not wish to be compared with the evolutionary speculators (like his grandfather). He wanted to have a full scientific theory with evidence to establish it. Getting that evidence was part of what he was doing during those twenty years since he first conceived the general outline of his theory of evolution by natural selection. He was still at it when Wallace forced his hand and made him go to press with his incomplete theory.

Darwin's Attack on the Design Argument

If Darwin had to "rush" into print before he was ready, he would not be able to produce a complete explanation in the accepted form of Newton's *Principia* (which was itself in the form of Euclid's *Elements*) with a complete theory that laid out the mechanism whereby species are produced and evolve. There were too many holes in the explanation, too much based on induction from observations, too little understanding of the mechanics of life processes. Whether he would have articulated the problem this way or not, Darwin somehow understood this. Faced with the need to go public, Darwin would have to lay out the general framework of his theory (in his 500-page "outline"), indicate how he thought the processes worked, and produce the mountains of empirical evidence that supported his view, acknowledging where he didn't know the exact process. It was bound to be controversial, and Darwin hated controversy.

The best course available to Darwin was going to be to show that his theory was plausible—it could account for evolution if the details could be worked out—*and* it was a better explanation than any of the alternatives. However, before he could do this, the first step was to show that his explanation was even possible. He had to take on the revered Design Argument, the basis of the doctrines of Natural Theology, and show that the logic was faulty. Whether he was consciously thinking of the logic of the Design Argument and was deliberately seeking to puncture it is hard to say. In any case, he did it well.

The history of the Design Argument goes back at least to Ancient Greek philosophy. To Aristotle, it was manifest that there was purpose in nature; so much so that no account of a thing that did not identify its purpose could count as an explanation. A

Chapter Twenty Two
Evolution by Natural Selection

corollary for Aristotle was that only events with purposes will repeat themselves. Accidents remain isolated events and can add up to nothing. As a result, they can be ignored.

Empedocles' Man-Faced Ox Progeny

A fascinating application of Aristotle's reasoning occurred in his *Physics* where Aristotle attempted to refute the idea of the world being organized by a blind mechanism instead of purposefully. He cited a speculation by the philosopher Empedocles—the originator of the concept of the four elements, earth, air, fire, and water—on the origin of species. Empedocles had speculated that originally various bodily parts existed independently and unjoined to particular animals. These then randomly ran into each other, attached, and formed bodies. Where those bodies made viable animals, they survived. Where they made monsters, they perished—the example being the "man-faced ox progeny." In a crude and bizarre way, Empedocles had actually proposed the principle of natural selection. What is interesting is how Aristotle disposed of it.

> Why should nature work, not for the sake of something, nor because it is better so, but just as the sky rains, not in order to make the corn grow, but of necessity? What is drawn up must cool, and what has been cooled must become water and descend, the result of this being that the corn grows. Similarly if a man's crop is spoiled on the threshing-floor, the rain did not fall for the sake of this…in order that the crop might be spoiled…but that result just followed. Why then should it not be the same with the parts in nature, e.g. that our teeth should come *of necessity*—the front teeth sharp, fitted for tearing, the molars broad and useful for grinding down food—since they did not arise for this end, but it was merely a coincident result; and so with all other parts in which we suppose that there is purpose? Wherever then all the parts came about just what they would have been if they had come to be for an end, such things survived, being organized spontaneously in a fitting way; whereas those which grew otherwise perished and continued to perish, as Empedocles says his "man-faced ox-progeny" did.[9]

What was wrong with Empedocles' argument, according to Aristotle, was that the random processes that Empedocles mentioned would not happen again and again, *because* they were random. Only things that repeated predictably could have consequences that went beyond the immediate result. In consequence, it was a waste of time to try to study

---

9. Aristotle, *Physics* Book II Ch 8, 198b 17-33, in *The Basic Works of Aristotle*, ed. Richard McKeon, 249 (New York: Random House, 1941).

things that happened by accident. Science, therefore, concerned regularities, not exceptions.

One can see that this was the only practical approach to take in classical times. It was hard enough to figure out what was going on with things that repeated again and again. A philosopher trying to find a pattern in random events would have gotten nowhere. As it was, the Greeks made most progress with making some sense out of planetary astronomy, where the number of objects of study was down to a handful and the events studied had a regular pattern.

To Aristotle, it is manifest that the "parts in nature," for example, bodily organs, can only exhibit the design that they do because they were made with purpose in mind. A design implies a designer. For Aristotle, that designer could be nature itself.

For 19th century Britain, the design was the "Effect of Choice," as Newton put it, and the designer was the God of creation. All those "Bridgewater Treatises" were written with this logic in the background. Nature is wonderful, because God made it so. For proof, you need only look at anything really complicated, like the hand, and you see divine intervention. If this logic is airtight, evolution by natural selection is impossible.

The Logical Structure of the Design Argument

Consider the logic of the Design Argument as a syllogism:

Let H_D be the hypothesis that God designed every species in its present form.

Let $T_1, T_2, T_3, \ldots$ be test implications—evidence of designs in nature, for example, the complex species and their superb adaptations for complex tasks.

The hypothesis accounts for the test implications:

$$H_D \text{ implies } T_1, T_2, T_3, \ldots$$

The Design Argument takes it as given that if there is design, there must be a designer:

$$T_1, T_2, T_3, \ldots \text{ implies } H_D.$$

Chapter Twenty Two
Evolution by Natural Selection

Since the test implications can be seen to be true (there is obvious design in nature), this argument has the valid conclusion that there must be a designer.

H_D is true.

It's quite a simple structure. The line about design implying a designer was the point argued by Aristotle in his refutation of Empedocles' natural selection argument. So long as this line is accepted as inviolable, the existence of God the creator is established.

Compare this to the logic of Darwin's explanation of evolution by natural selection:

Let H_E be the hypothesis that species evolved into their present forms via the process of Natural Selection.

Let $T_1, T_2, T_3, \ldots$ be test implications—evidence of designs in nature, just as in the Design Argument.

Darwin's hypothesis accounts for the test implications in just the same way as does the Design Argument:

H_E implies $T_1, T_2, T_3, \ldots$

But that other line is missing. Darwin cannot claim that just because there are designs in nature that those designs imply evolution:

$T_1, T_2, T_3, \ldots$ do not imply H_E.

If Darwin can show that the assertion that design implies a designer is invalid, then he puts his explanation on the same logical footing as the Design Argument: Namely that either a Designer (God) or Natural Selection *could* produce the designs that are evident in nature. Notice that the assertion of the Design Argument is *not* that it is highly improbable that living things were formed solely according to simple physical laws. Instead, it asserts that *it is impossible* for them to have done so. It is impossible because it would require many coincidences, and things that happen as a result of coincidence or chance do not happen with regularity. Since living things exhibit marvelous regularity in their structures and fit so well together in the order of nature, they could not have resulted from coincidental events that took place for no purpose and were merely the result of random physical motion.

No amount of evidence can assail this argument until the fallacy in it is seen. For me, the greatest achievement of *On the Origin of Species* was to reveal that fallacy. The fallacy is that the Design Argument fails to take account of time. Random causes, such as whatever caused the variations in plants and animals that Darwin observed, produce random effects in the *first* instance. But if one of those random variations, however infrequent, has a greater tendency to be preserved than other variations, then in successive repetitions of the generating process that variation becomes more and more frequent until its occurrence becomes the regular, expected effect.

The Eye

In Sir Charles Bell's Bridgewater Treatise *The Hand*, the staggering complexity and intricacies of structure of hand-like appendages in different animals that make them so perfectly adapted to their functions provide all the evidence of design necessary for Bell to conclude that only God could have made them so; they could certainly not have arisen as chance artifacts of natural processes.

To demonstrate a logical fallacy, it is only necessary to produce a *counterexample*, an implication of the logical construction that is false. This is the principle of *Modus Tollens*, the most useful of all the syllogistic forms for scientific reasoning. If Darwin can find a single example of a complex, purposeful design in a living being, then the argument of Bell loses its force. Bell had written about the hand. Darwin chose an even more mysterious organ for his counterexample, the eye. In the passage below from *The Origin*, notice that the thrust of Darwin's argument is *not* to prove that the eye evolved from natural causes, but only to defeat the logic of the design argument by showing that it is *possible* that it did evolve naturally:

> To suppose that the eye, with all its inimitable contrivances for adjusting the focus to different distances, for admitting different amounts of light, and for the correction of spherical and chromatic aberration, could have been formed by natural selection, seems, I freely confess, absurd in the highest possible degree. Yet reason tells me, that if numerous gradations from a perfect and complex eye to one very imperfect and simple, each grade being useful to its possessor can be shown to exist; if further, the eye does vary ever so slightly, and the variations be inherited, which is certainly the case; and if any variation of modification in the organ be ever so useful to an animal under changing conditions of life, then the difficulty of believing that a perfect and complex eye could be formed by

Chapter Twenty Two | 435
Evolution by Natural Selection

> natural selection, though insuperable by our imagination, can hardly be considered real. How a nerve comes to be sensitive to light, hardly concerns us more than how life itself first originated; but I may remark that several facts makes me suspect that any sensitive nerve may be rendered sensitive to light...

Darwin goes on to give some examples of light-sensitive nerves in various crustaceans, showing that eye-like contrivances exist in many degrees of complexity and relative perfection. This is to establish the plausibility of the idea that something like an eye evolved from a mere light-sensitive nerve. Then Darwin continues:

> With these facts, here far too briefly and imperfectly given, which show that there is much graduated diversity in the eyes of living crustaceans, and bearing in mind how small the number of living animals is in proportion to those which have become extinct, I can see no very great difficulty...in believing that natural selection has converted the simple apparatus of an optic nerve merely coated with pigment and invested by transparent membrane, into an optical instrument as perfect as is possessed by any member of the great Articulate class.
>
> He who will go thus far, if he find on finishing this treatise that large bodies of facts, otherwise inexplicable, can be explained by the theory of descent, ought not to hesitate to go further, and to admit that a structure even as perfect as the eye of an eagle might be formed by natural selection, although in this case he does not know any of the transitional grades. His reason ought to conquer his imagination; though I have felt the difficulty far too keenly to be surprised at any degree of hesitation in extending the principle of natural selection to such startling lengths.[10]

In both Darwin's analysis of the eye and Bell's analysis of the hand, the chief source of evidence is comparative anatomy of eyes and hands of living creatures. Bell's argument is that this great complex diversity proves the existence of a purposeful agent of design. Darwin's argument is that it doesn't prove that, because it is *possible* that these marvelous effects are the result of gradual evolution by natural selection over enormous periods of time.

Time as a factor in a scientific theory was not new with Darwin. The uniformitarian geology that guided Darwin's thoughts depended on the accumulation of small causes to produce large effects. But while the arrangement of geological things in the world was noticeably different from ages past, the uniformitarians did not claim that the level of order

---

10. Charles Darwin, *On the Origin of Species* (London, 1859), pp. 186-188.

in the present was different from that in the past. The nebular hypothesis required a vast amount of time in its explanation of how the apparent order of the solar system arose from a chaos of gas. But the order that was to be explained was the direction of orbit and rotation of the planets, and that was already present in the rotating nebula that was its former state. Similarly, Lamarck's theory of evolution held that species slowly changed over long periods of time, but the changes were effected by the individuals in each generation, adapting themselves to the environment and then passing on that adaptation to the next generation. Hence, the chief phenomenon to be explained, from the point of view of the Design Argument, the *orderedness* of living things, was given to them as a capacity from the start.

In Darwin's use of time, the resulting order is far greater than that of its causes, and the random causes are unrelated to the ordered effects. Darwin, of course, did not point out the fallacy in the Design Argument in this abstract fashion; he presented his case in terms of concrete biological phenomena. Because he argued in concrete terms, the truth or falsity of his conclusions depended on the facts. Because he did not have all the facts, he could not prove he was right, as he readily admitted. Nevertheless, his logic stands, independent of the theory of evolution.

The importance of the logical argument is that it not only cleared away the obstacle of the Design Argument from the advance of biological theory, it also opened new vistas for many subjects in which long sequences of events might produce unexpected results. The great popularity of Darwin's theory in the late nineteenth century assured that both his conclusions and his arguments would get careful scrutiny. From direct analysis of Darwin's works and from that indirect process through which bits and pieces of the thought of an influential person get assimilated into the common notions of an era, an evolutionary approach has become part of the analysis of many problems that have nothing to do with biological species.

An example is the work of Darwin's son, Sir George Darwin on lunar evolution. The younger Darwin argued that the reason that the Moon is always turned the same way toward the Earth, instead of rotating on its axis, is due to the long-term effect of the tides on Earth. It is the Moon that causes the tides in the first place, but since the Earth rotates, the gravitational pull of the Earth produces a slight drag on the Moon. That drag, argued

George Darwin, could suffice to slow down and eventually stop a rotation of the Moon on its own axis. The subject matter was not biological, but the reasoning followed the pattern of his father's logic, namely to show that the known effect *can* follow the proposed cause, given sufficient time. Then, anyone who wishes to oppose the suggested theory cannot do so by claiming it is impossible.

Weight of Evidence

If you accept Darwin's argument that complex designs can arise by random forces and the process of natural selection, then the clinching implication of the Design Argument is invalid and both explanations have the form:

$H_{(D \text{ or } E)}$ implies $T_1, T_2, T_3, \ldots$

$T_1, T_2, T_3, \ldots$ are true.

Therefore $H_{(D \text{ or } E)}$ is possible (but not necessary).

Once the theory of evolution and the design argument are on the same logical footing and neither is in itself conclusive, deciding among them becomes a matter of which explanation is more plausible, given the evidence. Darwin's work is primarily a reporting of a vast amount of evidence showing that the theory of evolution by natural selection accounts for the diversity of nature in a straightforward and logically simple manner.

What it does not do is *prove* that it is the correct interpretation; it can only claim to be simplest explanation, consistent with what else is accepted. This is true of all scientific theories. They all "save the phenomena" by showing how they make sense within a rational explanation. Newton's theory of universal gravitation is not proven, nor are his laws of motion, but if you accept them, you can explain a great deal. This is what a scientific explanation is. We accept it because it is superior to any other explanation we know of.

The theory of evolution by natural selection is no different from any other scientific theory in this regard, but because it relies upon a huge amount of evidence and because the Design Argument still has appeal for reasons which lie outside of science, the theory of evolution has sometimes been regarded as less scientific than theories of physics, or chemistry.

Part IV
What is Life?

Darwin on Man

Darwin portrayed as an ape in a cartoon that appeared shortly after his publication of *The Descent of Man* in 1871.

When *The Origin* was published, there was an immediate storm of protest in the public press. Darwin was called a fool, an atheist, a dangerous man. There can be no doubt that the reason that the public suddenly took such great interest in his theory of evolution is that it threatened them in some way. Evolutionary theories were popular topics of discussion throughout the 19th century, but they were easily dismissed as flights of fancy or at least bad science. The Design Argument still held sway, and this implied that there had been special creations for all species, especially the human species. Darwin's work was too carefully written to be dismissed as fantasy and his demolishing of the Design Argument pulled the rug out from under those who were clinging to the comfort of special creation.

Darwin had anticipated the reaction his work would get, and this is probably part of what held him back from publishing for so many years. Whether he addressed human evolution directly or not, it was bound to be what the public would fix upon. Back in 1857, before Wallace had dropped his bombshell on Darwin—but after they had begun to correspond—in answer to Wallace's question as to whether he was going to discuss human evolution in his upcoming work, Darwin replied:

> You ask whether I shall discuss "man." I think I shall avoid the whole subject, as so surrounded with prejudices; though I fully admit that it is the highest and most interesting problem for the naturalist.[11]

True to his word, *The Origin* does avoid human evolution altogether. Darwin slipped one sentence in suggesting that "Light will be thrown on the origin of man and his history." But of course, though he may have avoided it, his public critics did not. Neither did his public supporters. Darwin did his best to avoid having to discuss and defend his ideas in public, but others did, notably Thomas Henry Huxley, who so made a career of defending

---

11. Darwin to Wallace, December 22, 1857. In Appleman, ed., *Darwin*, p. 66.

Darwin from his critics that he earned the nickname "Darwin's Bulldog." The most famous confrontation that Bulldog Huxley made was at a famous meeting of the British Association for the Advancement of Science at Oxford in 1860. The Bishop of Oxford Samuel Wilberforce had turned to Huxley in the audience and sarcastically asked if Huxley claimed his descent from the apes through his grandfather or his grandmother. After the laughter had died down, Huxley rose and said that he would not be ashamed to have an ape for his grandfather, but he would be ashamed

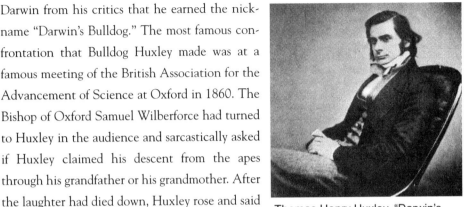

Thomas Henry Huxley, "Darwin's Bulldog."

to have to acknowledge as ancestor a man of such great influence and intelligence who, nevertheless uses those high endowments to introduce prejudice and ridicule into a sober scientific debate.

Darwin did have a lot to say about *human* evolution in particular. In 1871 he finally published a book specifically on human evolution, *The Descent of Man*, in which he established a relationship between humans and the primates. Darwin also made extensive use of embryological evidence, showing that at the earliest stages of development, human embryos bear close resemblances to embryos of other mammals. Embryological investigation in support of evolutionary theory was a fruitful area of research that was followed up by others with considerable impact in the next few decades.

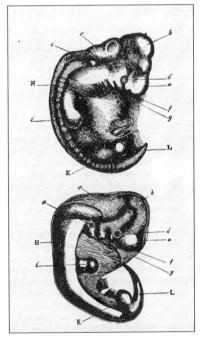

Darwin illustrated the close relationship between human embryos (at the top) and dog embryos (below) in *The Descent of Man.*

Part IV
What is Life?

The theory of evolution by natural selection has a special place in the history of science. It established many precedents in science: first, it introduced time as an important factor in scientific explanations. Galileo's and Newton's physics, for example, used time, but in the sense of time intervals. It didn't really matter when something started and when it finished, only how much time elapsed—for an object to fall or a planet to orbit, for example. But Darwin's use of time is a one-way chronology that affects the entire cosmos. The sort of changes he wrote about could not be transplanted to another time and place because the environment would be different. From Darwin onward, this use of historical time has been found to have great explanatory value.

Second, Darwin clarified the concept of a scientific theory as the best explanation that we have for its phenomena. The theory of evolution is a complex umbrella theory with many details unresolved. Nevertheless, the general framework is so superior to other possible explanations that we regard it as effectively proved. But that does not mean we regard it as absolutely true, nor would it mean that there is something basically wrong with scientific method if the theory was someday completely replaced by another explanation altogether. This was not that well understood in the 19th century. In the 20th century, physics had to face the major revision of Newton's system on both the large scale, with relativity, and the small scale, with quantum mechanics, but this does not mean that Newton's system was wrong and pointless just because better explanations were found in certain contexts.

Finally, Darwin gave biology a direction for research and a context in which to work. The tremendous variety of life forms was not going to be explained by physics and chemistry alone, without taking into account the processes over vast periods of time that Darwin had shown were necessary.

Darwin revolutionized biology.

Chapter Twenty Two
Evolution by Natural Selection

For More Information

Alioto, Anthony M. *A History of Western Science,* 2nd ed. Englewood Cliffs, NJ: Prentice-Hall, 1987. Chapters 20.

Appleman, Philip, ed. *Darwin.* A Norton Critical Edition. New York: Norton, 1970.

Barzun, Jacques. *Darwin, Marx, Wagner: Critique of a Heritage.* 2nd ed. Garden City, NY: Doubleday, 1958.

Bell, Charles. *The Hand: Its Mechanism and Vital Endowments as Evincing Design,* London, 1833.

Bowler, Peter J. *Evolution: The History of an Idea*, Rev. ed. Berkeley: University of California Press, 1989.

Burnet, Thomas. *The Sacred Theory of the Earth.* London, 1691. Reprinted with an introduction by Basil Willey. London: Centaur Press, 1965.

Chambers, Robert. *Vestiges of the Natural History of Creation.* London, 1844. Reprinted with an introduction by Sir Gavin De Beer. Leicester: Leicester University Press.

Cohen, I. Bernard. *Revolution in Science.* Cambridge: Harvard University Press, 1985. Chapter 19.

Coleman, William. *Biology in the 19th Century: Problems of Form, Function and Transmutation.* New York: Wiley, 1971.

Darwin, Charles. *Journal of Researches into the Geology and Natural History of the Various Countries Visited by H.M.S. Beagle.* London, 1839. Reprinted Brussels: Culture & Civilization, 1969.

_____. *The Descent of Man and Selection in Relation to Sex.* 2 vols. London, 1871. Reprinted Brussels: Culture & Civilization, 1969.

_____. *On the Origin of Species.* A Facsimile of the First Edition, with an introduction by Ernst Mayr. Cambridge: Harvard University Press, [1859], 1964.

Darwin, Erasmus. *Zoonomia; Or the Laws of Organic Life*, 4th ed. Philadelphia, 1818.

Eiseley, Loren. *Darwin's Century: Evolution and the Men Who Discovered It.* New York: Doubleday, 1958.

Greene, John C. *The Death of Adam: Evolution and Its Impact on Western Thought.* Ames, Iowa: The Iowa State University Press, 1959.

MacLachlan, James. *Children of Prometheus: A History of Science and Technology,* 2nd ed. Toronto: Wall & Emerson, Inc., 2002. Chapter 16.

Ronan, Colin A. *Science: Its History and Development Among the World's Cultures.* New York: Facts on File, 1985. Chapter 9.

Wall, Byron E., ed. *The Nature of Science: Classical and Contemporary Readings.* Toronto: Wall & Emerson, Inc., 1990.

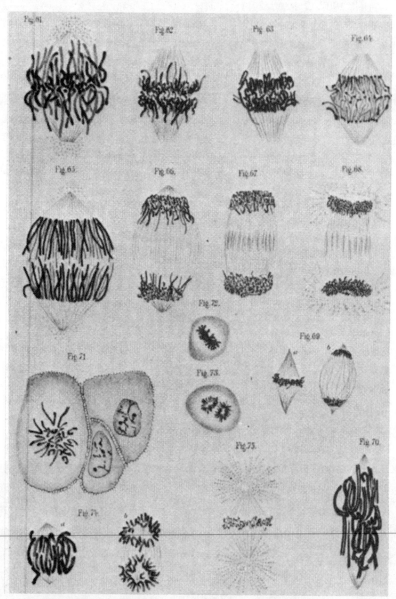

Chromosomes during cell division. From Walther Fleming, *Zellsubstanz, Kern und Zelltheilung*, 1882.

Chapter Twenty Three

Finding the Units of Life

While the naturalists were outdoors collecting specimens of living things and classifying them, a totally different approach to understanding life was developing indoors, in the scientific laboratory. This was the search for the minute structure of life, the building blocks out of which every living thing was made. The key to this work was the development of a tool—*the* scientific research tool for biology. This was the *microscope*. With the microscope, attention became focused on minute parts of living things, not on the whole individual. Researchers with microscopes tried to understand life processes in themselves, not just as they were exhibited by certain kinds of plants and animals.

The Microscope

The invention of the telescope was of tremendous importance for astronomy. Galileo showed that the sky had much more in it than could be even detected with the naked eye, and the detail that could be seen enabled him to draw conclusions that were important to establishing the Copernican worldview. Ever since, astronomy has relied upon telescopes of ever increasing sophistication to see farther and farther into the cosmos.

The telescope was invented at about the beginning of the 17th century. It was 1609 when Galileo started working with telescopes. The microscope dates from about the same time. There was an early version of the microscope invented in 1590 by a Dutch spectacle maker, Zacharias Janssen, which worked very poorly and was soon forgotten. One of the reasons it worked so poorly was that it was not much more than a telescope turned around. Early microscopes had two basic designs: *simple* and *compound*. Janssen's microscope was of the compound type, with two lenses, one positioned some distance over the other.

Microscopes of this type had one very serious limitation. Because light was refracted through the lenses, it separated into the different colors of the spectrum. The image that was produced of anything that was fairly small was so distorted by color fringes as to be nearly undecipherable. This is the problem of *chromatic aberration*. Telescopes have the same problem, as was noted by Newton, who got around the problem by inventing a telescope that worked entirely by reflection on a parabolic mirror. (See Chapter 15.) A similar solution for microscopes did not present itself.

Anton van Leeuwenhoek

Anton van Leeuwenhoek.

Anton van Leeuwenhoek's microscope.

The other kind of microscope was the *simple* design, which had one, nearly spherical lens. These microscopes gave much better resolution and kept the chromatic aberration problem to a minimum. Unfortunately, they were very difficult to use. It took great patience and practice to see through them and interpret what one saw. So, though this was a valuable tool, its use was restricted to very few people.

The person who did most of the useful research with the simple microscope was a Dutch lens grinder named Anton van Leeuwenhoek, who lived in the later 17th and early 18th centuries. Van Leewenhoek discovered a process for grinding lenses that gave far superior images. He was also a very secretive worker who sought to keep his grinding and observational techniques to himself. Nevertheless, he published detailed drawings of what he saw with his microscope, and the results were astounding. He got magnifications up to 2000 times and with that discovered an entire realm of life that was totally unexpected. He called these *animalcules*. They include most of the single-celled creatures that we call *amœbas*. These are found on almost all samples of living matter. It was a completely new and unexpected idea that such tiny forms of life existed, and existed on almost every surface that one examined.

Chapter Twenty Three
Finding the Units of Life

One of van Leeuwenhoek's more interesting drawings was made from the examination of scrapings from between his own teeth! These showed tiny little darting blobs later identified as bacteria, though their function was totally unknown in van Leeuwenhoek's lifetime. Later, Van Leeuwenhoek also discovered very strange looking beasts with roundish heads and long tails with which they could propel themselves very quickly. These are the *spermatozoa* found in seminal fluid. This was particularly important as it gave a clue to the reproductive process.

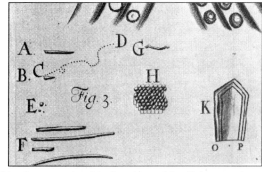

Van Leeuwenhoek's drawings of tiny living creatures that he found in scrapings from his own teeth.

Cells

Provocative as van Leeuwenhoek's discoveries were, they had limited impact on scientific research so long as only van Leeuwenhoek was able to operate his microscope. Meanwhile, the compound microscope had been improved over the original models, though still suffering from the problem of chromatic aberration.

One of the important developers of the compound microscope was Newton's arch enemy and rival Robert Hooke. Hooke's microscope did not have the powers of magnification of van Leeuwenhoek's, but its relative ease of use meant that observations made with it could be readily verified and then discussed. It was Hooke who gave biology the name for its basic building block. Hooke had taken a thin slice of cork and put it under the microscope to study its structure. What he saw was a tiny latticework that divided the cork up into small units, like the honeycombs made by bees. These he called *cells* and remarked that plants seemed to be made up of assemblies of these. These were relatively large and prominent features of cork, which is why he was able to see them with his comparatively weak instrument. Though he called these cells, what he was seeing was what we would consider the cell walls, which remain after the plant dies. Nevertheless, the name stuck and drew scientists' attention to the possibility of a unit of life smaller than a whole organism.

The Achromatic Compound Microscope

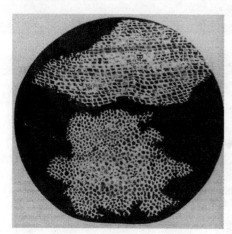

A section of cork as seen by Robert Hooke with his compound microscope.

Research might have remained very limited had not a remarkable improvement been made to the compound microscope that solved the problem of chromatic aberration. Aberration was caused by the different angles of refraction of light of different wavelengths as it passed through a lens. However the refractive angles varied with the kind of glass used. In the 1820s, a new way of making the lenses for compound microscopes was invented, one that used several different kinds of glass in layers and thereby corrected for the color fringes. With this new *achromatic microscope*, biology had a powerful research tool. By 1830, the new microscopes were standard equipment in universities everywhere.

Cell Theory

An entire new breed of scientists arose very quickly in the mid 19th century who did all their research work looking through these new instruments. They were sometimes called *microscopists*. Among them, for example, was the Scottish botanist Robert Brown, after whom Brownian motion is named. (Brownian motion is the irregular sudden movements of particles visible on any slide of a fluid seen under a microscope. Brownian motion was interpreted by Einstein in 1905 as being due to the impact of molecules in random motion striking a visible particle. See Chapter 18.) In 1833, Brown, studying plants with the new microscope, discovered that there was a dense area occupying part of the space in each cell. This seemed to be a regular feature of all plant cells and it seemed pretty important. In a very short time, microscopists were concentrating their research on the interior—i.e., the living—part of cells and less attention was given to the rather regular cell walls.

Chapter Twenty Three
Finding the Units of Life

Schleiden and Schwann

A few years later, two German microscopists declared that *all* life consisted of aggregates of cells. First Matthias J. Schleiden, a botanist, asserted in 1838 that all plant tissue was composed of nucleated cells, in other words, that plants were really just aggregates of cells. Then, the following year, Theodor Schwann, a zoologist, was stimulated by Schleiden's work to look at animal tissue more closely. Unlike plants, animals do not form the thick cellulose structures that Hooke originally called "cells." But Schwann found that on close examination, animals are also built up as aggregates of cells. He concluded then that all living things develop from the formation of cells.

Schwann believed that the key to understanding life was to figure out what chemical processes were taking place in the cells. He saw the cells as being "combustion chambers," where energy from food was converted into forms useful for the body. For these processes, he coined the term *metabolism*.

Omnis cellula e cellula

The usefulness of this conception was shown in 1858 by Rudolf Virchow, a German physician and pioneer of pathology. Virchow, using the microscope, studied the cells of diseased patients and found that there were consistent abnormalities in cells associated with specific diseases. This was contrary to the prevailing theory that disease was an imbalance of the humors of the body, dating back to Hippocrates. Virchow urged that medicine turn away from trying to cure the whole patient to addressing the abnormalities in the individual cells.

His doctrine, as expressed in Latin, was "omnis cellula e cellula"—all cells come from cells. In other words, not only are cells the building blocks of the body, they are the units of life itself.

In the field, naturalists continued to look at whole individuals and concerned themselves with groupings and transformations of entire species. In the laboratory, research had narrowed down to individual cells of the body.

Part IV
What is Life?

Vitalism versus Reductionism

Schleiden and Schwann were not just independent and open-minded scientists reporting what they saw. They represented one side in a growing debate about the nature of life. There were those who believed that life was so different from non-life that everything alive was imbued with a living force, an animal spirit of some kind. This group can be called *vitalists*. Opposite to them were those who, like Schleiden and Schwann, were certain that life would turn out to be understood and accounted for by the natural laws alone. This group can be called *reductionists*, because, in effect, they sought to reduce biology to very complex processes of chemistry and physics. Schleiden and Schwann saw cells as small machines (hence the "combustion chamber" analogy) that obeyed strictly mechanical laws.

Vitalists and reductionists appeared on all sides of the issues in biology. Their different explanations of the same phenomena only served to keep everyone looking more closely at all the facts and all the implications of any explanation. Sometimes it was surprising how observations and theories were interpreted in the light of these viewpoints.

Darwin's theory of evolution by natural selection is an interesting example. Darwin held back from publishing his theory because he could not find the mechanisms to explain every feature of evolution that he had envisaged. Natural selection was certainly a mechanistic notion, and Darwin certainly tried to propose a viable mechanism for inheritance, which would in principle fit the reductionist viewpoint. Darwin was particularly anxious that his work not be confused with the evolutionary ideas that had latched onto the idea of progress and inherent perfectibility since he viewed these as unscientific. Nevertheless, Darwin's worst fears were confirmed.

One of the most prominent biological scientists of the 19th century was the French physiologist Claude Bernard, whose life span was nearly the same as Darwin's. Bernard was a founder of experimental medicine and conceived the idea of the internal environment of the body and the principle of *homeostasis*. He was one of the most eloquent spokesmen for the reductionist viewpoint: all physiology was ultimately going to be explained by the principles of chemistry and physics. Bernard's view of Darwin was that he was just another armchair natural philosopher whose work did not qualify at all as science.

Compare this to the view of the German zoologist Ernst Haeckel, who made a career of promoting Darwin in Germany. Haeckel was 35 years old when *The Origin* was published in 1859. He seized upon it immediately and saw in it the much sought for principle of *Naturphilosophie*, the unity of nature, a particular German preoccupation since the time of Goethe. Evolution was the manifestation of that vitalist principle of perfectibility—the very principle with which Darwin did not want to be associated. Haeckel confidently proclaimed that all biology would be completely made over by the theory of evolution. Since embryological studies had shown that animal fetuses bear very close resemblances to each other at the earliest stages, but then differentiate into the different species during the gestation period, Haeckel proclaimed that the embryo was actually going through all the steps of evolution for each species before its birth. This was called *recapitulation theory*; its slogan was "ontogeny (the development of the individual) recapitulates phylogeny (the evolution of species)." To Haeckel, Darwin solved the "mystic problem of creation." Haeckel's version of Darwin was certainly different from Darwin's own theory, but Haeckel's version was what was best known in Germany.

Ernst Haeckel.

Inheritance

What's going on when an oak tree produces acorns from which other oak trees grow? What happens when a hen lays an egg from which a chick hatches? And for all mammals, how is it that whole newborn infants emerge from their mothers after a gestation period?

Inheritance is one of the main issues in understanding life. How does one generation of living things conceive and give birth to another generation? And how is it that the next generation turns out to be the same species? Biology has not gotten very far if it can't account for reproduction. Darwin would have had a far easier time trying to explain how species change if he had had a workable theory of how they stay the same. That was coming, but it would not reach a point where it would have been useful to Darwin until long after his death.

The Main Issues

We speak of creatures being alive or dead. What does that mean, exactly? Putting aside the question of death, what do we mean by birth? At some point, there is life where there was not life before. Is life created, as it were, from "nothing," or was life there all along in some other form? Does life arise from matter that was not itself alive, and if so, must something other than physical matter be added to it to make it alive? Or is life something immaterial altogether that is just housed in bodies but can come and go?

A perennial test case for these issues is spontaneous generation. At times, new living things arise where there was no sign of life before. Examples would be parasites, fungi, molds, even vermin in rotting food. Are these examples where life came from non-life or are there lurking parents that have not been detected?

Certainly mating has something to do with it, but what? If a pregnancy occurs after sexual activity, what actually happened to make the new life form begin to grow? If whatever it is that makes life is already in the parents, waiting to be set in motion by mating, where is it? Is it all in the male? All in the female? Or is there some combination of the two that is mixed together?

These are the major questions. They have been around since speculation about life first began. There have been answers proposed for the same amount of time. Here are some of the major theories.

Preformation Theories

The easiest way to get around the conundrum of explaining how an individual life began is to dodge the question altogether and say that it has always been there. Or at least say it has always been there since the Creation, which lumps all the creation problems into a single one and avoids the issue as a question of biology. Answers such as this have a long history; they can be classified as *preformation theories*—since they assume that every individual living thing already existed "preformed" before it was born. However, the next question would be, "Where?" If it is granted that the sex act is what starts the process

Chapter Twenty Three
Finding the Units of Life

of generation going, then the obvious choices are that the preformed life had to be either in the father or the mother. And in either case, what is the role of the other parent?

Those that took the view that the essential preformed life resided in the mother are called *ovists*. The idea was that the father's role was to contribute some sort of stimulus to the growth of the latent fetus which lay waiting in the mother. This would be the role of the seminal fluid for animals or of pollen for plants.

In the 17th century, one of the early researchers with the microscope was the Italian physician Marcello Malpighi, who did extensive investigations of the embryos of chickens in eggs at various stages of development. Malpighi traced the development of individual organs in the chick embryo and saw that the development followed a definite sequence on a definite schedule. His drawings of observations with the microscope did much to establish the idea that life has a growth pattern before birth that is even more complex and structured than after birth.

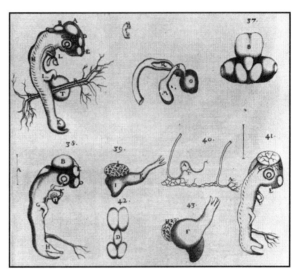
Marcello Malpighi's drawing of stages of development of the chick embryo.

The other obvious viewpoint is that all of the structure of life resides in the male and is transmitted to the female by mating. The female uterus is then the environment where the form from the father gets the nutrients and proper conditions to grow. This view fit nicely in with van Leeuwenhoek's discovery of spermatozoa in the male seminal fluid. The idea was that a small, fully formed individual, a *homunculus*, resided inside the sperm cell and once transmitted to the mother began to grow.

In the 17th century, the microscopist Nicolaas Hartsoeker drew this picture showing the human spermatozoa as a tiny homunculus, a complete folded-up human being which would grow into a baby when planted in the mother's uterus.

The logic of preformationism leads to some inevitable conclusions. If all life existed at the Creation and is being played out generation by generation, then there must be a limit to the number of generations that can be born. Life must come to an end once all the preformed beings have been born. An example of this thinking was the 17th century ovist Jan Swammerdam, who, thinking of the human species, said that if the ovum that grew into a fetus was already present in the mother, then that ovum must also have been present in the mother's mother within the ovum that became the mother. Therefore, eggs within eggs within eggs, etc., and the very first woman—Eve—must have had within her all the generations of humanity that ever would be. When the last generation nested inside Eve is finally born, that will be the end of humanity.

Idioplasm versus Germplasm

One of the leaders of the vitalist camp in the 19th century was Karl Nägeli, a German botanist and cell theorist. He had also studied philosophy under Hegel. Nägeli proposed that there were two kinds of substances in every living thing: a *tropoplasm*, which builds the bodily structure and dies when the organism dies, and an *idioplasm*, which directs the form that the organism takes. The idioplasm was immortal because it did not die with the individual. It was passed on from parent to offspring (how is another matter) and left the body upon death. It was too fine to be detected by instruments. As a biological explanation, preformationism is a dodge. It simply does not answer the question about how life comes to be. But it was a popular way out because the alternatives raised so many difficulties. All theories of inheritance other than preformationism have to account for the appearance of life where no life was there before. The theories proposed were inevitably tied up with the metaphysical issues of vitalism and mechanism, and also the question of whether life is some kind of continuous force or has discrete units. In the 19th century, what researchers believed about these issues often influenced what they saw through their microscopes and conceived in their theories.

Chapter Twenty Three | 453
Finding the Units of Life

Notice the Ad Hoc quality of the idioplasm. It provided all the direction, but it could not be detected. Nägeli was writing before Ernst Mach articulated the principles of Positivism: that science should use no entities that cannot be independently detected (see Chapter 18). The idioplasm had the major role in passing on inheritable traits and in making the organism grow into an individual of the correct species, but how it did so could not be investigated because it could not be detected.

In reproduction, Nägeli held that the idioplasm of both parents was mingled together in the fertilized ovum. This is similar to Darwin's pangenesis explanation, but the idioplasm of each parent is not derived from actual bodily parts as Darwin's gemmules were, so the problem of parents with amputated limbs does not arise. Nägeli was an evolutionist, but not of the Darwin school. His idea of evolution was much influenced by the German idealist school and by *Naturphilosophie*. To Nägeli, evolution was a striving to perfection, seen in his conception as the automatic perfecting process of the idioplasm. This was a theory that preserved the idea of life coming in forms that could be ranked on a scale, with each species striving to move up the ladder. To Nägeli, monkeys will be humans some day.

Nägeli was a strong voice for those who believed that life was essentially different from non-life and had to have a different explanatory model. By postulating the existence of a substratum that was beyond detection, he seemed to take the study of inheritance out of the mainstream of empirical science altogether and back to a speculative level. But this was not entirely true. His distinction between the tropoplasm and the idioplasm was insightful. Trying to understand how individuals reproduced themselves by looking at the entire organism only confused matters and led to the logical absurdities that Darwin brought on with pangenesis—e.g., the case of the children of amputees. Nägeli separated the body of the organism from the controls that made it grow as it did.

The views diametrically opposed to vitalism were those of the reductionists. In one of those curious twists of direction in the history of science, one of the leaders of the reductionist movement took Nägeli's vitalist model and turned it into an explanation entirely in terms of matter and mechanisms. The German zoologist August Weismann proposed a theory of inheritance that, like Nägeli's, divided the substances of life into two categories. Weismann's were both cells. *Soma cells* were the general cells of bodies which formed their structure. These had nothing to do with reproduction. They were comparable to Nägeli's tropoplasm, though soma cells were clearly individual, discrete units, in keeping with cell theory. The other kind of cells constituted the *germ plasm*, which, like Nägeli's idioplasm carried inheritance and directed the growth of the body. But the germ plasm had a material existence. Germ plasm was, like idioplasm, in a sense immortal, but not because it passed in and out of living bodies but for the more straightforward reason that it consisted of cells that reproduced by the process of cell division, and those cells were passed from parent to offspring during conception.

Soma cells derive their structure from germ cells, not vice-versa. So as far as life is concerned, the germ cells are far more important than the body cells. The continuity of a species is determined by the continuity of the germs cells. In embryological terms, Weismann said, a chicken is just nature's way of making another egg. Evolution occurs in the germ cells. What happens in the soma cells is immaterial. In Darwinian terms, natural selection is the selection of which germ plasm gets passed on.

And then one critical issue: germ plasm is *particulate*. It is a collection of discrete bits of matter that have separate roles, not an undifferentiated indivisible, and undetectable fluid like the idioplasm. The Nägeli-Weismann theories were yet another example of the conceptual differences between continuous and discrete views of reality. (For more discussion of this see Chapter 19.)

Chromosomes

The first cells in living matter were the cork cells seen by Robert Hooke in the 17th century with his fairly primitive microscope. Even after the vast improvement with the introduction of the achromatic microscope in the 1820s, most cell research continued to be with plant rather than animal cells. That was because plants grow sturdy and visible cell walls while animal cells are very hard to see on a microscope slide. Late in the 19th century, the German chemical industry came to the rescue of biological research with the development of some dyes that could be used on microscopic specimens. The dyes were particularly useful because some of the matter on the slides absorbed the dyes well while others did not, providing a sufficient contrast that differences could readily be seen.

Using these new dyes, microscopists discovered that there were certain long string-like structures in the nuclei of cells that took stains very well. In 1888, they were given the name *chromosomes*, meaning "color bodies."

Once chromosomes had been identified their extraordinary behavior began to be noticed. Whenever cells started the process of dividing in two, the chromosomes duplicated themselves exactly and then separated, moving apart to opposite ends of the nucleus. In the daughter cells, each one contained a complete set of chromosomes, just like the parent cell did before division.

Weismann expressed the view that the germ plasm that he had postulated must be carried by the chromosomes. Thus, chromosomes were seen to be important to the understanding of the mechanisms of inheritance, and they began to get a lot of attention.

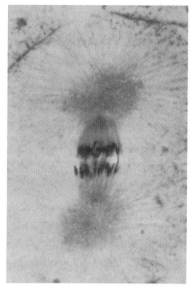

Chromosomes during cell division.

Gregor Mendel

The next logical step in the search for the mechanisms of inheritance would be the investigation of the properties of chromosomes and the discovery of their precise relation to inheritable traits. And this did happen. But before getting to that, we must pause in the recounting of this episode in biological laboratory research and go back to the world of the naturalists. The naturalists were those who tried to understand life by looking at it in its natural setting, studying whole organisms and how they behaved together. Microscopy was going just the other way, looking at smaller and smaller bits of the body and seeking the physical laws that operated there. By all reasonable expectation, the lab scientists were on the right track and the naturalists had reached the limit of what their methodology could discover. But there was at least one more very important lesson to be learned from looking at whole organisms.

Part IV
What is Life?

A biographical sketch of the life and work of Gregor Mendel might sum him up as a good-natured, reasonably intelligent, and patient man, who had a career as a high school science teacher and a monk, and who pursued science as a hobby. He was serious enough about science to do some experiments and report them at a meeting of other amateur science enthusiasts in his city. He would be remembered, if at all, as the Abbot of his monastery in Silesia, a position he held for the last sixteen years of his life. And so it would be had not that amateur experiment that he reported to his local society gotten the attention of some cell biologists years after Mendel's death. When finally, mainstream biologists read and understood what the Austrian monk had been doing, Mendel came to be viewed as the founder of a whole new branch of biological science, genetics.

Gregor Mendel.

Here are the basic details of his life and work: Johann Mendel was born in 1822 in Austrian Silesia of a peasant family. The Austro-Hungarian Empire had established public schools throughout the empire to enable all children (well, male children) with ability to get a good fundamental education. Young Johann Mendel attended one of these state-run schools and showed enough ability that his teachers urged him to continue his studies and become a teacher himself. He did so, and trained to be a teacher of basic sciences to young adolescents. He entered the monastery of the Order of St. Thomas in Brünn, Moravia (now Brno in the Czech Republic) in 1843 and became a monk in 1847, at which time he took the name Gregor.

Though Mendel was already teaching science subjects, his monastery sent him to the University of Vienna for further training. At the university Mendel did not cope well with what was expected of him, and after failing to pass a required examination, he was sent home without the credential he had sought. So, life as a monk and middle school science teacher seemed to be his destiny. However, Mendel wanted to do some real scientific work. He decided that he would pursue an interest he had as a child on his father's farm: he would study the process of plant hybridization. Hybridization is a breeding technique used by farmers to improve crops. Farmers will pay close attention to the plants they get from certain batches of seeds. When they find that the plants have properties they desire, they will choose these plants to fertilize other plants and develop new strains of their crop. Farmers did this by rule of thumb, knowing what worked and what

Chapter Twenty Three
Finding the Units of Life

did not, but they lacked any theoretical understanding of what was happening. Mendel realized that there were interesting issues about inheritance involved in this work, so he undertook to find out what he could about it, by doing some controlled planting and fertilizing of his own and tracking the results.

Mendel thought the whole project through and, since no one was expecting him to do anything anyway, took his time to do it all properly. He decided that he would investigate the inheritance patterns in the common garden pea (*Pisum sativum*). This was such an ideal choice for his experiment that it seems clear that he must have known some of its properties before he did his experiment and knew what he might discover. The pea plants, he knew, had some easily identified and sharply different characteristics that could be displayed by the plants in different combinations when they are crossed with each other. The plant he chose was an *inbreeder*, that is, the seeds are fertilized from the pollen of the same plant, because the petals of the flowers are such that they envelope the stigma that receives the pollen and the anthers that produce it. This gives the plants a very regular inheritance pattern compared with most living beings.

The model of a scientific experiment that came down from Galileo dictated that first one must decide what natural feature is to be studied, and then that must be isolated from other influences as much as possible. Galileo achieved this with highly polished inclined planes and very smooth metal balls so he could study the principles of free fall. Mendel could not create an artificial laboratory situation, but he could choose a plant that isolated itself from complications and made it easier to study a few principles in isolation. Even so, Mendel chose only those features that would be clear and distinct. As he described it later,

> The various forms of peas selected for crosses showed differences in length and color of stem; in size and shape of leaves; in position, color, and size of flowers; in length of flower stalks; in color, shape, and size of pods; in shape and size of seeds; and in coloration of seed coats and albumen. However, some of the traits listed do not permit a definite and sharp separation, since the difference rests on a "more or less" which is often difficult to define. Such traits were not usable for individual experiments; these had to be limited to characteristics, which stand out clearly and decisively in plants.[1]

---

1. Quoted in Colin Tudge, *The Engineer in the Garden* (New York: Hill and Wang, 1995), p. 12.

He took two years to prepare his experiment, eight years to run it, and another two years to analyze his results. At some point during his experiment, he wrote to Karl Nägeli to discuss his work. Nägeli replied, condescendingly, that Mendel was on the wrong track altogether as his work was "empirical" rather than "rational," but if Mendel wanted to do some useful scientific work, he could help Nägeli with one of his investigations. Mendel did not follow Nägeli's advice and pressed on.

He began with true breeding strains—those which, when kept in isolation, produced only the same characteristics year after year. Then he bred plants with a certain characteristic with those of the opposite characteristic. For example, he bred plants that all grew tall (over 2 meters) with those that always bred short plants (less than ½ meter). The result was that in the next generation all of the plants were tall. Then he took that second generation and let it self-pollinate. In this third generation he got both tall and short plants. He counted them. Having run the experiment with thousands of plants to get reliable numbers, his counting task was considerable.

After eight years of breeding, interbreeding, breeding characteristics in different combinations, and counting characteristics produced each time, he had a mass of data that he hoped would show him the pattern of heredity. It must have been fairly early on that Mendel found that in the second generation—the first one that he let self-pollinate, that the ratio of the two characteristics was sharply weighted in one direction. In the case of the tall and short plants, the talls outnumbered the shorts by a ratio that was very close to three to one. This same ratio came up again and again, and at some point, Mendel must have become confident that this was indeed the pattern, because the numbers reported in his data are so close to the theoretical three-to-one ratio that statisticians say his numbers are too good to be true and must have been calculations rather than real counts.

Nevertheless, Mendel found a startlingly simple relationship: the pea plant had seven different contrasting pairs of characteristics. A plant could be tall or short, have green peas or yellow peas, have round seeds or wrinkled seeds, etc., and could have the characteristics in any possible combination. However, on average, there were always three times as many of one of the pair of characteristics as of the other. He called the

Chapter Twenty Three
Finding the Units of Life

characteristic that appeared most often the *dominant* characteristic and the one that appeared less often the *recessive* characteristic.

So far so good, but what did it mean? It took more than years of patient plant breeding to make sense out of this. Charles Darwin was said to have known about the three-to-one ratio of characteristics in the pea plant, but made nothing of it.[2] To understand why the obscure monk Mendel was able to see a pattern that the great observer Darwin did not and was able to conceive the importance of a long and extensive experiment on plant breeding that a biological researcher like Nägeli could not even understand, it might be helpful to recall Mendel's extraordinary situation and compare it with that of others who might have stumbled onto the same phenomenon.

Mendel had grown up on a farm where he helped his father breed hybrids of plants. He had a close familiarity with plants and plant breeding. Darwin, for example, also had an extensive knowledge of plant breeding as his extensive studies of orchids showed. Both Mendel and Darwin were keen observers of nature, capable of recognizing patterns in visible characteristics, as they both did. Mendel, like Nägeli, was especially interested in the question of inheritance and had read all the literature available to him on existing theories of inheritance and possible explanations. Nägeli, however, was committed to the idea that microscopic research in the laboratory testing preconceived ideas was going to make progress while fooling around with growing thousands of plants was a waste of time.

But there was another difference that may have been important too: Mendel was a middle school science teacher. Day in and day out he was thinking through and explaining the basics of physics, chemistry, and mathematics to his pupils. He necessarily kept an integrated perspective on science while both the naturalist Darwin and the cell biologist Nägeli thought in biological concepts only.

What would cause a three-to-one ratio? There was more: not only did the third generation appear with the chosen characteristic pair in the three-to-one ratio, further generations of breeding with those same plants revealed that one-fourth of the third generation plants would breed nothing but the dominant characteristic if self-pollinated, another

---

2. See Ruth Moore, *The Coil of Life* (New York: Alfred A. Knopf, 1961).

fourth would breed nothing but the recessive characteristic and the remaining half would breed as the second generation had done, in the three-to-one ratio.

The vital insight that Mendel had that others had missed was that if every inheritable trait is determined not by one, but by two factors (Mendel called them *Anlagen*), and if one of those factors always outweighs the other, causing the organism to exhibit the trait associated with that factor, then this would produce the three-to-one ratio that was observed along with the true breeding recessive and dominant strains in subsequent generations as he had seen also.

Mendel devised the system, still used today in much the same form, that indicated dominant traits with capital letters and recessive traits with short letters. Using *T* for tall and *s* for short, and with each characteristic determined by a combination of two factors, one from each parent (in this case the "male" anthers and the "female" stigma), then that second generation of plants will all have one *T* and one *s*, which will make the plant tall. The third generation, the one self-pollinated from the second generation, will have an even distribution of *T* and *s* in every combination, namely *TT*, *Ts*, *sT*, and *ss*. All three groups containing one *T* will be tall plants, while only the *ss* group will be short. This arrangement is just what his experiments showed.[3] We have become accustomed to doing calculations of this sort in biological studies (especially in genetics), but in Mendel's time, it was unheard of. Biology was biology and had nothing to do with physics, let alone mathematics.

In 1865, Mendel was finally finished analyzing his results and was ready to make a formal presentation of them. He asked to be put on the program of the next meeting of the Brünn Society for the Study of Natural Science. When the time came, it was a cold

---

3. In short, what Mendel had done was applied some mathematical analysis, of the sort that he would have been teaching his students in elementary algebra, to the data he was getting.
   To give a single example of this, think of *T* and *s* as two variables in an algebraic formula. Think of the fertilization process that matches one factor from each parent with one from the other as a sort of "multiplication." Then take an equal number of *T* and *s*, call it k of each, and let them self-pollinate. Mathematically you can model that as performing the calculation $k(T+s)^2$. This is equal to $k^2(T^2 + 2Ts + s^2)$ and it is easy to see that there are three times as many plants carrying a tall factor as those carrying only a short factor, and that the number of true breeding talls and shorts are one-fourth each of the total.

and snowy February night, but enough people had heard of Mendel's long experiment that they wanted to hear his report. As Ruth Moore reported,

> As Mendel read, this curiosity gave way to incomprehension. The several botanists in the society were as much confused by the report on invariable hereditary ratios in peas as were the other members—a chemist, an astronomer, a geologist, and an authority on cryptograms. Mendel spoke for the hour allotted to him, and then announced that at the next meeting he would explain why the peculiar and regular segregation of characteristics occurred.
>
> The next month, as Mendel presented his algebraic equations, attention wavered. The combination of mathematics and botany which Mendel was expounding was unheard of. And the idea that lay behind it, that heredity was a giant shuffling and reshuffling of separate and invisible hereditary factors, stood in such diametrical contrast to all that had been taught that it probably could not be grasped. No one rose to ask a question. The minutes recorded no discussion.
>
> A number of the members spoke to Mendel afterward about his experimental work. "I encountered various views," Mendel told his friends. "No one undertook a repetition of the experiment."
>
> The editor of the *Proceedings* of the society extended the usual invitation to Mendel to prepare his paper for publication in the society's journal. "I agreed to do so," said Mendel, "after I had once more looked through my notes relating to the various years of the experiment without being able to discover any sort of mistake."[4]

Mendel presented his paper for publication. It appeared in the *Proceedings* of the Society, and copies made it to all the major scholarly libraries in Europe. In addition to that, Mendel sent offprints of the paper to a number of scientists whom he thought might be interested. No one was interested. No one understood what the significance was of discovering that the ratios of observable traits can be accounted for by assuming that each hereditary trait is determined by a pair of factors, one from each parent, and that these traits operate on an all or nothing basis. There were no medium-sized plants, only tall and short. There were no pink flowers, only white and purplish-red. There was therefore no blending going on between the traits of the parents. A recessive trait could be carried for generations and then finally appear when both parents contribute the recessive factor for that trait to the offspring.

---

4. From Ruth Moore, *The Coil of Life* (New York: Alfred A. Knopf, 1961).

This arrangement of distinct, discrete, factors that are always present in pairs and do not become washed out would help to make sense of a lot of issues in inheritance. But in Mendel's time, no one could understand what he was telling them. Mendel wrote *"Meine Zeit wird schon kommen."* ("My time will surely come.") What came for Mendel was that three years after he gave his famous paper, the Abbot of his monastery died and Mendel was elected to succeed him. Mendel spent the rest of his life doing administrative work for the monastery. He no longer had time for science.

Genetics

In 1900, there were three papers published in the same volume of the *Proceedings of the German Botanical Society*. The authors of the papers were Hugo de Vries, Carl Correns, and Erich von Tschermak. Each of these authors had pursued the trail that cell biology was forging and had come to the conclusion, independently of each other, that it must be the case that inheritance takes place through the contribution of discrete, particulate factors by each parent. No other possibility made sense. They each then undertook to find out what had been written about the subject in the scientific literature in hopes of finding some experimental work that would help confirm their hypothesis. They each were shocked to find that their entire thesis had been anticipated, and the critical experiments to establish its general conclusions had been done and published 34 years before, but no one had noticed at the time. Though each of these scientists had been planning to take the next significant step in understanding inheritance themselves, they found themselves instead in the position of being the ones to tell the scientific world that it had been asleep while the critical work had gone on in a monastery garden. The research into inheritance that their works launched became known as *Mendelian Genetics*.

The Gene

Mendel worked from the outside, as it were, and did not know what the internal processes in the plant might be. His analysis showed that the complex arrangement of characteristics shown by his plants throughout his experiment could be accounted for by postulating the existence of "factors" which controlled inheritance. The analytic part of his work was a straightforward Platonic exercise in "saving the phenomena." *If* such enti-

Chapter Twenty Three | 463
Finding the Units of Life

ties existed and behaved as he said, *then* the phenomena—his experimental results—would be as he found them.

This was good detective work and put science on the right track, but it did not get them to the "factors" that controlled heredity. What were the physical things that filled the role of the factors that Mendel had detected? Work on chromosomes had led scientists to believe that these factors were conveyed in the body by the chromosomes, but how that occurred was not known.

In 1906, at a meeting of the International Congress of Botany, British biologist William Bateson introduced the term *Genetics* to name "a new and well developed branch of Physiology."[5]

And then, three years later, Wilhelm Johannsen introduced the word *gene* to represent what these new geneticists were looking for. He wanted a word that would be as free of preconceptions as possible so that no particular theory of inheritance would inevitably come to mind.

> The word "gene" is completely free from any hypotheses; it expresses only the evident fact that, in any case, many characteristics of the organism are specified in the gametes by means of special conditions, foundations, and determiners which are present in unique, separate, and thereby independent ways—in short, precisely what we wish to call genes.[6]

Another two years after that Johannsen sought to make sure that it was understood that the "gene" remained an abstract term, meaning much the same as Mendel's term "factor" and that it would be premature to give it any physical meaning:

> The "gene" is nothing but a very applicable little word...it may be useful as an expression for the "unit factors"...demonstrated by modern Mendelian researches... As to the nature of the "genes," it is as yet of no value to propose any hypothesis; but that the notion of the "gene" covers a reality is evident in Mendelism.[7]

---

5. William Bateson, "The Progress of Genetic Research." In *Third Conference on Hybridization and Plant Breeding* (London, 1906), pp. 90-97.
6. Wilhelm Johannsen, quoted in Evelyn Fox Keller, *The Century of the Gene* (Cambridge: Harvard University Press, 2000), p. 2.
7. *Ibid.*

Thomas Hunt Morgan.

The Search for the Gene

Discovering the nature of the gene became a major activity of experimental biology in the 20th century. The leader of early genetic research was the American Thomas Hunt Morgan, a professor at Columbia University in New York. Morgan was the key figure in genetic research, because most of the important early research was done in his laboratory under his supervision, and for a good half-century all of the important geneticists in America were Morgan's former students.

Now that more was known about the cell and where one might reasonably look for more clues to the nature of the gene, more experiments of the sort that Mendel had undertaken would be useful. Mendel's choice of the garden pea plant was a most fortunate one for him because the plant was a self-breeder, had a simple set of inheritable characteristics, and those characteristics were all independent. But the choice of the pea plant led Mendel to state a general principle of heredity that was incorrect. Mendel's *Principle of Independent Assortment* asserted that all heritable characteristics can be inherited independently of each other, in other words, in any combination whatsoever. That was true of the garden pea, but was not generally true in nature, as Morgan discovered soon enough.

The pea plant would also be a poor choice for a research laboratory where people did not have years to spend waiting for their test organisms to reproduce. A different species was needed. Morgan found the ideal candidate for his work in the ordinary fruit fly, *Drosophila melanogaster*. Unlike pea plants, *Drosophila* had a generation time of just a few weeks. Several generations could be studied in one academic term. But like the pea plant, *Drosophila* had a small number of visible characteristics that could be easily identified and differentiated from each other. Another feature that was very important in the search for the gene concerned the chromosomes. Since there was good reason to think that genes and chromosomes were closely related, an ideal laboratory organism would be one that had chromosomes that were easy to study. *Drosophila*, as it turned out, had rela-

Chapter Twenty Three | 465
Finding the Units of Life

tively huge chromosomes that could be examined under the microscope and then compared with the visible characteristics of the fly.

At first Morgan's work with fruit flies served most to confirm that Mendel's methods and conclusions applied equally well to other species. But then Morgan discovered something new, that showed that Mendel's Principle of Independent Assortment was not entirely correct. In his various breeding combinations, Morgan produced some fruit flies with white eyes—red-eyes were the dominant trait. When he then mixed these in with a group of red-eyed flies and let them breed, he found a surprising result: all of the white-eyed fruit flies were male. The characteristic of sex and the characteristic of eye-color were linked, not independently sorted as per Mendel's principle. This is the principle of *linkage* of traits. Some inheritable traits always occur together. It was an important clue to determining the nature of the gene.

The connection between chromosomes and hereditable traits began to become clearer when Morgan and his associated began studying the chromosomes of the flies they were breeding and correlating the dark and light bands seen on the chromosome with the visible traits of the flies. Careful study enabled them to identify certain portions of certain chromosomes with different hereditable traits. For example, the trait of eye color was, sure enough, found to be correlated with a particular section of the chromosome of the fly that determined its sex.

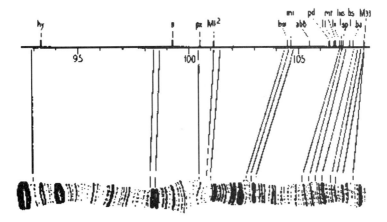

Morgan's map of a *Drosophila* chromosome correlated with different visible characteristics.

It was still not clear what a gene was, or whether it really existed as an entity. In 1933, Morgan remarked that "There is no consensus opinion amongst geneticists as to what genes are—whether they are real or purely fictitious."[8] Whatever genes would turn out to be, their close correlation with places on chromosomes turned the attention of all geneticists firmly onto the chromosome as the key to understanding heredity.

◆

How did the work of the geneticists in the 20th century reflect upon the theories of the 19th century? The emerging notion of the gene as a place, or closely related to a place, on the chromosomes, provided a physical confirmation of Mendel's theoretical factors. What Mendel called factors appeared in every respect to be the same as what the 20th century was calling genes. The fact that chromosomes come in pairs, one contributed from each parent, dovetailed perfectly with Mendel's notion of pairs of factors—one from each parent—being the determinant of the visible characteristics. Mendel was seen to be wrong about the Principle of Independent Assortment, but that was because he was misled by the particular species he chose to investigate. And a good thing that he did choose a species with independent assortment or he might not have found his simple ratios. Mendel's idiosyncratic mathematical analysis of inheritable traits proved itself as an essential component of this kind of research in order to track the distribution of traits through a population with statistical analysis.

At first glance, both Mendel and 20th century genetics seemed to indicate a fatal flaw in Darwin's theory of evolution by natural selection. Mendelian traits formed a fixed set of factors. They could be shuffled and reshuffled, but it was still the same group of variations. No evolution from one species to another was going to emerge from different combinations of the same characteristics. Darwin required that there be continually changing variations, from which those which were favored by natural selection would survive and others perish.

But 20th century genetics did provide an answer in the form of mutations. The characteristics didn't always remain the same. The characteristics exhibited by an organism

---

8. *Ibid.*

depended on the detailed sequence of pieces of chromosomes. But in the conception and the process of cell division, the chromosomes come apart and then reconfigure themselves. Where the reconfiguring does not turn out the same, the characteristics change. These are *mutations*. They happen naturally but can be stimulated by subjecting the flies to doses of radiation. Morgan and his colleagues used X-rays in their work and found that they could produce new characteristics, often injurious ones, in their fly populations. If only some of the mutations were advantageous, the species would have a direction in which to change by natural selection. Mendel and 20th century workers such as Morgan had found a physical interpretation of variation for Darwin.

For More Information

Alioto, Anthony M. *A History of Western Science,* 2nd ed. Englewood Cliffs, NJ: Prentice-Hall, 1987. Chapters 21 & 26.

Cohen, I. Bernard. *Revolution in Science.* Cambridge: Harvard University Press, 1985. Chapter 21.

Coleman, William. *Biology in the 19th Century: Problems of Form, Function and Transmutation.* New York: Wiley, 1971.

Keller, Evelyn Fox. The Century of the Gene. Cambridge: Harvard University Press, 2000.

Lewontin, Richard. The Triple Helix: Gene, Organism, and Environment. Cambridge: Harvard University Press, 2000.

MacLachlan, James. Children of Prometheus: A History of Science and Technology, 2nd ed. Toronto: Wall & Emerson, Inc., 2002. Chapter 21.

Moore, Ruth. The Coil of Life. New York: Alfred A. Knopf, 1961. Ronan, Colin A. *Science: Its History and Development Among the World's Cultures.* New York: Facts on File, 1985. Chapters 9 & 10.

Tudge, Colin. *The Engineer in the Garden: Genes and Genetics—From the Idea of Heredity to the Creation of Life.* New York: Hill and Wang (Farrar, Straus and Giroux), 1995

Wills, Christopher. *Exons, Introns, and Talking Genes: The Science Behind the Human Genome Project.* New York: Basic Books, 1991.

Francis Crick and James Watson on the grounds of Cambridge University.

Chapter Twenty Four

DNA, the Key to the Mystery

As geneticists got closer and closer to identifying genes as sequences on chromosomes, the question became more and more pressing: how could some snippet of a tiny string-like piece of the nucleus of a cell tell the body how to grow? And how could one set of these strings—chromosomes—make matter grow into a fruit fly while another set makes flesh grow into a human being?

Genes had to be information in coded form. In 1935, a physicist named Max Delbrück addressed this issue. Delbrück, a former student of Niels Bohr, had become interested in biology and made a career change, putting his mind to the issues raised by the reductionist point of view. He published a paper entitled "On the nature of gene mutation and gene structure" in which he suggested that if the genes conveyed information to the body, it had to be via the arrangement of the individual molecules of the gene. Somehow, genes had to be complex arrangements of molecules that could embody complex codes that the body could "read" to find out how to grow.

Delbrück's work was an important change in thinking, but his writing was somewhat labored and the paper was not widely available. However, the same idea occurred to one of the top physicists of the day, Erwin Schrödinger, who wrote a similar analysis in a short book called simply, *What is Life?* in 1944. Schrödinger's book was much more widely read and influenced several important researchers to turn their mind to this problem.

The Phage Group

Delbrück had decided to give up physics for biology and went to do post-doctoral work at the California Institute of Technology ("Cal Tech") on *phages* (bacterial viruses) to study their replication. Phages are among the simplest life forms and can be studied at a much more fundamental level than either pea plants or fruit flies. Phages infect bacterial hosts and immediately begin to multiply. Within a half-hour, a single phage will have produced one hundred identical copies of itself. The phage suggested itself as an ideal subject of study to try to discover the nature of the gene. While at Cal Tech, Delbrück met two other young scientists, Salvador Luria and Alfred Hershey, who were both interested in the work on phages. They decided to keep in contact with each other and share their research. Together they formed the *Phage Group* in 1943, to study the nature of the gene via research on phages and similar organisms. At this point Delbrück was at Vanderbilt University in Nashville, Luria was at Indiana University, and Hershey was at Washington University in St. Louis. (In 1969, the three of them shared a Nobel prize in medicine for their work on phages.) Having gone their separate ways, they began to draw other people, primarily their graduate students, into their interests.

The Rise of the Multidisciplinary Laboratory

People like Max Delbrück were becoming more common in biological research in the mid 20th century, that is, people with considerable training in other disciplines, particularly physics and chemistry. The reductionist program to reduce biology to physics and chemistry would get nowhere without them. But how many Delbrücks could there be? And, even if large numbers of physical scientists suddenly turned their attentions to biology, they would not be expert biologists. All the sciences were becoming so complex in the 20th century that important new insights depended on a mastery of many specialties, too much for any one individual.

Consider just the case of all the separate scientific issues that had to come together to find the physical nature of the gene. First there was the *heredity problem* itself. Cell biology had identified the importance of the sperm and egg cells. This is what was known: The nucleus of the sperm cell joined with the egg in the fertilized egg. *Mitosis* (cell division) had been studied; chromosomes had been identified and tracked through cell division and the fertilization processes. Everything pointed to the cell nucleus as the location of activity. The work of Mendel, Morgan, and others had established that the gene must be a discrete entity, located, somehow, on the chromosomes.

Then there were the developments in *Organic Chemistry*. A curious issue that had puzzled science, since Aristotle, is that animals (at least, "warm-blooded animals") produce body heat. Why should that be? This phenomenon, called *animal heat*, had been one of the mysteries that the vitalists, who sought an explanation for life outside natural physical causes, would take as evidence that life had rules of its own and factors that came

Chapter Twenty Four
DNA, the Key to the Mystery

to it from outside of the physical world. However, this view gave way when work in cell biology and organic chemistry established that heat is produced by exothermic chemical reactions in the cells. This is the function that Schwann named *metabolism*. (See Chapter 23.) The cell theorists' view was that the important life processes, those which can be studied through physics and chemistry, must occur in the cells. An attention to the details of organic chemistry (what we would now call biochemistry) would reveal much of the mechanics of bodily processes.

As opposed to organic chemistry, a new discipline that arose in the 20th century, *Physical Chemistry*, concerns the actual configuration of atoms within molecules and the forces at play between them that hold molecules together, determine their shape, and account for how they form compounds and break apart. Niels Bohr's model of the atom with its electron shells helped chemists a great deal to picture how molecules were arranged and held together. Bohr's model, despite its faults for atomic physics, enabled chemists to imagine the actual shape of a molecule and to better understand how it formed compounds and which compounds it could form. Quantum mechanics provided the analytic tools to determine the strength and configuration of chemical bonds.

Physics itself provided the research tool of X-rays to investigate structures. Very soon after the discovery of X-rays by Röntgen in 1895, their uses as a research tool became apparent. Medicine began using X-ray machines for diagnostic purposes almost right away. But there were uses in physics itself. Materials that formed crystals when they solidified could be studied by bouncing X-rays off them and analyzing the pattern of shadows cast. Because of the regular structure of crystals, X-rays aimed at them would be reflected at the same angles off every crystal which would then cast shadows and light patterns that would be strong enough to be read on a photographic plate, and from that inferences could be made about the structure of the crystal. Thus arose within physics a new specialty called *Crystallography*, which used what are called *X-ray diffraction* techniques. Knowing the physical structure of a molecule helps to determine its actual molecular configuration and how the individual atoms are bound together. If the gene were an arrangement of molecules on the chromosomes, it would be vital to know its actual configuration to break the code.

All these strands were complex and required specialization to understand. No one could be expected to keep up with each field. Moreover, academic departments that put researchers with like backgrounds together made it even more difficult to know what was going on outside one's own specialty. The solution was to create deliberately multidisciplinary laboratories that would collect people from a variety of different areas of expertise, put them together, and set them to solve some of the difficult intractable problems.

Among the best of these multidisciplinary laboratories were the Cavendish Laboratories at Cambridge University in England. The laboratories worked best when they identified "hot problems" that needed multidisciplinary expertise to solve and which, if solved, would be very important. Among the hot problems being looked at in the early 1950s at the Medical Research Division of the Cavendish was DNA.

DNA

DNA turned out to be the key to understanding the gene, the nature of inheritance, and the organization of living matter itself. It also finally made dominant the reductionist viewpoint with a mechanistic explanation of the mysteries of reproduction and development of the individual. It was the discovery of the physical structure of the DNA molecule in 1953 that was the breakthrough that transformed biology. It also happens to be one of the most interesting scientific adventures on record. To appreciate the steps in the discovery of the structure of DNA, it is necessary to know a bit about what was already known about DNA and who was working on related issues.

The Components of DNA

In the 1860s, the Swiss chemist Friedrich Miescher discovered that cell nuclei contained acids not found elsewhere. He called these *nucleic acids*. By 1900, biochemists had established that nucleic acids all contained: four nitrogenous bases, a five-carbon sugar, and a molecule of phosphoric acid. Any nucleic acid could be built up from units that each contained one molecule of one of the bases, one molecule of the appropriate sugar, and one molecule of phosphoric acid. These building block units were called *nucleotides*.

Chapter Twenty Four
DNA, the Key to the Mystery

By the 1920s, a nucleic acid was seen as a chain of nucleotides, or a *polynucleotide* for short. Any polynucleotide contains only one kind of sugar. Most sugars have six carbon atoms in their structure. The sugars in nucleic acids are anomalies containing only five carbon atoms. There are two basic designs. One is called *ribose*. The other one has an almost identical chemical structure, except that it is missing a single atom of oxygen. It is therefore named *de-oxyribose*. They each form nucleic acids, so there are two basic kinds of nucleic acids, *ribonucleic acid* or RNA for short, and *deoxyribonucleic acid*, better known as simply DNA. DNA has four possible bases. Two are *Purines*: *Adenine* and *Guanine*. Two are *Pyrimidines*: *Cytosine* and *Thymine*. The components of RNA are similar, except that in place of the Pyrimidine Thymine, RNA contains another very similar Pyrimidine, *Uracil*. In the late 1920s, it was discovered that DNA is found almost exclusively in the chromosomes, while RNA (despite being a "nucleic" acid) was found mostly outside the nucleus, in the cytoplasm of the cell.

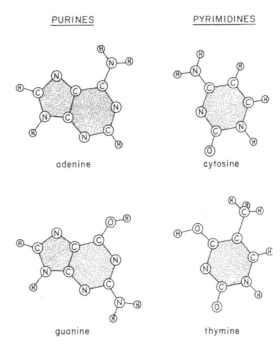

The chemical structure of the four bases in DNA, as conceived in 1951. (From Watson, *The Double Helix*.)

Mechanical Models

One of the tools developed for the study of physical chemistry was mechanical models. Actual ball and stick constructions of a molecule would be built to scale, so as to get the angles and distances corresponding to physical theory. This would help determine the actual structure of the molecules under study. Linus Pauling at the California Institute of Technology was a leader in this work.

His book *The Nature of the Chemical Bond* was the standard text in the field. In 1951, Pauling discovered the basic structure of many protein molecules (polypeptides) by building such 3-dimensional models. His alpha-helix (α-helix) model showed that these proteins were strings of amino acids arranged as building blocks. He discovered that they were bound at points by a large turn in the string that stuck to itself, forming the shape of the Greek letter α, hence the name. Pauling's work demonstrated the power of model building for figuring out these complex structures. But not all scientists agreed that this was real science.

Linus Pauling with a rather elaborate model of a polypeptide chain.

Chargaff's Rules

Erwin Chargaff.

Erwin Chargaff, an Austrian-born chemist at Columbia University in New York, had been studying DNA with standard chemical analyses. He found that in any sample of DNA, the bases can appear in any order and be in different proportions to each other, except that:

The amount of Guanine = the amount of Cytosine.

The amount of Thymine = the amount of Adenine.

These are called Chargaff's rules. Chargaff discovered these one-to-one ratios by doing analyses of the DNA from several species. Though the total amounts of the bases varied from species to species, and even from individual to individual, Chargaff discovered a remarkably close matching of the amount of Guanine with the amount of Cytosine and the amount of Adenine with the amount of Thymine.

The Gene: Protein or Nucleic Acid?

Genes had been identified as places on the chromosomes. Chromosomes were made of protein and DNA, with a greater amount of protein. Since protein had been seen to be

Chapter Twenty Four | 475
DNA, the Key to the Mystery

the basic building structure of life, the general view was that the genes would be found to be made of protein. In 1944, Oswald Avery fed a purified sample of DNA from donor bacteria to recipient bacteria. Some of the recipients of the DNA transformed and began to function like the donor bacteria. Therefore, Avery concluded that the DNA had transmitted hereditable information. However, his methods were disputed; the possibility existed that more than DNA had been transferred, and hence, his results were not widely accepted.

In 1952, Martha Chase and her supervisor at Washington University in St. Louis, Missouri, Alfred Hershey—one of the original members of the Phage Group—did more experiments and showed that *only* the DNA of a phage infects a bacterial host, leaving its protein outside. This confirmed Avery's results. DNA was therefore much more strongly indicated as the likely carrier of the genes.

Crystallography – X-Ray Diffraction

Two physicists, W. H. Bragg and W. L. Bragg, father and son, had invented the discipline of crystallography in 1912. They used it to study the structure of many simpler crystallized structures. Britain was the center of crystallography in the twentieth century. W. L. Bragg, the son, was the head of the Medical Research Division of the Cavendish Laboratories in the 1950s, which was one of the centers of crystallography. Another was King's College at the University of London. At King's, the head crystallographer was Rosalind Franklin, who was studying the structure of DNA using X-ray diffraction techniques.

Rosalind Franklin.

Watson and Crick

The pivotal figures in the search for the structure of DNA were James D. Watson and Francis Crick. James Dewey Watson was born in Chicago in 1928 and attended the University of Chicago, where he completed his bachelor's degree at the age of 19, with a degree in biology and a desire to learn what a gene was. Salvador Luria—another one of the original Phage Group—took on the young Watson for graduate studies at Indiana

Part IV
What is Life?

University. Watson completed his Ph.D. in 1950. Luria invited him to join the Phage Group, which would help him keep abreast of the work that members of the group were doing, trying to find the secret to inheritance by studying simple organisms.

Watson's training had been that of a traditional biology student of the time, with only minimal physics and chemistry. He approached the gene problem as a biological one, and expected to solve its mysteries without having to become a physical scientist. Watson described his first encounters with DNA research as follows:

> My interest in DNA had grown out of a desire, first picked up while a senior in college, to learn what the gene was. Later, in graduate school at Indiana University, it was my hope that the gene might be solved without my learning any chemistry. This wish partially arose from laziness since, as an undergraduate at the University of Chicago, I was principally interested in birds and managed to avoid taking any chemistry or physics courses which looked of even medium difficulty. Briefly the Indiana biochemists encouraged me to learn organic chemistry, but after I used a bunsen burner to warm up some benzene, I was relieved from further true chemistry. It was safer to turn out an uneducated Ph.D. than to risk another explosion.[1]

Watson continued his studies with a post-doctoral research position at the University of Copenhagen, where, alas, he was expected to learn biochemistry. Even then he found a way to dodge the chemistry and continue to do experimental work with phages. While still officially doing research in Copenhagen, Watson attended a conference in Naples, Italy, and heard a talk by Maurice Wilkins of King's College, London, in which Wilkins showed the audience a detailed X-ray diffraction photograph of DNA. Watson became very interested in this and sought a way to learn more about crystallography and X-ray diffraction. He heard about the work being done at the Medical Research Division of the Cavendish Laboratories at Cambridge, and with Luria's help got an appointment there for more post-doctoral research, arriving in 1951.

Francis Harry Compton Crick, born in 1916, had been trained as a physicist, and worked as such for the British military during World War II. The war, especially the dropping of the atomic bomb, disillusioned Crick about a career in physics and he, like several others, turned away from physics to biology where he believed his training in the physical

---

1. James D. Watson, *The Double Helix: A Personal Account of the Discovery of the Structure of DNA*, p. 17.

Chapter Twenty Four | 477
DNA, the Key to the Mystery

sciences would enable him to do good work. Because of the war, Crick had not been able to complete his studies, so at the age of 35, he was still working on his Ph.D. at the Cavendish Laboratories when Watson arrived in 1951. Crick was known for his brilliance and exuberance. When he and Watson met and each learned of the other's interest in DNA, they became immediate friends and tried to find ways to work together on DNA investigations, despite the fact that the official reasons for both of them being at the Cavendish Labs were different. Watson was there to learn X-ray diffraction techniques. Crick was doing research for his thesis on the theory of protein structures.

Watson summed up his first impressions of his new colleague Crick as follows:

> Often he came up with something novel, would become enormously excited, and immediately tell it to anyone who would listen. A day or so later he would often realize that his theory did not work and return to experiments, until boredom generated a new attack on theory.
>
> There was much drama connected with these ideas. They did a great deal to liven up the atmosphere of the lab, where experiments usually lasted several months to years. This came partly from the volume of Crick's voice: he talked louder and faster than anyone else and, when he laughed, his location within the Cavendish was obvious. Almost everyone enjoyed these manic moments, especially when we had the time to listen attentively and to tell him bluntly when we lost the train of his argument.[2]

The facilities of the Cavendish made the perfect environment for Watson and Crick to draw from the expertise of researchers in many different disciplines and bring that all to bear on the DNA problem. Watson was a biologist; Crick was a physicist. Close by at the Cavendish were chemists, mathematicians, crystallographers, and other scientists to whom they could go for help at any time and use as sounding boards for their ideas. Since they themselves were so differently trained, they tended to see the same problem different ways and avoid getting stuck in intellectual blind alleys.

The Structure of DNA

One of the most entertaining and eye-opening books written about science is James Watson's personal account of his and Crick's work together in discovering the structure of

---

2. *Ibid., pp. 9-10.*

Part IV
What is Life?

the DNA molecule. This is *The Double Helix,* from which the above two quotations have been taken. Watson's irreverent account of their work, the personalities of their colleagues, and the race to get their result before anyone else pushed the limits of propriety too far in the view of some of his colleagues. Harvard University Press had undertaken to publish Watson's book, but after Watson had circulated it in manuscript to several of the people mentioned in it, asking for their comments, Harvard changed its mind. The Press took the unprecedented step of refusing to publish it after receiving protests from some scientists, including Francis Crick himself.

Immediately thereafter the book was published by the Atheneum Press in 1968 and has been a steady best-seller ever since.[3] It is one of the few books ever written by a scientist that recounts all the false starts, missteps, and blunders along the way to a major discovery, as well as the motives and personality quirks of the people involved.

In a much more mundane format, here are the elements of Watson and Crick's breakthrough discovery.

A Double Helix

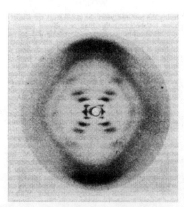

An X-ray diffraction photograph of DNA taken by Rosalind Franklin showing the tell-tale cross sign of a helical structure.

Like many complex organic compounds, DNA has a helical structure. That means that it is in the shape of a long strand that is twisted around on itself, which makes it more compact. Watson and Crick had suspected that DNA had a helical structure throughout most of their research, and the X-ray diffraction pictures they had seen suggested that this was correct. However, what was not clear was the number of strands that were twisted together. The main possibilities were either two or three strands. For a time they leaned toward the three-strand idea but were not able to make it work with everything else they knew about DNA. When they focused on the two-strand idea, the rest came together fairly quickly.

---

3. Walter Sullivan, "A Book That Couldn't Go to Harvard." In James D. Watson, *The Double Helix*. Gunther S. Stent, ed. New York: Norton, 1980., pp. xxiv-xxv.

Chapter Twenty Four
DNA, the Key to the Mystery

A Sugar-Phosphate Backbone on the Outside

DNA could be thought of as a polynucleotide, meaning that it was composed of an assembly of "nucleotides," each of which consisted of a sugar molecule (deoxyribose), a phosphate molecule, and one of the four bases found in DNA. There were many ways to imagine how these nucleotides could be attached to each other to make a polynucleotide. One possible single-strand arrangement was proposed by Alexander Todd and his group, who were working in a laboratory near Watson and Crick.

When Watson and Crick began to think of a double helix, the main question

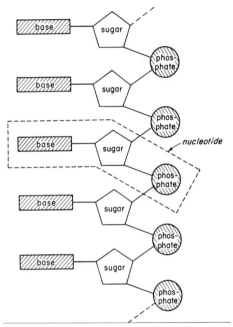

Alexander Todd's single-strand model of DNA with a sugar-phosphate backbone, proposed in 1951.

was whether the sugar-phosphate backbone would be on the outside of the helix or on the inside. If the DNA molecule somehow stood for the genes, then it had to be capable of embodying in its structure the complex code that the body could "read" for instructions on how to grow. The only possibility for a complex code in the DNA lay with the bases, not with the constantly repeating sugars and phosphates. So the "code" was in the arrangement of the bases. Therefore, for this to do any good, the bases needed to be exposed where they could influence the chemical reactions taking place in the cell.

Hence, it seemed necessary that the backbone be on the inside and the bases on the outside. But then it would be difficult for the DNA molecule to crystallize, since its outer edge would not be regular. Yet, DNA did crystallize, which is why it was possible to take X-ray diffraction photographs of it. The

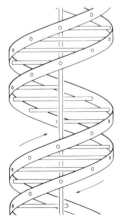

A double-helix model of DNA with the bases on the inside.

photographs taken by Rosalind Franklin led her to the opinion that the backbone had to be on the outside and the bases on the inside. So be it, but then the question remained of how this arrangement of bases could be of any use as information.

Accounting for Chargaff's Rules

Erwin Chargaff's finding that the amount of guanine always equals the amount of cytosine while the amount of adenine equals the amount of thymine surely had to have some physical meaning. Such one-to-one ratios do not happen as coincidences. However, what meaning his rules had evaded Watson and Crick so long as they continued to think that the strands of DNA were identical copies of each other. If they were not, then how could they function during cell division when the chromosomes break apart and make identical copies of themselves?

By this point, Watson and Crick had decided that the key to finding the structure of the molecule was to use Pauling's model-building approach. So they directed the shop at the Cavendish to make metal models of all the component molecules of DNA for them; they would then try to assemble them in different ways until they got an arrangement that satisfied all the requirements of chemical bonding and fit in with what else was known about DNA.

Thymine bonded to Adenine.

Cytosine bonded to Guanine.

Watson didn't wait for the metal jigs and instead cut out some cardboard models to work with. After much fruitless fiddling, he had the inspiration to try bonding the adenine to the thymine and the guanine to the cytosine. Amazingly, they appeared to work, and when he checked the idea out with a chemist at the Cavendish, he got a nod of approval. The adenine-thymine combination occupied the same amount of room as the guanine-cytosine combination, and both combinations presented outer edges that could bind with sugar phosphate backbones. Moreover, the combinations could be flipped over and still worked (e.g., adenine on the right instead of on the left). That meant that there were four different ways that the base pairs could be fit between the backbones.

Chapter Twenty Four
DNA, the Key to the Mystery

There was one other necessary detail: the sugar-phosphate backbones had to run in opposite directions, one up, one down. Chargaff's rules would be accounted for and a model could be constructed with two sugar-phosphate helical backbones running in opposite directions, held together by the base pairs in between.

April 25, 1953

And then it all came together very fast. Crick made the calculations that verified that all the bond distances and angles would work, that there was room for everything else that was supposed to fit in (for example, quite a few water molecules had to fit around the bases and the backbones). Maurice Wilkins and Rosalind Franklin came down from King's College, London, and inspected the model, giving it their approval. Much of the data that finally led Watson and Crick to their correct model had come from the Rosalind Franklin's crystallographic work at King's College.

It was decided that three papers would be published simultaneously in *Nature* on April 25, 1953. The first was Watson and Crick's description of their model. The second was by Maurice Wilkins and his associates, and the third was by Rosalind Franklin and an associate of hers. The second and third papers provided the supporting data to confirm Watson and Crick's model. As is so often the case with publications that completely revolutionize a subject, the wording is very restrained. The paper by Watson and Crick begins:

> We wish to suggest a structure for the salt of deoxyribose nucleic acid (D.N.A.). This structure has novel features which are of considerable interest.[4]

There are few dates in the history of science that can rival April 25, 1953 for changing the course of scientific research. There were enough people studying DNA and enough others who realized its importance that the papers published in *Nature* did not go unnoticed. All that Watson and Crick had settled was the general structure of the DNA molecule. Everything else remained to be solved: How did the molecule reproduce itself; how did the sequence of bases trapped in the middle of the molecule capture and then

---

4. J. D. Watson and F. H. C. Crick, "Molecular Structure of Nucleic Acids," *Nature*, no. 4356 (April 25, 1953), p. 737.

convey information to the body as a gene must do; what, if anything, did DNA have to do with the other nucleic acid, RNA?

These were just the first urgent questions that had to be answered. Once the general framework was in place, research had a clear direction. What Watson and Crick had found was that the mystery entity, the gene, was actually a sequence of bases within the DNA molecule. The answers to the questions of inheritance were going to be found in the mechanist realm of biochemistry.

Watson and Crick had made use of every insight available about DNA and every technique that could be used to find its structure. They knew of the work of cell biologists who had identified the location of DNA as part of the chromosomes; they knew that the chromosomes were the likely location of the genes; they relied heavily on work of the crystallographers, especially Rosalind Franklin, to get the general shape and structure of the molecule; they incorporated Chargaff's insight about the one-to-one ratios in the base pairs; they used Linus Pauling's model-building techniques to fit all the possibilities together; and, they had all the expertise in other scientific subjects around them at the Cavendish to help them with special issues as they arose.

The discovery earned them the Nobel Prize in Medicine or Physiology in 1962, which they shared with Maurice Wilkins of King's College. Rosalind Franklin might have been a more appropriate person than Wilkins to be honored along with them, but unfortunately she had died prematurely in 1958 and Nobel prizes are only awarded to the living.

Molecular Biology

According to Gunther Stent, one of the leading DNA researchers in the second half of the 20th century, the era of *molecular biology* began exactly on April 25, 1953. The term "molecular biology" had been introduced years before by W. T. Astbury, one of the early crystallographers to get interested in the structure of proteins and nucleic acids. Nevertheless, the name did not catch on and people doing similar work did not think of themselves as molecular biologists—until April 25, 1953, when suddenly all such work seemed to hang together as a new discipline in biology.

Chapter Twenty Four
DNA, the Key to the Mystery

The first task of molecular biology was to find out how DNA actually worked. This can be divided into two topics: (1) how the molecule reproduces itself, preserving all that incredible detail in the ordering of the base pairs, and (2) how that order is used as information by the body. The standard answers to these questions is called, believe it or not, the *Central Dogma of Molecular Biology*.

The Autocatalytic Function

DNA is a double helix. It reproduces itself by uncoiling into separate strands and then making two molecules where there had been one. But this is where Watson, Crick, and everyone else had gone off the rails. Because it was necessary that the DNA reproduce itself in cell division so as to make a complete set of DNA for every cell of the body, they had imagined that the two strands would be identical and then just make copies of themselves when they came apart. But in fact DNA strands make complements of themselves instead of identical copies.

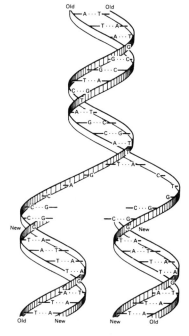

The Autocatalytic Function.

A molecule of DNA that is reproducing itself during cell division uncoils from one end by breaking the bonds between the bases down the middle. Then each exposed strand attracts to itself the complementary components of the other side. For example, a strand that exposes the bases Adenine, Cytosine, and Guanine, in that order, will attract the sugar and phosphate molecules necessary and with them the bases Thymine, Guanine, and Cytosine, in that order. These processes are all controlled by enzymes that the body secretes to start and control the process. When the entire DNA chain has unraveled and attracted its complement to each side, the result will be two new DNA helices, identical to the original DNA molecule. This is called the *Autocatalytic Function*.

Part IV
What is Life?

Here is a place where mutations can occur. When the DNA uncoils and replicates itself, it can make mistakes. This function is controlled by enzymes in the cell nucleus. Problems with the enzymes due to chemical imbalance could lead to the wrong bases becoming attached, or the DNA can break apart and recombine in a different order. Moreover, the base pairs are so similar to each other in chemical formula that a very slight change, caused perhaps by radiation, can swing the alignment of the atoms in these molecules enough to turn them into another. It does not take much to change Cytosine-Guanine into Thymine-Adenine. If that happens, the DNA does not reproduce exactly and a mutation occurs. This provides the opportunity for new variations and therefore gives natural selection some new material to work with

The Heterocatalytic Function

The DNA molecule contains the coded information that represents genes in the arrangement of the base pairs. However, this is of little use unless there is a way for this arrangement to influence processes in the body. It does so with the *Heterocatalytic Function*. From time to time as the body requires a new supply of a given protein in a certain kind of cell, an enzyme is secreted into the cell nucleus which causes the DNA molecule to open up a bit at a specified place, breaking the bonds between the purines and pyrimidines. At the place where the DNA is open, enzymes cause a backbone of ribose and phospate to form and attract to it the purines and pyrimidines that are the complements of the exposed bases on the DNA. This forms a piece of RNA (which is single stranded).

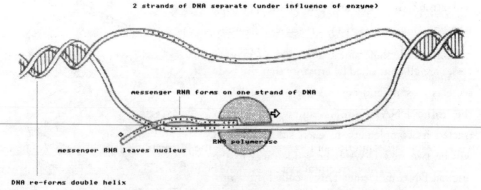

The formation of messenger-RNA.

Chapter Twenty Four
DNA, the Key to the Mystery

The piece of RNA that has formed and copied the sequence of bases onto its own molecule then migrates out of the nucleus into the cytoplasm, where it becomes the template for protein synthesis. This piece of RNA is called messenger-RNA or mRNA for short.

Codons

There are four different bases that form the sequences in DNA. Think of this as an alphabet with four letters A, C, G, and T. The sequence of these "letters" on a stretch of DNA is transferred to messenger-RNA. Actually, the complement of the sequence is transferred, with Uracil substituting for Thymine. In any case, it is still a sequence written in four letters. Proteins are made up of strings of amino acids. There are twenty amino acids that go into proteins. The sequences of bases on the mRNA determine which amino acid goes next in the sequence on a protein when it is being formed. To make four "letters" point to 20 different amino acids, they are grouped in threes. Each group of three bases is called a *codon*. Since there are 4 bases to choose from for each "letter" of the codon "word," there are 64 possible codons (because 4 x 4 x 4 = 64). In 1961, a young biochemist names Marshall Nirenberg read a paper to the International Congress of Biochemistry in Moscow explaining how he had devised an experiment that enabled him to find out what the meaning of a particular codon was. That meant that in principle there was a way that all 64 codons could be identified and matched with their corresponding amino acid. Crick, who was at the meeting, wrote that when he heard this he was "electrified."[5] Identifying all the codons would "break the genetic code." A race ensued to work out all the correspondences, and in a few years it was done. Each codon points to a particular amino acid; since there are 64 codons and only 20 amino acids, several codons point to the same amino acid, and a few others have special meanings, such as "stop the sequence."

Protein Synthesis

The chief function of DNA is to provide the instructions on how to make proteins, which are the chief structural materials in the body. If the DNA is the "library" of instructions for a given organism, the section of DNA that opens up in the presence of certain

---

5. Quoted in Watson, *The Double Helix*, Stent edition, p. xxi.

enzymes is the blueprint for a particular protein. That blueprint is transferred to the messenger-RNA where the sequence of bases may be notionally grouped in threes—the codons—each of which represents a particular amino acid. Proteins are simply sequences of amino acids. The messenger-RNA carries the codons that specify a certain order of amino acids. That mRNA travels out of the nucleus of the cell and into the cytoplasm where, in a complicated process, it is "read off" by units in the cytoplasm that then assemble the desired protein amino acid by amino acid.

One-Way Process

In the general course of DNA-body interactions, information flows from the DNA to the body, not vice-versa. If so, there is really no mechanism here to support the inheritance of acquired characteristics. Changes in the environment of an individual would not affect that individual's DNA. The DNA therefore is much like Weismann's germ plasm. And so it seemed that the matter was settled. Natural selection was possible. The inheritance of acquired characteristics was not. But that was before *reverse transcriptase* was discovered in 1970, with which the process from DNA to RNA to protein is reversed and the DNA is rewritten by the cell itself. Science, it seems, never ever settles a question forever.

DNA Technology

The complexity of DNA has made it very difficult to study its particular sequences in detail. Even a virus can have as many as 5000 base pairs. A human has more like 100,000 base pairs in its DNA. Breakthroughs in research came in the mid-1970s with the development of two techniques for working with DNA: *cleaving enzymes* and DNA *ligases*.

In the 1970s, it was found that when samples of DNA were placed in a solution of certain enzymes, they fell apart in several places that were later identified as being where a particular sequence of bases appeared on the DNA. For example, the enzyme ECORI cuts DNA at the sequence GAATC. With the discovery of other cleaving enzymes, scientists had a technique for breaking DNA apart in a number of predictable places. Then, after a time of working with the cleaving enzymes, another remarkable discovery was made of other enzymes that will cause pieces of DNA to rejoin. These are the ligases. Thus, DNA research had the "scissors" and "paste" tools necessary to manipulate DNA and study the results of experiments. Thus began the technology of Recombinant DNA—the disassembling of DNA and reassembling of it with different components.

Cloning

Cloning is the process of producing a strain of DNA and then inserting that DNA into a host where it will replicate. A phage can be used to convey the DNA. The replicated DNA is called a clone. Cloning as a technique has many uses. For example, it can be used to replicate rare hormones and proteins such as insulin and interferon that have much medical usage

Insulin is a protein hormone produced in the pancreas that the body uses to regulate blood sugar concentrations. Diabetics have lost the ability to produce insulin and must have an outside source of it. In the 1920s, insulin from cows and pigs was isolated and made available to humans with diabetes, even though it is not identical to human insulin. But supply was a major concern since the number of diabetics was on the rise. Cloning insulin became an ideal usage for recombinant DNA technology.

In 1978, Herbert Boyer and colleagues at the University of California in San Francisco created a synthetic version of human insulin using recombinant DNA technology. The DNA sequence representing the instructions on growing insulin was separated and then inserted into the bacterium E. coli. The E. coli then produced prodigious amounts of human insulin.

Boyer set up a company to manufacture and sell the products of recombinant DNA technology. His company, Genentech, began manufacturing recombinant human insulin. Genentech now manufactures a variety of synthetic hormones for the treatment of cancer, heart disease, immune system disorders, and other problems. A large industry with many companies in many countries has followed.

Another industrial application of recombinant DNA is to insert genes from one species into another species altogether to make a new hybrid that has desired characteristics. One of the biggest applications of this today is genetically modified foods. Crops may be modified for a number of reasons: to provide natural resistance to insects, which would make insecticides unnecessary, or to provide tolerance for herbicides so that they may then be sprayed on the crops to kill weeds without killing the crop.

Cloning is also being pushed to its logical limits. Whole organisms have been reproduced from DNA taken from other bodies. In 1997, the sheep "Dolly" was cloned from an adult sheep. Dolly is an exact replica of its "mother"—the animal from which the cell was taken.

Dolly, the fully cloned sheep.

The Human Genome Project

One of the largest scientific projects ever undertaken has been the decoding of the entire gene map of the human species. This project once fully complete offers the possibility of researchers finding the causes and perhaps cures for genetic diseases of all sorts. Indeed, for finding out what diseases are genetically determined in the first place.

It has been controversial from the start. In 1985, Robert Sinsheimer, Chancellor of the University of California at Santa Cruz (UCSC), who was himself a distinguished molecular biologist, had been looking for a way to bring the university to greater prominence. He thought that if a major biological research project could be based at UCSC, it would bring funding to the university and establish it as a research center. He convened a meeting of prominent geneticists to discuss a project to determine the exact sequence of every gene in the human body, what is called the human genome. Though some individual genes had been located within narrow ranges on particular chromosomes, their exact sequence of bases had not been determined. A few genes, those important in certain diseases, such as Huntington's disease and muscular distrophy, had been mapped, but these were individual cases and there were not many of them. Sinsheimer's plan was to map the entire genome. Every position on a DNA molecule is characterized by one of four base-pairs (Adenine-Thymine, Guanine-Cytosne, Thymine-Adenine, or Cytosine-Guanine). There are approximately 3 billion positions on the DNA of a single human genome. To figure out their precise sequence would be a very large project indeed. When Sinsheimer

Chapter Twenty Four
DNA, the Key to the Mystery

convened his meeting in 1985, a good molecular biologist could work out the sequence of about 50 to 100 base-pairs a day, at a cost that would work out to about $10 per base-pair. That would make the cost of the whole project about $30 billion and would take the entire working careers of around 25,000 molecular biologists working full time on this and nothing else. It was preposterous. Yet, the benefit to science to have a complete map of the human genome could not be denied, and doing it systematically instead of going after one gene at a time would clearly be more efficient.

The next year, it was suggested by Nobel Laureate Renato Dulbecco, who was unaware of Sinsheimer's meeting, that sequencing the human genome would be the ideal means of understanding the genetic causes of all diseases, such as cancer. Still, the project seemed insurmountable. All the funding for scientific research available in the United States and many other countries would scarcely be enough to underwrite the project, and even if it did, it would leave very little funding for any other scienctific research at all.

The crucial support for the project came from an unexpected quarter. The U.S. Department of Energy—the former Atomic Energy Commission—happened to have a number of laboratories devoted to research on the biological effects of radiation and a considerable expertise in biotechnology, much of it devoted to determining the genetic effects of nuclear weaponry. Their research laboratories had already completed some significant work in gene mapping. They had the mandate to learn as much as possible about genetic effects on humans of atomic energy, and one way to do that was to map the human genome. After considerable debate within the scientific community and in hearings before Congress, the project was approved. It would be a joint effort between the Department of Energy and the National Institutes of Health. James Watson was appointed Associate Director at the National Institutes of Health, in charge of genome research, thus giving the project a very high profile. (Watson is said to have remarked that it made a perfect complement for him to discover the structure of DNA at the beginning of his career and oversee the mapping of the human genome at the end of it.)

Other countries were brought into the research, including Canada, Japan, Germany, Great Britain, France, and Italy. Research was conducted in laboratories all over the world in a systematic plan. Work proceeded slowly at first, but then much of the sequencing was automated, and the process sped up. A "draft sequence" of the entire human genome was published in February 2001. The goal was to complete the project by 2003, fifteen years after commencement.

It's called the Human Genome Project because its main goal is to map the genes of the human species, but practically speaking, it is worthwhile to map the genomes of some simpler species along the way. Bacteria, with far simpler genetic structures, were among the first other species to have their genomes mapped. In 2000, the genome of the old workhorse of genetic research, *Drosophila melanogaster*, was finally mapped, all 13,601 genes spread over 250 million base-pairs on five chromosomes.

◆

We appear to be at a stage in the history of science where the bulk of scientific research is very expensive. In physics, for example, the project to determine the ultimate physical particle with research in high energy accelerators is so expensive that Congress in the U.S. had to consider which of the two projects—the genome project or the accelerator—it could support. In astronomy, the Hubble orbiting telescope is a major expenditure, and space exploration can only be undertaken by the richest of countries working in concert. In the Scientific Revolution, and even into the late 19th century, significant scientific research could be conducted by individuals who were independently wealthy or were pursuing a serious hobby. In the last hundred years, it has become professionalized and supported by government agencies. Since the public purse is the source of funding, projects have to justify themselves before they get support. Research has therefore become goal-directed: "Please give me the money I need to investigate this problem. I hope to find an answer in this amount of time (specified) and at this cost (specified)." This can be very effective and efficient. Unfortunately, the most extraordinary discoveries are those we did not expect to make.

Science is just beginning to reap the rewards of compiling this encyclopedia of the human species. As it does, it finds that new and unexpected questions are beginning to arise about the nature of genes, of life, of the boundary between the species. These we will also investigate. And it will go on and on. In the physical sciences we are learning more and more about the fine structure of matter and the vast reaches of the universe. Every-

Chapter Twenty Four | 491
DNA, the Key to the Mystery

time we learn something new, many more facets of nature appear dimly on our horizons. As always throughout the history of science, once some aspect of reality has been brought into focus, another, previously unsuspected one is glimpsed off in the distance.

For More Information

Access Excellence and the National Health Museum. *Introduction—The Human Genome Project.* http:/www.accessexcellence.org/AB/IE/Intro_The_Human_Genome.html.

Alioto, Anthony M. *A History of Western Science,* 2nd ed. Englewood Cliffs, NJ: Prentice-Hall, 1987. Chapter 26.

Cook-Deegan, Robert Mullan, *Origins of the Human Genome Project.* http:/www.fplc.edu/risk/vol5/spring/cookdeeg.htm.

Keller, Evelyn Fox. *The Century of the Gene.* Cambridge: Harvard University Press, 2000.

Lewontin, Richard. *The Triple Helix: Gene, Organism, and Environment.* Cambridge: Harvard University Press, 2000.

MacLachlan, James. *Children of Prometheus: A History of Science and Technology,* 2nd ed. Toronto: Wall & Emerson, Inc., 2002. Chapter 21.

Ronan, Colin A. *Science: Its History and Development Among the World's Cultures.* New York: Facts on File, 1985. Chapters 9 & 10.

Tudge, Colin. *The Engineer in the Garden: Genes and Genetics—From the Idea of Heredity to the Creation of Life.* New York: Hill and Wang (Farrar, Straus and Giroux), 1995.

U. S. Department of Energy. *Human Genome Project Information.* http://www.ornl.gov/hgmis/.

Wall, Byron E., ed. *The Nature of Science: Classical and Contemporary Readings.* Toronto: Wall & Emerson, Inc., 1990.

Watson, James D. *The Double Helix A Personal Account of the Discovery of the Structure of DNA.* Ed. by Gunther S. Stent. A Norton Critical Edition. New York: Norton, 1980.

Wills, Christopher. *Exons, Introns, and Talking Genes: The Science Behind the Human Genome Project.* New York: Basic Books, 1991.

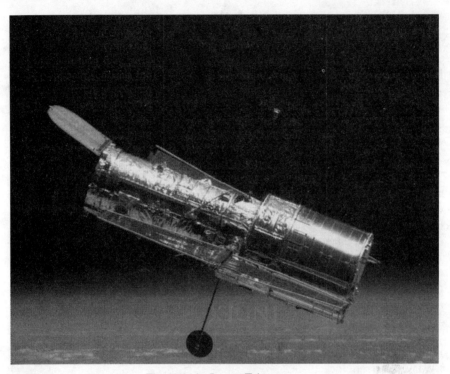

The Hubble Space Telescope.

INDEX

A

Absolute
  brightness 363
  space 309–311, 323
  time 309–311, 323
  zero 295
Academy, Plato's 52, 60, 77–79, 96, 112
Acceleration and Gravity 327
Achilles and the Tortoise 31–32
Actualism 387
Ad Hoc explanation 158, 160, 198, 232, 304, 315–316, 319, 332, 372, 429, 453
Adenine 473, 474, 483
Adriatic Sea 190
Æther 299–307
Age of the Earth 383
Agricola 142, 284
Alberti, Leon Battista 142
Alexander the Great 78–79, 96, 135
Alexandria 96, 111, 113
Algebra 234
Alpha-helix 474
Alps 191
Analytic geometry 234
Anaximander 24–25
Anaximenes 25–27
Andromeda Nebula 368
Animal heat 470
Animalcules 444
Anlagen 460
Annus mirabilis
  Einstein 318
  Newton 242, 248
Antichthon 42
Antiperistasis 87–88, 232
Apeiron 24

Apollo Lykaios, Temple of 78
Apollonius 175, 236
"April 25, 1953" 481–482
Aristarchos of Samos 116
Aristotle 2, 19, 37, 43, 76, 77–91, 95–96, 127, 186, 200, 209, 227, 257, 309, 350, 379, 380, 383, 394, 431, 432
  Final Cause 380
  Flees Athens 79
  Metaphysics 43
  On the Heavens 215
  Physics 431
  Universe 85–90
Armadillo 418
Arrow, Flying 32
Artificial selection 427
Astbury, W. T. 482
Aswan 112
Atheneum Press 478
Athens 79
Atmospheric pressure 284
Atomic Energy Commission 489
Action at a distance 277
Autocatalytic Function 483
Avery, Oswald 475
Axiomatic system, Euclid 102

B

Babylonia 12
Babylonian number system 12–15
Bacteria 445
Barberini, Cardinal Maffeo 202
Barnacles 420
Barrow, Isaac 260
Bateson, William 463
Becquerel, Henri 340

Bell, Charles
  The Hand 412–414, 434
Bern, Switzerland 318
Bernard, Claude 448
Bessel, Friedrich 362
Beta = Eratosthenes 112
Bible 199–201
Big Bang 371–374
Big Crunch 376
Billiard ball universe 347
Biringuccio, Vanocchio 142
Black Body Radiation 344
Blueshift 369
Boethius 39
Bohr Atom 348–351
Bohr, Niels 337, 348, 469, 471
Bologna 143
Boulton, Matthew 287
Boyer, Herbert 487
Bragg, W. H. 475
Bragg, W. L. 475
Brahe, Tycho 169–172
Bridgewater Treatises 412–414, 432
Bridgewater, Earl of 412
British Admiralty 415
British Association for the Advancement of Science 439
Brown, Robert 319, 446
Brownian motion 319, 446
Brunelleschi, Fillippo 142
Brünn Society for the Study of Natural Science 460
Brünn, Moravia 456
Buckland, William 391
  Geology and Mineralogy Considered with Reference to Natural Theology 412
  Relics of the Flood 391

Buffon, Georges 397
  Histoire Naturelle 397
  Theory of degeneration 397
Bunsen, Robert 349
Burke, Bernard 374
Byzantine Empire 135

C

Calculus 172, 242, 248—253
California Institute of Technology 469, 473
Cambridge 242
Cambridge University 341, 415, 472
Canon 143
Cat paradox 356
Catastrophism 388—389
Cathode Rays 339
Catholic Church 383
Catholic University of Louvain 371
Cattle breeders 420
Causes, The Four 83—85
Cavendish Laboratories 341
  Medical Research Division 472, 476
Celestial sphere 127
Cell theory 446, 447
Cells 445
Central Dogma of Molecular Biology 483
Cepheid Variables 363—364, 368
Cepheus 363
Chalcis 80
Chambers, Robert
  Vestiges of the Natural History of Creation 409
Change 28—29
Chargaff's rules 474, 480—481
Charlemagne 137
Chase, Martha 475
Chicago 475
Chicago, University of 367, 475—476
Chick embryo 451

Christina, Queen 238
Chromatic aberration 444—446
Chromosomes 454—455, 463, 466, 470, 474, 482
Circle
  circumscribed 167
  equation of 237
  inscribed 167
Circular motion perfect 89
Cleaving enzymes 486
Clockwork universe 270
Cloning 487—488
Coal 283, 292
Codons 485
Cogito ergo sum 231, 234
Color 243
Columbia University 464, 474
Columbus 117
Commensurability 44—48
Common notions 101
Comparative anatomy 389
Cone 175
Conic section 175—176
Conservation of Energy 290
Constantinople 135
Continuous variation 422, 427
Continuous viewpoint 337—338
Copenhagen interpretation 354
Copenhagen, University of 476
Copernican System 144—160
Copernicus, Nicholas 141—162, 177, 350
  On the Revolutions of the Heavenly Spheres 144—162, 165
Cork 445
Correlation of parts 389
Correns, Carl 462
Cosmic Background Radiation 372—374
Coulomb, Charles 305
Council of Trent 142
Counterexample 434

Cracow, University of 143
Crick, Francis 475—482
Crookes, William 339
Crystal spheres 73
Crystallography 475
Cube 167
Curie, Marie 341
Curie, Pierre 341
Curtis, Heber 367
Curvature of Space 334
Cuvier, Georges 389
Cyrene 112
Cytoplasm 473
Cytosine 473—474, 483

D

Dark Ages of Europe 135
Dark Matter 375—376
Darwin, Charles 394, 403, 414—440, 448, 459, 466
  Descent of Man 438—439
  Journal of Researches 416
  On the eye 434—436
  Origin of Species 426—435
  Weight of evidence criterion 437
Darwin, Erasmus 407, 414
  Zoonomia 407, 408
Darwin, George
  on lunar evolution 436
Darwin's Bulldog, T. H. Huxley 439
de Broglie, Louis-Victor-Pierre-Raymond 351—352
de Broglie, Maurice 351
de la Mettrie, Julian Offroy
  Man a Machine 278
de Laplace, Pierre Simon
  Celestial Mechanics 278
  No need for God 279
de Vries, Hugo 462
Deferent 127
del Monte, Guidobaldo 187
Delbrück, Max 469, 470
Democritus 337
Deoxyribonucleic acid *See* DNA

Deoxyribose 473, 479
Descartes René 227–238, 299, 338
    Discourse on Method 238
    Geometry 237
    Le Monde 238
    Principles of Philosophy 229–232, 263
Design Argument 410–415, 430–438
    Logical structure 432–434
di Lasso, Orlando 182
Diagonal of the Square 45–46
Dicke, Robert 373–374
Diffraction 301
Diluvialism 391
Discrete viewpoint 337–338
Divided Line, The 53–58, 95
DNA 472–489
    components of 472–473
    technology 486
Dodecahedron 167
Doge of Venice 189
Dolly, the sheep 488
Doppler effect 369
Doppler, Christian 369
Down, Village of 420
Drake, Stillman 198
Drosophila melanogaster 464–466
Dualism, Cartesian 231
Duke of Tuscany 192

E

E. coli 487
Earth 113, 155, 181
    size of 111–115
Earthshine 191
Easter 141
Ecuador 418
Eddington, Arthur 332, 371
Edentates 418
Edinburgh, University of 415
Effect of Choice 410–411, 432
Efficient cause 84

Egypt 7
Egyptian arithmetic 9
Eiffel Tower 352
Einstein, Albert 317–334, 337, 346, 446
    "On the Electrodynamics of Moving Bodies." 320
    Elevator Experiment 328
    Photon theory 352
Elea 28–33
Electromagnetism 299–307
Electrons 327–342
Ellipse 175
Empedocles 431
Empiricism 80–82
Energy 283–293, 307
    Available 292
    Conservation of 290
    Quantum of 345–346
    Unavailable 292
Enlightenment 277, 395
Ephesus 18, 27
Epicycle-Deferent system 127–130
Equant 130–131, 145
Eratosthenes 111–115
Etna, Mount 392
Euboea 80
Euclid 96–111, 118, 258, 415
    axiomatic system 102, 258
    Elements 94–111, 136, 249, 257, 263, 314, 430
Eudoxus 72–74, 90, 96, 115, 127, 350
    Spheres of 72–74
Everett, Hugh 357–358
Eye, The 434–436

F

Factors 460, 463, 466
Faraday, Michael 306
Featherless bipeds 395
Ferdinand, Archduke 172

Fertile Crescent 12
Feynman, Richard 337
Figurate numbers 41
Final cause 84
Finches 419
Fitzroy, Robert 415
Florence 202, 205
Forced motion 87, 209
Formal cause 83
Forms 54, 55
Fossils 388, 389
Four Causes 83–85
Frames of reference 322
Franklin, Benjamin 305
Franklin, Rosalind 475, 480–482
Frauenberg 143–144
Frederick, King 171
Freiberg, Saxony, 385
Fresnel, Augustin 302
Fruit fly 464–466

G

Galápagos Islands 416–419
Galaxy 365
Galen 132
Galilean relativity 322
Galileo 185–205, 209–224, 227, 257, 284, 443, 457
    Dialogue on the Two Chief World Systems 202–205
    Law of Free Fall 211–216, 266
    Starry Messenger 190–199
    Theory of the tides 187–189
    Two New Sciences 218–224
Gamow, George 373
Garden pea 457
Gdansk 143
Gedanken Experimenten 321
Gemmules 429, 453
Gene 462–466, 474, 482
    search for 464–466
Genentech 487

General Theory of Relativity 326–332
Genesis 372, 383, 385, 397
Genetics 463
Geography 112
Geometrical and Military Compass 186
Geometrical harmony 40–41
Germ plasm 454, 486
German Botanical Society, Proceedings of 462
Glasgow, University of 286
Glaucon 95
Gnomon 113
Goethe 449
Grand Duchess of Tuscany 200
Gravitational mass 328
Gravity 271–273
Gray, Asa 426
Graz, University of 166
Great Chain of Being 395
Great Debate on the Nature of the Universe 367
Great Translation Project 137
Greece, Ancient 19–33
Greek modes 38
Gregorian calendar 161
Guanine 473–474, 483

H

Haeckel, Ernst 378, 449
Hale, George 367
Halley, Edmund 261–262
Harmony, geometrical 40–41
Harvard University Press 478
Hawking, Stephen 260
Heat 292
    as an element 288
    Mechanical Equivalent of 289
Heat Death of the Universe 296
Heisenberg Uncertainty Principle 353–355
Heisenberg, Werner 354–355
Helium 369

Hellenic period 97, 111
Hellenistic period 96–97, 116–117, 135–136, 174
Henslow, John 415
Heraclitos 27–28
Heredity 470
Hermes Trismegistus 147
Hermetic traditions 147, 259
Hershey, Alfred 470
Heterocatalytic Function 484
Hieratic 7
Hieroglyphics 7
Hipparchus 15, 116, 129
Homunculus 451–452
Hooke, Robert 261, 263, 445
Hooker, Joseph 426
How, not Why 216
Hoyle, Fred 372
Hubble constant 370, 377
Hubble telescope 490
Hubble, Edwin 360, 367–370
Hubble's Law 370–371
Human evolution 439
Human Genome Project 488
Huntington's disease 488
Hutton, James 387, 393
    Theory of the Earth 387, 409
Huxley, Thomas Henry 438
Huygens, Christiaan 299
Hven 170
Hybridization 456
Hydrogen, Spectral Lines of 349–350
Hyperbola 175
Hypotheses non fingo 272

I

Iceland spar 303
Icosahedron 167
Idea of Progress 407
Idioplasm 452–454
Imperial Mathematician
    Galileo 200
    Johannes Kepler 172
    Tycho Brahe 171

Inclined Plane 212
Incommensurability 44–48
Indiana University 470
Induction 82
Industrial Revolution 277, 283, 384, 389, 407
    beginning of 287
Inertial mass 327–328
Inertial motion 299
Inheritance 449–455
Inheritance of acquired characteristics 399–402
Inquisition 204–205
Instantaneous Velocity 249–253
Institute for Advanced Study 317, 333
Insulin 487
Intelligible world 54
Interferometer 313, 316
Interferon 487
Invariants 291, 323
Ionia 20–21, 28
Ionian mode 38
Islam 136
Istanbul 135

J

Janssen, Zacharias 443
Jardin du Roi 397–398
Johannsen, Wilhelm 463
Joule, James 289
Joule's Churn 290
Julian calendar 141–142, 144, 242
Julius Caesar 141
Jupiter 170, 181
    satellites 191
Justinian 52

K

Kaiser Wilhelm Institute 333
Kant, Immanuel 277
Kelvin, Lord (William Thomson) 296

Index

Kepler, Johannes 165–182, 198, 217, 227, 257, 350
  Cosmographical Mystery 166–169
  Harmony of the World 178–182
  Laws 174-180, 259, 266
  New Astronomy 174, 176
Kinematic model 128
King's College, London 475, 476, 481
Kirchhoff, Gustav 344, 349
Koran 136

L

Lamarck, Jean 397–403, 407, 412–413, 419
  Inheritance of acquired characteristics 399–402
language 179
Laplace, Pierre Simon de 366, 411
Law of contradiction 29
Leaning Tower of Pisa 186, 211
Leavitt, Henrietta 364–366
Leibniz, Gottfried 248, 260
Leipzig 169
LeMaître, George 371–372
Leonardo da Vinci 134, 142
Light 243, 244-248, 299–302
Light quanta 347
Lincolnshire 242
Linkage 465
Linnæus (Linné, Carl von) 395–397
  The System of Nature 396
Linnean Society, London 426
Lockyer, Norman 368
Lodestone 305
Logarithms 172
Logic 82–83
Longitudinal waves 302
Lorentz, Hendrick Antoon 316
Lorentz-FitzGerald contraction 316, 323
Luminiferous æther 304
Lunar evolution 436
Lunar Society 414
Luria, Salvador 470, 475
Luther, Martin 199
Lutheran University of Copenhagen 169
Lyceum 78–79
Lyell, Charles 392–394, 398–403, 413, 424–425
  Principles of Geology 393–394, 401–403, 413, 418, 423

M

M.I.T. 371, 374
Macedonia 77, 96
Mach, Ernst 317, 319, 453
Magellanic clouds 364
Magnetism 305
Malpighi, Marcello 451
Malthus, Thomas
  Essay on Population 421, 424
Man-Faced Ox Progeny 431
Many Universes Interpretation 357–358
Mars 176, 181
Marx, Karl
  Das Capital, 279
Mass and energy 325
Mästlin, Michael 165, 171
Material cause 83–84
Mathematical elegance 157
Mathematics & music 38–40
Matter Waves 351
Maxwell, James Clerk 294, 306
  Demon 294–295
  Wave Equations 306–307
Mayer, Julius 291
Mechanical models 347, 473
Mechanist philosophy 338
Mendel, Gregor 455–463
  Independent Assortment 464–466

Mendeleev, Dmitri 350
Mendelian Genetics 462
Mercury 181
Mersenne, Marin 238
Mesopotamia 12
Mesopotamian number system 12
messenger-RNA 485, 486
Metabolism 447, 471
Meyer, Julius 350
Michelson, Albert A. 311
Michelson-Morley Experiment 311, 312, 313, 323
Microscope 443–446
  achromatic 446
  compound 443–446
  simple 443, 445
Microscopists 446
Middle Ages 383
Miescher, Friedrich 472
Miletos 21, 24–25, 27
Milky Way 192, 365–366, 375
Mitosis 470
Modus Tollens 314–315, 434
Molecular biology 482–486
  central dogma 483
Momentum, conservation of 291
Moon 182, 191, 195, 258
Morgan, Thomas Hunt 464–466
Morley, Edward 312
Mount Everest 191
Mount Wilson observatory 368
mRNA 485–486
Muhammad 136
Multidisciplinary Laboratory 470
Muscular distrophy 488
Muséum d'histoire naturelle 389
Museum in Alexandria 97, 111, 174
Music 38–40
Mutations 484

N

Nägeli, Karl 452–454, 458–459
Naples, Italy 476
National Academy of Sciences 367
National Institutes of Health 489
Natural motion 87, 209
Natural selection 426–427, 432, 448
Natural Theology 411–412, 430
Naturphilosophie 306–307, 326, 449, 453
Nebulae 278
Neptunism 384–385
Netherlands 189
Newcomen, Thomas 285
Newton, Isaac 241-275, 299, 338
    Axioms of Motion 264
    Crucial Experiment on Light 245
    De analysi 248, 260
    Effect of Choice 410, 411, 432
    Falling apple 253
    Frames no hypotheses 272
    Hypotheses non fingo 272
    Letter to the Royal Society of London 246
    Lucasian Professor of Mathematicks 260
    Mathematical Principles of Natural Philosophy See *Principia*
    Prism experiment 243–248
    Reflecting Telescope 261
Nichomachus 77–78
No vestige of a beginning, — no prospect of an end 388
Noah's Ark 397
Nova 170
"November 24, 1859" 426
Nucleic acids 472
Nucleotides 472
Numbers 43
    Egyptian 7
    figurate 41
    origin of 4–6

O

Octahedron 167
"October 23, 4004 B.C" 383
Odd-Number Rule 214–215
Oersted, Hans Christian 306
Olbers, Heinrich 365
    Paradox 365
Omnis cellula e cellula 447
"On the Electrodynamics of Moving Bodies" 320
Orchids 420
Order of St. Thomas 456
Organic Chemistry. 470
Orlando di Lasso 182
Osiander, Andreas 161
Ottoman Turks 136
Oxford University 391

P

Padua 143, 187
Pangenesis 428–429, 453
Parabola 175
    equation of 236
Paracelsus 142
Parallax 331
Parallel postulate 100–101
Parasitic cones 392
Paris Basin 389
Paris, University of 351
Parmenides 28–30, 337
Particulate structure 454
Pauling, Linus 473, 480, 482
Peebles, P. J. E. 373–374
Pendulum 213
Penzias, Arno 373–374
Perihelion of Mercury 333
Periodic Table of Elements 350–351
Perpetual motion machine 292
Phaestis 77
Phage Group 469–470, 475
Phages 469
Philosophy of nature 306
Photoelectric Effect 319, 346–347
Physical Chemistry 471
Pigeon fanciers 420–422
Pineal gland 234
Pisa 187
    Leaning Tower of 186
    University of 185
Pisum sativum 457
Plague 242, 243
Planck, Max 345
Planck's constant 345, 348
Planets, problem of the 70–74
Plato 19, 37, 50–70, 77–79, 95, 117, 167, 201, 231, 335
    Divided Line 231
    Republic 53, 95
    Timaeus 167
Platonic Solids 166–169
Playfair, John 388
Plutonism 386
Poitiers, University at 228
Polarization 303
Polonium 341
Polynucleotide 473, 479
Polypeptides 474
Polytechnic Institute in Zurich 318
Positivism 317, 323, 453
Postdiction 333
Postulates 100
Precession of the equinoxes 116
Preformation theories 450–452
pre-Socratics 19–33, 37–48
Princeton University 357, 373
Princeton, New Jersey 317, 333
Principia 256–273, 277–279, 299, 309, 327, 410, 430
Projectile motion 217, 235
Protein 474

Index

Protein Synthesis 485
Protestant Reformation 199
Prutenic Tables 161
Ptolemaic Universe 110
Ptolemy Soter 97
Ptolemy, Claudius 15, 117–132, 148, 190
    Almagest 117–132, 136
    Geography 117
    Mathematical Composition 117
Purines 473
Pyrimidines 473
Pythagoras & Pythagoreans 37–48, 51, 165, 168, 180, 335
Pythagorean Theorem 42–43, 104–109, 191

Q

Q.E.D. 264
Quantum mechanics 471
Quantum of Energy 345–346
Quod erat demonstrandum 264

R

Radiation 338–341
Radioactivity 340–341
Radium 341
Railroads 287
Raphael, *School of Athens* 80–81
Recombinant DNA 487
Redshift 369–370
Reductionism 448–449
Reductionists 448–449
Regular Solids 166–169
Relative brightness 363
Relativity
    Galilean 322
    General Theory 326–332
    Special Theory 320–326
Renaissance 138, 383
Res cogitans 231–233, 278
Res extensa 231–232
Reverse transcriptase 486

Ribonucleic acid *See* RNA
Ribose 473
RNA 473, 482, 484–485
Robertson, Howard 369
Roman Empire 135
Rome 202
    Fall of 132
Röntgen, Wilhelm 339, 471
Royal Society of London 261, 332, 412
Rudolph II, Holy Roman Emperor 171
Rules of Reasoning 263
Rumford, Count (Benjamin Thompson) 289
Rutherford Experiment 342–343
Rutherford model of the atom 343
Rutherford, Ernest 342–343, 348

S

Sagredo 202–204
Salviati 202–204
Samos 37
Saturn 170, 182
    rings of 198
Saving the phenomena 111, 117, 437, 462
Schleiden, Matthias J. 447
Schrödinger, Erwin 337, 353–356, 371
    What is Life? 469
Schrödinger's Cat 355, 356
Schwann, Theodor 447
Scientific Revolution 138, 257, 273, 309, 350, 380, 383, 395, 407
Sedgwick, Adam 415
Sensible world 54
Sexagesimal number system 13–14
Sexual selection 427
Shakespeare, William 185
Shapley, Harlow 366–367
Simplicio 202–204

Simultaneity 322
61 Cygnus A 362–364
Skåne 169
Slide rule 172
Smith, Adam
    Wealth of Nations 279
Socrates 19, 51–53, 95
Sodium 349
Solvay conference 351
Soma cells 453
Sosigenes 141
Sound waves
    longitudinal waves 303
South America 416, 418, 421
Special Theory of Relativity 320–326
Spectral Lines of Hydrogen 349–350
Spectroscopy 344
Speed of light 323
Speed vs. velocity 249
Spermatozoa 445, 451
Speusippus 78
Spheres of Eudoxus 72–74
Spy glas 189
Stade, defined 114
Stadium, paradox of the 30
Stagirite, The 77
Steam Engine 284–288
    Newcomen Atmospheric Engine 285
    Savery steam pump 285
    Watt-Boulton 286
Steamships 287
Stellar parallax 158–161, 361–363, 368
Stent, Gunther 482
Straits of Magellan 418
Strata 384, 385, 386, 388, 389
Sublunar World 85, 86, 87, 88
Suction pump 284
Sumeria 12
Sumerian number system 12
Sun spots 198
Supernova 170
Swabia 165

Swammerdam, Jan 452
Syene 112–113
Syracuse 78

T

Table of Chords 116
Tartaglia, Niccolo 210–211, 218
Taxonomy 394
Telescope 189, 443
Temperature
    Absolute Zero 296
    versus Heat 293
Tetractys 41–42
Tetrahedron 167
Thales 21–24
Thermodynamics 293
    First law 293, 345
    Second law 293
    Third law 296
Thompson, Benjamin (Count Rumford) 289
Thomson model of the atom 342
Thomson, J. J. 341
Thomson, William (Lord Kelvin 296
Thought experiments 321
Thymine 473–474, 483
Time
    in Darwin's theory of evolution 435
    measuring 212
Times-Squared Law 215
Todd, Alexander 479
Torun, Prussia 143
Train Station Experiment 321–322
Transverse Waves 302–303
Tropoplasm 452–453
Tübingen, University of 165
Tuscany 199
Twin Paradox 324
Tycho Brahe 169–172, 367

U

U.S. Dept of Energy 489
Ulm, Germany 317
Ultraviolet Catastrophe 343–345
Uncertainty Principle 353–355
Uniformitarianism 392–394
Universal Gravitation 271–273
Universe
    Age of 377
    Critical Mass 376
    Open or Closed 374–377
    theories of 361–377
University of California at Santa Cruz 488
University of Zurich 319
Uppsala, University of 396
Uracil 473
Uraniborg 170
Urban VIII 202–204
Ussher, Archpishop James 383

V

van Leeuwenhoek, Anton 444–445, 451
Vanderbilt University 470
Vatican 204
Venice 190, 199
Venus 181
    phases of 159–160, 316
Vienna, University of 456
Vincenzio Galilei 185
Violent Motion 87
Virchow, Rudolf 447
Virus 486
Visible world 54
Vitalism 448
Vitalists 448–449, 470
von Tschermak, Erich 462
Vortices 233
Voyage of the Beagle 415–419
Vulcanism 387

W

Wallace, Alfred Russel 423–426, 438
Washington University 470, 475
Water clock 213
Waterwheels 283
Watson, James D. 475–482
    Double Helix, The 478
Watt, James 286
Wedgwood, Josiah 415
Weismann, August 453–455, 486
Werner, Abraham 385–386
Wheeler, John 357
White light 244
Wilberforce, Samuel 439
Wilkins, Maurice 476, 481–482
Wilson, Robert 373–374
Wine barrels 173
Woolly mammoths 389
Woolsthorpe Manor 253
World War I 352
Wren, Christopher 262
Writing, origin of 5

X

Xenophanes 28
X-Rays 339–340, 471

Y

Young, Thomas 301
    2-slit experiment 301–302

Z

Zeitgeist 407
Zeno 30–33, 250
Zero, lack of 15
Zeus 42
Zürich, University of 353